COMPLEX
VARIABLES

COMPLEX VARIABLES

George Polya
Professor Emeritus
Stanford University

Gordon Latta
Professor
University of Virginia

JOHN WILEY & SONS, INC.
New York · London · Sydney · Toronto

Library of Congress Cataloging in Publication Data:

Polya, George, 1887–
Complex variables.

1. Functions of complex variables. I. Latta, Gordon, 1923–   joint author. II. Title.

QA331.P794   515'.9   73-14882
ISBN 0-471-69330-8

Printed in the United States of America

10 9 8 7 6 5 4 3 2 1

PREFACE

Take care of the sense, and the sounds will take care of themselves. *Alice in Wonderland*

After having lectured for several decades on complex variables to prospective engineers and physicists, I have definite and, I hope, not unrealistic ideas about their requirements and preferences.

Students are students. Since they are required to take several courses, they may study some subjects just for the examination, with the intention of forgetting what they have learned after the examination. Yet they may (and the more intelligent and purposeful students do) ask pertinent questions about the subject: Is it interesting? Can I use it?

These questions are fully justified. The instructor of a more advanced branch of mathematics—such as the theory of complex variables—who is a mathematician should try to put himself into the position of his students who are prospective physicists or engineers. Before going into heavy definitions and lengthy proofs, the student wants to satisfy himself that the subject is interesting and useful enough to expend his time and effort on definitions and proofs.

Having realized these points as I taught successive generations of students, I adapted my lectures to their standpoint. I evolved the following guidelines.

Start from something that is familiar, useful, or challenging—from some connection with the world around us, from the prospect of some application, or from an intuitive idea.

Do not be afraid of using colloquial language when it is more suggestive than the conventional precise terminology. In fact, do not introduce technical terms before the student can understand the need for them.

Do not enter too early or too far into the heavy details of a proof. First, give a general idea or just the intuitive germ of the proof.

Generally, realize that the natural way to learn is to learn by stages. First, we want to see an outline of the subject in order to perceive a concrete source or a possible use. Then, gradually, as we can see use,

v

connections, and interest, we accept more willingly the responsibility of filling in the details.

The ideas just stated influenced the organization of this book.

Whenever the mathematical context offers a natural opportunity, there are a few words inserted about concrete phenomena or connected general ideas. Before the introduction of a formal definition, the intervening ideas may be previously discussed by examples or in more colloquial language. The proofs emphasize the main points and may leave to the student, now and then, more intermediate points than usual. The most notable departure from the usual is to be found, however, in the "Examples and Comments" that follow most sections and each chapter. There are, of course, the examples of the standard kind that offer an opportunity to practice what has been explained in the text. Yet there is what is not usual—a definite effort to let the student learn the subject by stages and by his own work. Some problems or comments ask the student to reconsider the definitions and proofs given in the text, directing his attention to more subtle points. Other problems introduce new material: a proof different from the one given in the text, or generalizations of (or analogues to) the facts considered, encouraging the student toward further study. Moreover, even the simpler problems, insofar as is possible, are arranged to give the student an opportunity to face a variety of research situations that will awaken his curiosity and initiative.

I hope that this book is useful not only to future engineers and physicists but also to future mathematicians. Mathematical concepts and facts gain in vividness and clarity if they are well connected with the world around us and with general ideas, and if we obtain them by our own work through successive stages instead of in one lump.

The course presented here has been taught several times at Stanford by me and by my friend and colleague, Gordon E. Latta, who shares my pedagogical ideas. I am grateful to him for sharing the writing which, because of other interests and duties, I was not able to do alone.

We may have achieved less than we hoped for at various points and in various respects, yet we still think that this book is a modest concrete contribution to the widespread debate about the lines along which the instruction in the universities should evolve.

George Polya

Stanford, August 1974

HINTS TO THE READER

Section 4.3 means Section 3 of Chapter 4.

Subsection (b) of Section 4.3 is quoted in that section simply as (b), or Subsection (b), but in any other section as Section 4.3 (b).

Formula (2) of Section 4.3 is quoted in that section simply as (2), or Formula (2), but in any other section as Formula 4.3 (2). (In Chapter 1, the rules of basic importance are labeled with Roman numerals.)

We distinguish two kinds of exercises (examples, comments, and problems). Some are attached to a complete chapter and are printed at the end of the chapter, and others are attached to a single section of a chapter and are printed at the end of that section. The two kinds are labeled differently: the third exercise attached to Chapter 4 is numbered 4.3 and quoted as Ex. 4.3, but the third exercise attached to Section 4.2 is numbered 4.2.3 and quoted as Ex. 4.2.3.

As a rule, the exercises attached to whole chapters are more difficult than the exercises attached to single sections; but there are exceptions, and they are not very rare.

(S) is printed at the end of the problems *attached to sections*, the solution of which is given at the end of the chapter.

The solution may give just the result (for more straightforward problems) or just a sketch or a hint, and more details are given only when the problem is more important or more difficult. A serious understanding of the solution demands work from the reader—and the reader will profit more if he does a good part of the work before he looks at the solution.

Hints to facilitate the solution of problems are placed in square brackets [ ]—do not look at the hints too early. Problems related to each other are often printed near each other, and therefore a look at the surrounding problems may help.

Most problems are only numbered, but some (more important or more interesting) problems have titles.

Iff. The abbreviation "iff" stands for the phrase "if and only if."

Read with a paper and pencil in hand. Carry out intermediate steps of calculation, build up the figures yourself, and add one part after the other as they arise in the reasoning.

Note. The last two chapters are of particular interest to the reader who studies complex variables in view of their applications. He is introduced to physical applications by Chapter 8 and to analytical applications by Chapter 9. These chapters are somewhat more demanding. Especially, the reader should acquire from other sources a more complete understanding of the physical theories that can only be hinted in this text.

The student should study at least the first six chapters and should do, if not all, then a substantial part of the problems attached to the sections. He should do as many chapter problems as his time allows, and he should pick and choose. He should also select problems that are closer to his interest or to his major study. He may (and he will, we hope) come back to the chapters and the problems skipped—he may need them in later studies.

An instructor, weighing the available time and the interest and preparation of the students, may consider omitting some of the more demanding topics and a good part, or even almost all, of the more demanding problems.

Solutions to exercises attached to whole chapters are not provided (except in a few cases; then they follow the problem or the other solutions at the end of the chapter concerned). The instructor may find an opportunity to discuss some solutions, for example, in practice sessions. The student may consult more extensive textbooks on complex variables. Some exercises are not usually treated in these textbooks, as, for instance, at the end of the first chapter; information can be found in G. Polya and G. Szegö, *Problems and Theorems in Analysis* (Springer, Vol. 1, 1972, and Vol. 2, 1974*).

For instance, for exercise 1.18 on p. 26 of this book, see problem III 22 in Polya-Szegö Vol. 1, p. 107, of which the solution is on p. 301 of the same Vol. 1. It may be useful to look not only at the problem quoted but also at the neighboring problems.

Thus,

FOR EXERCISE	SEE PROBLEM (Polya-Szegö)
1.5	VI 1
1.13	II 80
1.15	II 90
1.17	III 18
1.18	III 22
1.19	III 23
1.26	VII 14
2.10.1	III 8

---

* As this goes to press, Vol. 2 is about to appear but is not yet in print. Both volumes are available in German.

FOR EXERCISE	SEE PROBLEM (Polya-Szegö)
2.10.2	III 9
2.24	III 31
2.33	III 34
3.4	III 58
3.5	III 55.4
3.21	III 122
3.22	III 106
3.36	III 124
5.8	III 169
6.7.9	III 280
6.2	III 174
6.30	III 194
6.34	III 192
6.39	III 209
6.42	III 120
6.43	III 175
6.44	III 176
6.45	III 275
7.4.1	III 174
9.4.1	II 202
9.4.2	II 205
9.5.3	II 204

G. P.

CONTENTS

COMPLEX VARIABLES

In introducing complex numbers, we stress their analogy with real numbers. Complex numbers are subject to the same algebraic laws of addition, subtraction, multiplication, and division as real numbers, and they are used similarly in describing geometrical and physical situations.

1.1 REAL NUMBERS

We make no attempt to explain what real numbers are but, instead, remind ourselves how they are used. From time to time we require some particular type of real number, such as a negative integer, a rational number, or an irrational number, and we shall assume that these terms are meaningful to the reader.

(a) Let us denote real numbers by the letters $a, b, c, \ldots, x, y, \ldots$. We can perform the operations of addition, subtraction, multiplication, and division using real numbers, and these operations obey the following rules

(I) $a + b = b + a$, the commutative law of addition.
(II) $ab = ba$, the commutative law of multiplication.
(III) $(a + b) + c = a + (b + c)$, the associative law of addition.
(IV) $(ab)c = a(bc)$, the associative law of multiplication.
(V) $(a + b)c = ac + bc$, the distributive law.
(VI) Given a, b, there is a uniquely determined x such that

$$a + x = b$$

the law of unique inverse of addition.
(VII) Given a, b, there is a uniquely determined x such that

$$ax = b,$$

provided that $a \neq 0$, the law of unique inverse of multiplication.

1

Figure 1.1

The absolute value of the number a is denoted by $|a|$; for instance, $|2| = 2$, $|-2| = 2$. We mention two rules dealing with absolute values:

(VIII) $|a + b| \leq |a| + |b|$

(IX) $|ab| = |a|\,|b|$

(b) Real numbers may be interpreted as marks on a straight line. As is usually done, we consider an infinite straight line (the x-axis) drawn from left to right in a horizontal plane. We select a point we call the origin, and a certain (arbitrary) unit of length (see Figure 1.1). Then x refers to the point on the line $|x|$ units from the origin, to the right if $x > 0$, and to the left if $x < 0$. Correspondingly, each point of the line yields a real number x. We refer to it as *the real line*.

(c) Real numbers can be used to describe vectors along a fixed straight line (such as the real line). Clearly this interpretation of real numbers is somewhat artificial, but it does pave the way for a very useful interpretation of complex numbers.

We recall that the notion of *vector* is obtained by abstraction from the notion of *displacement*. Displacement is the operation of transferring a particle from point A to point B. Such displacement has a *magnitude*, a *direction*, and an initial point (A). We now regard two displacements as equivalent (equal) if they have the same magnitude and the same direction (Figure 1.2). If the whole space is shifted as a rigid body, without rotation, all its points undergo equivalent displacements, and we call such a change of position a *vector*. Thus, a vector has a definite magnitude and a definite direction, but no specified initial point. A vector is represented by a directed line segment. Two such segments represent the same vector if they are parallel and have the same length and sense (direction). Accordingly, if x "is a point on the real line," we may (if it is convenient or to our advantage to do so) interpret x as a vector, of length $|x|$, parallel to the line, and pointing to the right if $x > 0$, or to the left if $x < 0$.

Figure 1.2

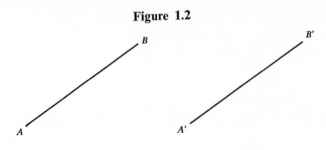

1.2 COMPLEX NUMBERS

If we apply the usual procedure for solving quadratic equations to the equation

$$x^2 + 1 = 0$$

we obtain the roots $\sqrt{-1}$ and $-\sqrt{-1}$. There being no real number whose square is -1, such roots (when they arose for example in constructing a certain triangle) were regarded as "impossible" or "imaginary" and were discarded.

The procedure for solving quadratic equations was essentially known in Babylonia, around 2000 B.C., but the mathematicians of those days did not speculate about the nature of imaginary roots. Such speculation started in the sixteenth and seventeenth centuries, and mathematicians gradually found that the rules of algebra may be applied to imaginary quantities, and that the introduction of such quantities may simplify certain questions. By the eighteenth century, imaginary numbers were being introduced so frequently that Euler found it convenient to introduce i as standing for $\sqrt{-1}$. This notation is almost universally used today (except in electrical engineering, where j, not i, is used).

Today, complex numbers play an important role in almost all branches of mathematics and mathematical physics. To the student who is not familiar with complex numbers, large parts of algebra, of trigonometry, of the integral calculus, and of the theory of ordinary and partial differential equations must remain inaccessible.

1.3 COMPLEX NUMBERS AS MARKS IN A PLANE

Consider a rectangular coordinate system in the usual position (Figure 1.3). To each complex number $x + iy$ we associate the point (x, y) and conversely. In this way, a one-to-one correspondence is set up between complex numbers $z = x + iy$ and points of the plane (x, y). We shall use the terminology "the

Figure 1.3

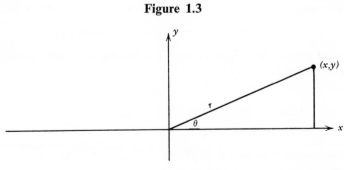

point z", or "the point $x + iy$" in place of the longer (but more precise) statement "the point (x, y) that corresponds to the complex number $z = x + iy$." The x-axis consists of the real numbers $x + 0i = x$, and the points on the y-axis are marked iy, y real. As a concession to history and tradition, we call numbers of the form iy *purely imaginary*. Also, we call the x-axis the *real axis*, the y-axis the *imaginary axis*, x the *real part* of $x + iy$, and y the imaginary part of $x + iy$. In symbols

$$x = \mathbf{R}z \qquad y = \mathbf{I}z$$

The distance from the origin to the point z is $\sqrt{x^2 + y^2}$, called the *absolute value* or *modulus* of z and denoted by $|z|$:

(1) $$|z| = \sqrt{x^2 + y^2}$$

In terms of the polar coordinates of $z = x + iy$,

(2) $$x = r \cos \theta \qquad y = r \sin \theta$$

we write

(3) $$z = x + iy = r(\cos \theta + i \sin \theta)$$

The terminology for complex numbers differs from analytic geometry; we call the distance r the absolute value of z

$$r = |z|$$

and we call the angle θ the *argument*, or the *phase* of z.

$$\theta = \arg z$$

It follows from (2) that

$$\frac{y}{x} = \tan \theta$$

$$x^2 + y^2 = r^2$$

Hence $r \geq 0$ is uniquely determined, while θ can have any one of an infinite set of values and still satisfy (2). In fact, if θ satisfies (2), then all of the angles

$$\ldots, \theta - 4\pi, \theta - 2\pi, \theta, \theta + 2\pi, \theta + 4\pi, \ldots$$

satisfy it just as well. As an agreement, or convention, we pick out the value between 0 and 2π,

$$0 \leq \theta < 2\pi$$

and call it the principal value of $\arg z$.*

---

* Other texts may define $-\pi \leq \theta < \pi$ as the principal-value range. Alternately, it may be advantageous to consider this range of arg z instead of the principal-value range.

A point is determined by its rectangular coordinates and determines them completely. If x_1, y_1, x_2, y_2 are real, and

$$(4) \qquad x_1 + iy_1 = x_2 + iy_2$$

then

$$(5) \qquad x_1 = x_2 \qquad y_1 = y_2$$

The single equation (4) between complex numbers is equivalent to two equations (5) between real numbers.

The corresponding relations in polar coordinates are less simple. If r_1, θ_1, r_2, θ_2 are real, r_1, $r_2 \geq 0$, we can infer from

$$r_1(\cos \theta_1 + i \sin \theta_1) = r_2(\cos \theta_2 + i \sin \theta_2)$$

that

$$r_1 = r_2$$

If $r_1 = 0$, we can say nothing about θ_1, θ_2. If $r_1 > 0$, then $\theta_1 - \theta_2$ is an integral multiple of 2π:

$$\theta_2 = \theta_1 + 2n\pi \qquad n = 0, \pm 1, \pm 2, \ldots$$

Comparing Figures 1.3 and 1.4, we survey the expressions for the rectangular and polar coordinates of the point z in terms of the complex number z:

$$(6) \qquad x = \mathrm{R}z \qquad y = \mathrm{I}z \qquad r = |z| \qquad \theta = \arg z$$

1.3.1   Find the absolute value, and the principal value of the argument, for the complex numbers

(a) $2 + 3i$ (S)   (b) $-5 + 12i$

(c) $-1 + i\sqrt{3}$ (d) $-1 - i$

1.3.2   For which points of the plane is $\mathrm{R}z = \mathrm{I}z$?

Figure 1.4

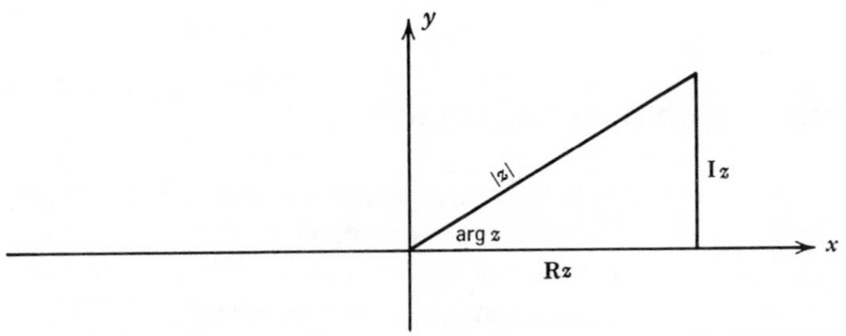

1.3.3 Regard the coordinate axes and the two lines bisecting the angles between the axes as mirrors, and list and plot all the points that can be obtained by successive reflections in these mirrors from the point $3 + 2i$.

1.3.4 Show that

$$\frac{\mathbf{R}z + \mathbf{I}z}{\sqrt{2}} \leqq |z| \leqq |\mathbf{R}z| + |\mathbf{I}z| \tag{S}$$

1.4 COMPLEX NUMBERS AS VECTORS IN A PLANE

Consider a vector in a plane containing a system of rectangular coordinates (Figure 1.5). The projections of this vector onto the x- and y-axes have

Figure 1.5

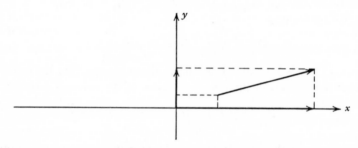

definite magnitude and direction (sense) and so may be regarded as vectors on separate lines. These projections (components) determine the magnitude and direction of the vector, but not its initial point. In the absence of any other requirement, it is convenient to choose the origin as the initial point of the vector; then its projections x, y on the x- and y-axes represent the *vector $x + iy$*, whose terminal point coincides with the *point $x + iy$*.

1.4.1 The initial point of a line segment is $-3 + 2i$, and its terminal point is $7 + 12i$. Find the vector represented by the segment, and obtain its absolute value and direction.

1.5 ADDITION AND SUBTRACTION

In the remainder of this text, we frequently interpret our results involving complex numbers in terms of points in a plane or in terms of vectors in a plane (and, occasionally, both ways). We are never forced to make these interpretations, but we do so when it is to our advantage.

Figure 1.6

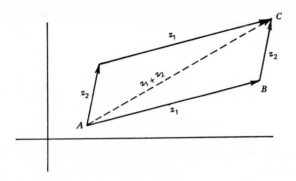

For instance, let us interpret the addition of complex numbers in light of the vector interpretation. If the vector $z_1 = x_1 + iy_1$ carries a particle (point) from A to B, and the vector $z_2 = x_2 + iy_2$ carries the particle from B to C, then we call $z_1 + z_2$ the vector carrying the particle from A to C (Figure 1.6). In terms of the projections of the vectors on the axes, we have

$$(1) \qquad (x_1 + iy_1) + (x_2 + iy_2) = (x_1 + x_2) + i(y_1 + y_2)$$

and verify the rule: *adding complex numbers, we add real parts to real parts and imaginary parts to imaginary parts.*

Recalling the properties of addition of real numbers from Section 1.1, we see that

$$(2) \qquad z_1 + z_2 = z_2 + z_1$$

which is also clear from the figure and emphasizes the "parallelogram law" of vector addition. Again

$$(3) \qquad (z_1 + z_2) + z_3 = z_1 + (z_2 + z_3)$$

is the associative law of addition. We readily see how this law follows from the properties of real numbers and from (1). We leave the proof of unique inverse of addition as an exercise.

As an example of the value of an interpretation, we note in Figure 1.6 that z_1, z_2 and $z_1 + z_2$ form the sides of a triangle. Since one side of a triangle cannot exceed the sum of the other two sides

$$(4) \qquad |z_1 + z_2| \leqq |z_1| + |z_2|$$

The modulus of a sum is not greater than the sum of the moduli. This result (Equation 4) is usually called the "triangle inequality" (see Problem 1.12).

1.5.1   Find $z_1 + z_2$ and $z_1 - z_2$ by computation *and* a figure

(a) $z_1 = 1 + i$ $z_2 = 1 - 2i$ (S)

(b) $z_1 = 2$ $z_2 = \dfrac{1 + i}{\sqrt{2}}$

1.5.2   Establish the associative law of addition

$$z_1 + (z_2 + z_3) = (z_1 + z_2) + z_3$$

by considering real and imaginary parts, and draw a figure to visualize its geometric meaning.

1.5.3   Describe geometrically the conditions imposed on z_1 and z_2 by the equation

(a) $|z_1 + z_2| = |z_1| + |z_2|$
(b) $|z_1 + z_2| = |z_1| - |z_2|$ (S)

1.5.4   Deduce that

$$\|z_1| - |z_2\| \leqq |z_1 - z_2|$$

first by interpreting the quantities involved geometrically, and then by an application of the triangle inequality.

1.5.5   If three sides of a quadrilateral, considered as vectors drawn in a suitable direction, are z_1, z_2, and z_3, show that the fourth side is given by $z_1 + z_2 + z_3$. Deduce that

$$|z_1 + z_2 + z_3| \leqq |z_1| + |z_2| + |z_3|$$

1.5.6   Generalize the result of Example 1.5.5 to a polygon of $n + 1$ sides.

1.6 MULTIPLICATION AND DIVISION

Rather than attempt to interpret the product of two complex numbers as a vector operation, let us fall back on the formal application of the laws of algebra

(1) $$z_1 z_2 = (x_1 + iy_1)(x_2 + iy_2)$$

$$= x_1 x_2 - y_1 y_2 + i(x_1 y_2 + x_2 y_1)$$

Figure 1.7

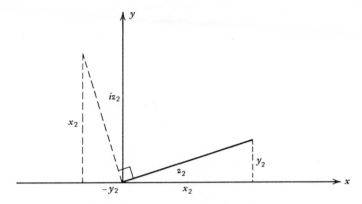

where we have expanded the bracketed terms and replaced ii by -1. This formula contains a number of special cases worth noting. If $y_1 = 0$, $x_1 = p > 0$, we find

(a) $pz_2 = px_2 + ipy_2$

which preserves the direction of z_2 but multiplies its length by p.

(b) $y_1 = 0$, $\qquad x_1 = -1$

$$-z_2 = -(x_2 + iy_2) = (-x_1) + \iota(-y_2)$$

Here the length of z_2 is preserved, but its direction is reversed. Combining (a) and (b),

$$-pz_2 = -px_2 + i(-py_2) \qquad p > 0$$

reverses direction and scales length.

(c) $\qquad\qquad\qquad x_1 = 0 \qquad y_1 = 1$

$$iz_2 = i(x_2 + iy_2) = -y_2 + ix_2$$

We observe that multiplication by i rotates the vector $z_2 = (x_2 + iy_2)$ through $90°$ in the positive (counterclockwise) direction. Two applications of this operation yield $i(iz_2) = -z_2$ in which the direction of z_2 gets reversed, as in (b).

In polar coordinates, Equation 1 assumes a particularly useful and important form

(2) $\qquad\qquad z_1 z_2 = r_1 r_2 (\cos \theta_1 + i \sin \theta_1)(\cos \theta_2 + i \sin \theta_2)$
$$= r_1 r_2 [\cos (\theta_1 + \theta_2) + i \sin (\theta_1 + \theta_2)]$$

In multiplying complex numbers, we multiply the absolute values and add the arguments.

It is instructive to reconsider the special cases (a), (b), and (c) in light of (2) and, at the same time, observe how the commutative and associative laws of multiplication follow from (2) and the corresponding laws for real numbers.

(d) We now examine division, or the unique inverse to multiplication. Consider the equation

$$(3) \qquad\qquad z_1 z = z_2$$

being given z_1, z_2, with z unknown. From (3) we obtain the two real equations

$$x_1 x - y_1 y = x_2 \qquad y_1 x + x_1 y = y_2$$

These equations are equivalent to

$$(x_1^2 + y_1^2)x = x_1 x_2 + y_1 y_2$$
$$(x_1^2 + y_1^2)y = -y_1 x_2 + x_1 y_2$$

which have unique solutions x, y, provided that $x_1^2 + y_1^2 \neq 0$. This unique inverse of multiplication exists unless $x_1^2 + y_1^2 = 0$; $x_1 = 0$, $y_1 = 0$; $z_1 = 0$. Excluding $z_1 = 0$, we find that

$$z = \frac{z_2}{z_1} = \frac{x_1 x_2 + y_1 y_2 + i(- y_1 x_2 + x_1 y_2)}{x_1^2 + y_1^2}$$

In Section 1.8 we shall find a more convenient way of computing the quotients of complex numbers. For the moment, we are concerned mainly with the verification that the laws of algebra are valid for complex numbers.

1.6.1 Perform the indicated operations, and reduce the numbers to the form $x + iy$.

(a) $(4 + 3i)(4 + 2i)(3 - i)(1 - i)$ $\hfill$ (S)

(b) $\dfrac{i}{(i - 1)(i - 2)(i - 3)}$

(c) $(\sqrt{3} - 1)^6$

(d) $\dfrac{1 + z}{1 - z}$ where $z = \cos\theta + i\sin\theta$

(e) $\sqrt{2}\,(\cos\alpha + i\sin\alpha)(\cos\beta + i\sin\beta)$ where

$$0 < \alpha < \frac{\pi}{2} \qquad 0 < \beta < \frac{\pi}{2}$$

and

$$\alpha = \arctan 2 \qquad \beta = \arctan 3$$

1.6.2   Establish the associative law of multiplication

$$z_1(z_2 z_3) = (z_1 z_2) z_3$$

by considering absolute values and arguments.

1.6.3   Establish the distributive law

$$z_1(z_2 + z_3) = z_1 z_2 + z_1 z_3$$

1.6.4   Show that if $z_1 z_2 = 0$, then either $z_1 = 0$ or $z_2 = 0$ (or both).

1.6.5   If $z_2 \neq -z_3$, show that

$$\left| \frac{z_1}{z_2 + z_3} \right| \leqq \frac{|z_1|}{||z_2| - |z_3||}$$

1.6.6   Show that

$$z^n = [r\,(\cos\theta + i\sin\theta)]^n$$
$$= r^n\,(\cos n\theta + i\sin n\theta)$$

for positive integral values of n. (The case $r = 1$ of this result is known as de Moivre's theorem).

1.6.7   Derive from de Moivre's theorem that

(a) $\cos 2\theta = \cos^2\theta - \sin^2\theta$ (S)
(b) $\sin 2\theta = 2\sin\theta\cos\theta$ (S)
(c) $\cos 3\theta = \cos^3\theta - 3\cos\theta\sin^2\theta$
(d) $\sin 3\theta = 3\cos^2\theta\sin\theta - \sin^3\theta$

1.6.8   Assuming, as usual, that

$$r\,(\cos\theta + i\sin\theta) = x + iy$$

show that

$$r^n\cos n\theta = x^n - \binom{n}{2}x^{n-2}y^2 + \binom{n}{4}x^{n-4}y^4 - \cdots$$

$$r^n\sin n\theta = \binom{n}{1}x^{n-1}y - \binom{n}{3}x^{n-3}y^4 + \cdots$$

1.6.9   Show that the formula of Example 1.6.6 holds also for the cases $n = -1, -2, -3$, and so on, for any negative integer n. [Show first that

$$z^{-1} = r^{-1}(\cos\theta - i\sin\theta)$$

and then apply Example 1.6.6.]

1.6.10 The complex numbers z_1 and z_2 mark the two endpoints of a straight-line segment. Find the complex number that marks (a) the midpoint of the segment, and (b) the point that divides the segment in the proportion $1:2$ and is closer to z_1 than to z_2. (S)

1.6.11 The complex numbers z_1, z_2, and z_3 mark the three vertices of a triangle. Find the complex number that marks the centroid of the triangle. (The centroid divides in the ratio $1:2$ a median that joins the midpoint of a side to the opposite vertex.)

1.7 SUMMARY AND NOTATION

From this point on, we shall use complex numbers of the form

$$(1) \qquad\qquad z = x + iy$$

and may speak of "the point z of the plane," or the "vector z." When necessary to refer to the various real numbers associated with z, we let

$$(2) \qquad x = \mathbf{R}z, \qquad y = \mathbf{I}z, \qquad r = |z|, \qquad \theta = \arg z$$

The laws of algebra, as outlined in Section 1.1, apply equally to the whole set of complex numbers. Accordingly, the term "number" will always mean complex number (with real numbers as special cases), although in such statements as $\mathbf{R}c = a, \mathbf{I}c = b$, it is understood that for this particular discussion, a and b are real.

As a special convention, of the last six letters of the alphabet

$$u, v, w, x, y, z$$

u, v, x, y have real, and w and z complex values, with

$$z = x + iy$$
$$w = u + iv$$

although there will be exceptions to this rule (but not without adequate warning).

Example 1. Write

$$a + ib = c, \qquad x + iy = z$$

and assume that a, b, x, y are real. Observe that

$$|z - c| = [(x - a)^2 + (y - b)^2]^{1/2}$$

is the distance between the points c and z. Now consider c as a constant (fixed) and z as a variable. Let $p > 0$ be another constant. The equation

$$|z - c| = p$$

states that each point z is at the same distance from the fixed point c. Therefore the equation represents a circle.

The inequality

$$|z - c| < p$$

states that z is inside the circle just discussed.

Example 2. Let z_1, z_2, z_3, z_4 denote four complex numbers linked by the equation

(3) $$z_1 - z_2 + z_3 - z_4 = 0$$

Consider the points z_1, z_2, z_3, z_4 in the plane. What is the geometric interpretation of Equation 3?

We write (3) in the form

$$z_1 - z_2 = z_4 - z_3$$

That is, the directed line-segment joining z_2 to z_1 has the same length and direction as the segment joining z_3 to z_4. Therefore, the points z_1, z_2, z_3, z_4, taken in this order, are the vertices of a parallelogram.

We write (3) in the form

$$z_1 - z_4 = z_2 - z_3$$

That is, two sides of the parallelogram, joining z_4 to z_1 and z_3 to z_2 respectively, which we have not mentioned before, are also parallel and equal in length.

We write (3) in the form

$$\frac{z_1 + z_3}{2} = \frac{z_2 + z_4}{2}$$

That is, the diagonal of the parallelogram joining the vertex z_1 to the opposite vertex z_3 has the same midpoint as the diagonal joining z_2 to z_4; see Example 1.6.10(a).

Again, we write (3) in the form

$$\frac{z_1 + z_3}{2} = \frac{1}{2}\left(\frac{z_1 + z_2}{2} + \frac{z_3 + z_4}{2} \right)$$

That is, the common midpoint of the diagonals we have just considered is also the midpoint of the segment joining the midpoints of two opposite sides. The endpoints of the first side are z_1 and z_2, and the endpoints of the other side are z_3 and z_4.

We can prove many other theorems of elementary plane geometry in the same way, that is, by algebraical manipulation and geometrical interpretation of equations between complex numbers. See, for instance, Examples 1.7.4 and 1.29.

1.7.1   Describe geometrically the regions of the z-plane determined by the following inequalities

(a) $|3z + 2| < 1$ (S)

(b) $|3z - 2| < 1$

(c) $0 < \arg z < \pi$

(d) $\dfrac{\pi}{2} \leqq \arg z \leqq \pi$

(e) $\mathbf{I} z \geqq 2$

(f) $\mathbf{R}(1 + i)z < 0$

(g) $\mathbf{R} z^2 > 0$

(h) $|z - 1| + |z + 1| \leqq 4$ (S)

(i) $\mathbf{R} z^2 + \mathbf{I} z^2 = 0$

(j) $|z - p| \leqq p + \mathbf{R} z$ $p > 0$

1.7.2   Assuming that $p > 0$ and $p \neq 1$, show that the locus of the points z for which

$$\left| \frac{1 - z}{1 + z} \right| = p$$

is a circle. What is the locus for $p = 1$?

1.7.3   Find the set of all points z for which

$$\mathbf{I}\left(z + \frac{1}{z}\right) = 0$$

1.7.4   On each side of a given (arbitrary) quadrilateral describe an exterior square. Join the center of each square to the center of the square described on the opposite side. Show, by complex numbers, that the two segments thus obtained are of equal length and are perpendicular to each other.

(Let $2a$, $2b$, $2c$, and $2d$ be the vertices of the given quadrilateral, encountered in this order by a person who walks counterclockwise along the perimeter, and express the four centers.)   (S)

1.7.5   Show, by complex numbers, that the line joining the midpoints of two sides of a triangle is parallel to the third side and half its length.

1.7.6   Show, by complex numbers, that the three medians of a triangle pass through the same point—the centroid—which divides each median in the ratio 2:1. (Let a, b, and c be the three vertices of the triangle; a median joins the midpoint of the side with endpoints b and c to the opposite vertex a.)   (S)

1.8 CONJUGATE NUMBERS

Two complex numbers are called *conjugate* if they are represented by points symmetrical to each other with respect to the real axis. The number conjugate to z is denoted by $\bar{z}$. If we use our standard notation and write

$$z = x + iy$$

then

$$\bar{z} = x - iy$$

In other terms, if we leave x as it is, but interchange y and $-y$, then we interchange z and $\bar{z}$. (Or leave x, y as they are, and interchange i and $-i$.)

We have (also standard notation)

$$z = r(\cos\theta + i\sin\theta)$$

$$\bar{z} = r(\cos\theta - i\sin\theta) = r[\cos(-\theta) + i\sin(-\theta)]$$

Observe that

$$z\bar{z} = x^2 + y^2 = r^2 = |z|^2$$

These formulas are often useful. For instance, they lead quickly to various expressions for $1/z$, the reciprocal of z:

$$\frac{1}{z} = \frac{\bar{z}}{|z|^2} = \frac{x - iy}{x^2 + y^2} = \frac{1}{r}(\cos\theta - i\sin\theta)$$

Let us write

$$z_1 = x_1 + iy_1 = r_1(\cos\theta_1 + i\sin\theta_1)$$

where x_1, y_1, r_1, θ_1 are real and r_1 positive, as the notation suggests. Then we can express the quotient z_1/z as follows:

(1) $$\frac{z_1}{z} = \frac{z_1\bar{z}}{|z|^2} = \frac{r_1}{r}[\cos(\theta_1 - \theta) + i\sin(\theta_1 - \theta)]$$

If we wish to write the conjugate of a composite expression we use longer horizontal lines. For instance, if we suppose, as usual that x and y are real,

$$\overline{z^2} = \overline{(x + iy)^2} = x^2 - y^2 - 2ixy.$$

With this notation, we may express concisely the most important rules concerning conjugate complex numbers:

$$\text{(X)} \quad \overline{a + b} = \bar{a} + \bar{b} \quad \text{(XI)} \quad \overline{ab} = \bar{a}\,\bar{b}$$

$$\text{(XII)} \quad \bar{\bar{a}} = a$$

The rules are easy to verify and easy to describe intuitively: *the symmetrical to a composite number is obtained by composing the symmetricals to the component numbers.* This statement takes a more usual but somewhat less-intuitive form is instead of "symmetrical to" we say "conjugate of." The statement remains valid even if the composite number is derived from t components, not by a single addition or multiplication, but by any combin tion of additions, subtractions, multiplications, and divisions. For instanc

$$\bar{a} - \frac{b}{c} = \overline{a - \frac{b}{\bar{c}}}$$

1.8.1   Verify that

$$\mathbf{R}z = \tfrac{1}{2}(z + \bar{z})$$

$$\mathbf{I}z = \frac{1}{2i}(z - \bar{z})$$

$$|\bar{z}| = |z|$$

$$\arg \bar{z} = -\arg z$$

1.8.2   Show that

(a) $\overline{z_1 z_2} = \bar{z_1}\bar{z_2}$ (b) $\overline{(z^4)} = (\bar{z})^4$

(c) $\overline{\left(\dfrac{z_1}{z_2}\right)} = \dfrac{\bar{z_1}}{\bar{z_2}}$ (d) $\overline{(i\bar{z})} = -iz$

1.8.3   Solve Example 1.6.9, starting from

$$\frac{1}{z^n} = \frac{\bar{z}^n}{(z\bar{z})^n}$$

1.8.4   Verify that

$$(1 + \bar{a}\,b)(1 + a\,\bar{b}) - (a + b)(\bar{a} + \bar{b}) = (1 - a\,\bar{a})(1 - b\,\bar{b})$$

1.8.5   Assuming that $|z_1| < 1$ and $|z_2| < 1$, show that (Example 1.8.4):

$$|z_1 + z_2| < |1 + \bar{z}_1 z_2| \qquad\qquad (S)$$

1.8.6   Express the reflection (the mirror image) of z in the lines

(a) $x = 0$
(b) $y = x$
(c) $y = kx$ k real

in terms of $\bar{z}$. [If w is the image point required by (c), and arctan $k = \alpha$,

$$w(\cos \alpha - i \sin \alpha) = \overline{z(\cos \alpha - i \sin \alpha)}]$$

1.8.7   Let z be a point inside the unit square with vertices at $0, 1, 1 + i, i$.

(a) Express the mirror images of z with respect to each of the four sides in terms of $\bar{z}$.

(b) Each image can then be reflected in the other sides, yielding an infinite set of reflections. Find these points for the special case in which z lies at the center of the square.

1.9 VECTORIAL OPERATIONS

We are given certain vectors in the plane, represented by directed line-segments. These vectors may have some physical significance as displacements, velocities, and forces, for example. We may represent these vectors by complex numbers $a, b, \ldots z$; in order to do so we must choose a system of coordinates. Physical phenomena do not depend, however, on our choice of the coordinate system. Therefore, we should pay particular attention to *operations with vectors whose results are independent of the choice of the coordinate system; we call* such operations *vectorial operations*.

For instance, addition is a vectorial operation. In fact, being given the vectors a and b, we find $a + b$ as the third side of a triangle whose other sides are a and b. In constructing this triangle, we make no use whatever of the coordinate system and could ignore its position altogether (Figure 1.6).

There are a few other simple vectorial operations. Let

$$\alpha = \arg a \qquad \beta = \arg b$$

Then

$$a = |a| (\cos \alpha + i \sin \alpha), \qquad b = |b| (\cos \beta + i \sin \beta)$$

$$\bar{a}b = |a| |b| [\cos (\beta - \alpha) + i \sin (\beta - \alpha)]$$

and, therefore,

(1) $$\mathbf{R}\bar{a}b = |a| |b| \cos (\beta - \alpha)$$

(2) $$\mathbf{I}\bar{a}b = |a| |b| \sin (\beta - \alpha)$$

The operations expressed by (1) and (2) are vectorial operations.

In fact, the angle $\beta - \alpha$ is the angle between the vectors a and b; if we wish to rotate vector a so that it coincides in direction with b, we must rotate it through the angle $\beta - \alpha$. Thus (1) is the product of the lengths of both vectors multiplied by the cosine of the included angle; it has a significance independent of the choice of the coordinate system. The expression has physical significance if the vector a is a force acting on a particle that moves along b. Observe that $|a| \cos (\beta - \alpha)$ is the projection of vector a on the direction of vector b, or the component of the force a acting in the direction of the displacement b. Multiplying this component by the distance through which the particle is moved, $|b|$, we obtain (1) that is, therefore, the *work* done when a particle is moved along the displacement b by a force a.

If we construct a parallelogram whose sides are vectors a and b, and we consider b as the base, the altitude of the parallelogram is $|a| \sin (\beta - \alpha)$. Thus (2) is the *area of the parallelogram*, and so it has a significance independent of the choice of the coordinate system. The expression (2) has physical significance if vector a is the velocity of a uniformly moving fluid layer whose thickness measured at right angles to the plane is unity. For the sake of concreteness, let our plane be horizontal, and let the fluid glide over it. Let us pay attention to those particles of the fluid that cross the line segment b in unit time; they fill a parallelepiped whose base is a rectangle with sides $|b|$ and 1 and whose altitude is $|a| \sin (\beta - \alpha)$. Therefore, the *volume of the fluid crossing in unit time* is just (2). This volume of fluid is called the *flux* across b.

We may condense the interpretation of expressions (1) and (2) just discussed into the formula

$$(3) \qquad \bar{a}b = \mathbf{R}\bar{a}b + i\mathbf{I}\bar{a}b = \text{work} + i \text{ flux}$$

which is not quite orthodox but is easy to remember.

If the reader is familiar with the elements of vector analysis (which is not required), he may observe that $\mathbf{R}\bar{a}b$ is the scalar product of the vectors a and b, and $\mathbf{I}\bar{a}b$ is closely connected with their vector product.

For the discussion of the sign of $\mathbf{I}\bar{a}b$, see Example 1.9.1 and 1.9.2.

1.9.1   Lay down vector a, take its endpoint as the initial point of vector b, and complete the parallelogram of which a and b so constructed are consecutive sides. The area of this parallelogram, taken with a certain sign, is $\mathbf{I}\bar{a}b$. On what geometric feature does the sign depend? See Figure 1.8.   (S)

1.9.2   If a and b are interchanged, $\mathbf{R}\bar{a}b$ remains unchanged, but $\mathbf{I}\bar{a}b$ changes sign. Visualize the geometric ground for this change: what is the difference between the two parallelograms one of which is so connected with $\mathbf{I}\bar{a}b$ as the other is with $\mathbf{I}\bar{b}a$?

1.9.3   Let a and b be complex numbers,

$$\arg a = \alpha \qquad \arg b = \beta$$

Figure 1.8

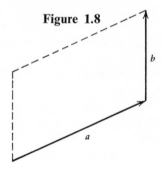

consider the triangle with vertices a, b, and 0, and show that the square on the side opposite to 0 is

$$(a - b)(\bar{a} - \bar{b}) = |a|^2 + |b|^2 - 2\,|a|\,|b|\cos(\beta - \alpha)$$

What theorem of elementary geometry is expressed by this equation?

1.9.4   How is the expression

$$(2a\bar{a}b\bar{b} - a^2\bar{b}^2 - \bar{a}^2b^2)/16$$

connected with the triangle considered in Example 1.9.3?   (S)

1.9.5   Prove that
$$(a + b)^2 + (a - b)^2 = 2a^2 + 2b^2$$

What is the geometrical interpretation?   (S)

1.9.6   Let $z_1, z_2, z_3, \ldots, z_n$ denote the vertices of a polygon, described in the positive sense. Show that the value of the expression

$$\tfrac{1}{2}I(z_2\bar{z}_1 + z_3\bar{z}_2 + \cdots + z_n\bar{z}_{n-1} + z_1\bar{z}_n)$$

remains unchanged by translations and rotations of the polygon. What is the geometric meaning of this value?   (S)

1.9.7   Consider the centroid of a triangle and the distances from the centroid to the vertices: let G stand for the sum of the squares of these three distances and S for the sum of the squares of the three sides of the triangle. Show that

$$S = 3G$$

(You are at liberty to choose the location of the triangle with respect to the coordinate system. What location may be the most advantageous?)   (S)

1.9.8 (continued)   We consider two mass distributions connected with the same triangle. Both have the same total mass and (as you should convince yourself) the same centroid.

1. The mass is distributed uniformly, with surface density 1, over the area of the triangle.
2. One third of the total mass is concentrated in each vertex of the triangle.

Let I stand for the polar moment of inertia in case 1, I^* in case 2. (The polar moment of inertia is computed with respect to an axis perpendicular to the plane of the triangle and passing through the centroid.) Show that

$$I^* = 4I$$

(Take for granted the expression $I = AS/36$ where A is the area of the triangle.) (S)

1.9.9   Let ρ satisfy the equation

$$\rho^2 + \rho + 1 = 0$$

Show that the quantity

$$Q = |a + \rho b + \rho^2 c|^2$$

remains unchanged by translations and rotations of the triangle with vertices a, b, c. (S)

1.9.10 (continued)   Express Q as a linear combination of A and I/A with numerical coefficients (notation of Example 1.9.8). (S)

1.9.11 (continued)   Check your result in the particular case of the equilateral triangle. (S)

1.9.12 (continued)   Of all triangles with a given area A the equilateral triangle has the minimum polar moment of inertia I. (S)

1.10   LIMITS

The reader should be familiar with the concept of limit in the real domain, at least to a certain extent. Then he can easily acquire the concept of limit in the complex domain, to the same extent, because (1) these concepts are closely analogous and (2) it is easy to reduce the case of complex numbers to that of real numbers.

(a) There are various sorts of limiting processes of which the simplest is concerned with an infinite sequence of numbers $z_1, z_2, z_3 \ldots$. These numbers, as the notation suggests, are assumed to be complex. We wish to understand the meaning of the sentence "z_n tends to z as n tends to infinity" or, what is the same thing, the meaning of the formula

(1)
$$\lim_{n \to \infty} z_n = z$$

The meaning is, roughly, that z_n becomes undistinguishable from z when n becomes large. More precisely, the difference $|z_n - z|$ is arbitrarily small when n is sufficiently large. Still more explicitly, the absolute value of the difference between a term of the sequence and the limit of the sequence, that is $|z_n - z|$, becomes and remains smaller than any preassigned positive quantity, however small, provided that n is sufficiently large.

We may express the same thing in geometrical terms if we represent all numbers concerned as points in the plane. What is the connection between a sequence of points, $z_1, z_2, z_3, \ldots$, and its limit point, z? The distance of a term of the sequence from the limit, $|z_n - z|$, becomes arbitrarily small when n is sufficiently large. In order to present this idea more clearly let us draw a small circle around the point z as center. If the points z_1, z_2, z_3, ... of the sequence have z as the limit point, then at most a finite number of them can lie outside the circle. An overwhelming majority of them, that is, an infinite number, must be inside the circle; and this is true for any circle about z, however small. Now let us suppose that we observe the points through a given means of observation such as a microscope; then the circle can be drawn so small that the infinite number of points remaining inside it become, for us, indistinguishable from the limit point z. If we wish to express intuitively the relative positions of the points z_1, z_2, z_3, ... and of their limit point z we may say that z_1, z_2, z_3, ... "cluster" around z.

(b) The relation expressed by (1) is equivalent to

$$\text{(2)} \qquad \lim_{n \to \infty} |z_n - z| = 0$$

and we can understand the meaning of Equation 1 if we are acquainted with complex numbers and with the concept of limit insofar as it is concerned with real numbers alone.

There is another way of explaining the relation (1) in terms of real quantities. We write

$$z_n = x_n + iy_n \qquad z = x + iy$$

with real x_n, y_n, x, y; then the one complex equation (1) is equivalent to the two real equations

$$\text{(3)} \qquad \lim_{n \to \infty} x_n = x \qquad \lim_{n \to \infty} y_n = y$$

and again, the understanding of these equations requires only a knowledge of limits as applied to real numbers.

Observe that

$$|z_n - z|^2 = (x_n - x)^2 + (y_n - y)^2$$

If the left-hand side of this equation is very small, both terms on the right-hand side must be very small, and the converse is also obvious. Thus we see that Equation 2 is actually equivalent to the two equations (3).

Example.   Since

$$(1 - z)(1 + z + \cdots + z^n) = 1 + z + \cdots + z^n - z - \cdots - z^n - z^{n+1}$$

$$= 1 - z^{n+1}$$

we have

$$\text{(4)} \qquad 1 + z + z^2 + \cdots + z^n = \frac{1 - z^{n+1}}{1 - z}$$

Let us examine whether the left-hand side tends to a limit as n tends to infinity. In fact,

$$1 + z + \cdots + z^n - \frac{1}{1-z} = -\frac{z^{n+1}}{1-z}$$

and the absolute value of the right-hand side is

$$\frac{1}{|1-z|} |z|^{n+1}$$

Now, this expression tends to zero when $n \to \infty$ *provided that* $|z| < 1$ since a high power of a proper fraction is very small. Using the definition of the limit as it was given in connection with (1), we may write our result in the form

$$\lim_{n \to \infty} (1 + z + z^2 + \cdots + z^n) = \frac{1}{1-z}$$

We mean the same thing when, using the more convenient notation of infinite series, as for real numbers, we write

(5)
$$1 + z + z^2 + z^3 + \cdots = \frac{1}{1-z}$$

we have supposed that $|z| < 1$.

   We can obtain further useful results from (5) by using the fact that this one complex equation is equivalent to two real equations. This principle can be applied to (5) just as it stands, but we obtain the result in a simpler form if we first transform slightly by multiplying both sides by 2 and then subtracting 1, getting

(6)
$$1 + 2z + 2z^2 + 2z^3 + \cdots = \frac{1+z}{1-z}$$

We now separate both sides (6) into their real and imaginary parts. To separate the left-hand side we write z in the form

$$z = r(\cos \theta + i \sin \theta)$$

Then by Example 1.6.6 we obtain

$$z^n = r^n(\cos n\theta + i \sin n\theta)$$

$$1 + 2r(\cos \theta + i \sin \theta) + \cdots + 2r^n(\cos n\theta + i \sin n\theta) + \cdots$$

(7)
$$= \frac{1 + r \cos \theta + ir \sin \theta}{1 - r \cos \theta - ir \sin \theta}$$

To separate the right-hand side into its real and imaginary parts we observe that it equals [compare Section 1.8 (1)]

$$\frac{(1 + r \cos \theta + ir \sin \theta)(1 - r \cos \theta + ir \sin \theta)}{(1 - r \cos \theta)^2 + r^2 \sin^2 \theta}$$

Now, equating the real and the imaginary parts in (7) and separating, we finally obtain

$$1 + 2r \cos \theta + 2r^2 \cos 2\theta + \cdots + 2r^n \cos n + \cdots = \frac{1 - r^2}{1 - 2r \cos \theta + r^2}$$

$$r \sin \theta + r^2 \sin 2\theta + \cdots + r^n \sin n\theta + \cdots = \frac{r \sin \theta}{1 - 2r \cos \theta + r^2}$$

These results hold provided $|z| < 1$, that is, provided $r < 1$. We could not obtain them with equal facility without using complex quantities.

1.10.1   Has the following sequence a limit and, if it has one, what is the limit?

(a) $1, i, -1, -i, 1, i, -1, -i, 1, \ldots$ (S)

(b) $1, \dfrac{i}{2}, -\dfrac{1}{3}, -\dfrac{i}{4}, \dfrac{1}{5}, \dfrac{i}{6}, -\dfrac{1}{7}, -\dfrac{i}{8}, \dfrac{1}{9}, \ldots$ (S)

(c) $\dfrac{1 + i}{2}, \left(\dfrac{1 + i}{2}\right)^2, \left(\dfrac{1 + i}{2}\right)^3, \ldots, \left(\dfrac{1 + i}{2}\right)^n, \ldots$

(d) $\dfrac{3 + 4i}{5}, \left(\dfrac{3 + 4i}{5}\right)^2, \left(\dfrac{3 + 4i}{5}\right)^3, \ldots, \left(\dfrac{3 + 4i}{5}\right)^n, \ldots$

1.10.2   For which values of z does

$$\lim_{n \to \infty} \left(\frac{z}{\bar{z}}\right)^n \text{ exist?}$$

1.10.3   Assuming the convergence of $\Sigma |a_n|$, prove the convergence of the five series

$$\Sigma(|a_n| + \mathbf{R}a_n), \ \Sigma(|a_n| - \mathbf{R}a_n), \ \Sigma(|a_n| + \mathbf{I}a_n), \ \Sigma(|a_n| - \mathbf{I}a_n), \ \Sigma a_n$$

Additional Examples and Comments on Chapter One

1.1   The mirror image of the face of a clock is so placed on the complex plane that the center of the face coincides with the origin, and when the hour hand points to z on the unit circle, the minute hand points to z^{12}. When both hands point in the same direction $z = z^{12}$ or, since $z \neq 0$,

$$z^{11} - 1 = 0$$

Express similarly, by an equation, that the two hands

(a) Point in opposite directions.
(b) Are perpendicular to each other.
(c) Include an angle equal to *any* multiple of 90°.

1.2 Compute $(2 + i)(3 + i)$ and prove that

$$\frac{\pi}{4} = \arctan \tfrac{1}{2} + \arctan \tfrac{1}{3}$$

1.3 Compute $(5 - i)^4(1 + i)$ and prove that

$$\frac{\pi}{4} = 4 \arctan \tfrac{1}{5} - \arctan \tfrac{1}{239}$$

(This relation, due to Machin, has been used to compute π with high accuracy: the series expansion of the first term on the right hand side is easy to evaluate, that of the second term converges very fast.)

1.4 Prove that

$$\frac{\pi}{4} = 3 \arctan \tfrac{1}{4} + \arctan \tfrac{1}{20} + \arctan \tfrac{1}{1985}$$

1.5 Derive from de Moivre's theorem (Example 1.6.6) the fact that $\cos n\,\theta$ is a polynomial of degree n in $\cos\theta$, and $\sin n\,\theta / \sin\theta$ is a polynomial of degree $n - 1$ in $\cos\theta$. (They are called Tchebycheff polynomials.)

1.6 Given z_1 and z_2 as fixed points, describe the locus of the point

$$z_1 + t(z_2 - z_1)$$

where t is variable and

(a) t is real
(b) $0 \leq t \leq 1$

1.7 Let z_1, z_2, and z_3 be the vertices of a triangle, and p_1, p_2, and p_3 nonnegative real numbers such that

$$p_1 + p_2 + p_3 = 1$$

Show that the point

$$z = p_1 z_1 + p_2 z_2 + p_3 z_3$$

lies in the interior or on the boundary of the triangle and, conversely, that any point z so situated can be represented in the given form with appropriate p_1, p_2, and p_3. (Example 1.6)   (S)

1.8   If, in Example 1.7, $p_1 = p_2 = p_3$, where is the point z?

1.9   Analogously to Example 1.7, characterize the interior of a convex quadrilateral with vertices z_1, z_2, z_3, and z_4.

1.10   The points represented by the complex numbers z_1, z_2, and 0 are noncollinear. Prove that it is possible to express any complex number z in the form

$$z = az_1 + bz_2$$

where a and b are real. Furthermore, this representation is unique.

1.11   If the points z_1, z_2, and z_3 are collinear, show that real numbers a, b, and c not all equal 0 exist so that

$$a + b + c = 0$$

$$az_1 + bz_2 + cz_3 = 0$$

1.12 THE TRIANGLE INEQUALITY

The fact

$$|z_1 + z_2| \leqq |z_1| + |z_2|$$

is obvious when we regard z_1, z_2, and $z_1 + z_2$ as three vectors forming a triangle. However, as there are occasions when such geometric arguments are less immediate, we must be prepared to deduce such inequalities by analytical means. We might well proceed this way:

$$
\begin{aligned}
|z_1 + z_2|^2 &= (z_1 + z_2)(\overline{z_1 + z_2}) \\
&= z_1\bar{z}_1 + z_1\bar{z}_2 + \bar{z}_1 z_2 + z_2\bar{z}_2 \\
&= |z_1|^2 + 2\mathbf{R}(z_1\bar{z}_2) + |z_2|^2 \\
&\leqq |z_1|^2 + 2\,|z_1\bar{z}_2| + |z_2|^2 \\
&= |z_1|^2 + 2\,|z_1|\,|z_2| + |z_2|^2 \\
&= (\,|z_1| + |z_2|\,)^2
\end{aligned}
$$

Try to see some of the simple geometry hidden by this algebraic approach.

1.13   The inequality

$$|\mathbf{R}z_1\bar{z}_2| \leqq |z_1|\,|z_2|$$

(cf. the derivation in Example 1.12) is equivalent to an inequality between real numbers (components of the vectors z_1 and z_2):

$$(x_1 x_2 + y_1 y_2)^2 \leq (x_1{}^2 + y_1{}^2)(x_2{}^2 + y_2)^2$$

(a) What is the geometric interpretation of this inequality?
(b) Could you extend it to vectors in three dimensional space?
(c) Could you extend it to n dimensions?

1.14   The inequality considered in Example 1.13 can be derived by consideration of the quadratic equation in t

$$(x_1 t + x_2)^2 + (y_1 t + y_2)^2 = 0$$

(a) How?
(b) Can you extend this derivation to n dimensions?

1.15   Prove the triangle inequality in n dimensions

$$[(x_1 + y_1)^2 + \cdots + (x_n + y_n)^2]^{1/2} \leq$$

$$[x_1{}^2 + \cdots + x_n{}^2]^{1/2} + [y_1{}^2 + \cdots + y_n{}^2]^{1/2}$$

1.16   Show that

$$|z_0 + z_1 + z_2 + \cdots + z_n| \geq |z_0| - |z_1| - |z_2| - \cdots - |z_n|$$

1.17   We consider the polynomial

$$P(z) = a_0 z^n + a_1 z^{n-1} + a_2 z^{n-2} + \cdots + a_n$$

and assume that $a_0 \neq 0$, $a_n \neq 0$. Show that

(a) the equation

$$|a_0| x^n + |a_1| x^{n-1} + \cdots + |a_{n-1}| x - |a_n| = 0$$

has just one positive root r, and

(b) $P(z)$ does not vanish in the circle $|z| < r$.

1.18   If the coefficients of the equation

$$a_0 z^n + a_1 z^{n-1} + a_2 z^{n-2} + \cdots + a_n = 0$$

are positive and nondecreasing, that is, $0 < a_0 \leq a_1 \leq a_2 \leq \cdots \leq a_n$ the equation has no root in the circle $|z| < 1$.

1.19   If the coefficients of the equation

$$a_0 z^n + a_1 z^{n-1} + a_2 z^{n-2} + \cdots + a_n = 0$$

are positive, and r and R denote the minimum and the maximum among the n ratios

$$\frac{a_1}{a_0}, \frac{a_2}{a_1}, \frac{a_3}{a_2}, \ldots, \frac{a_n}{a_{n-1}}$$

the equation has no root *outside* the annular region

$$r \leqq |z| \leqq R$$

(To remember this result, observe that this annular region contains the roots of all the equations

$$a_0 z + a_1 = 0, \ a_1 z + a_2 = 0, \ldots, a_{n-1} z + a_n = 0)$$

1.20 Describe the region formed by those points z for which the term $z^n/n!$ is greater in absolute value than the other terms of the infinite series

$$1 + \frac{z}{1!} + \frac{z^2}{2!} + \frac{z^3}{3!} + \cdots$$

1.21 Let A and B denote complex constants, and z a complex variable. Assume that $B \neq 0$, and show that $A + \bar{A} + B\bar{z} + \bar{B}z = 0$ is the equation of an (arbitrary) straight line.

1.22 Let A, B, and C denote complex constants, and z a complex variable. Show that

$$A + \bar{A} + B\bar{z} + \bar{B}z + (C + \bar{C})z\bar{z} = 0$$

is the equation of an (arbitrary) circle. (Degenerate cases, especially straight lines, are included.)

1.23 Show that the three points a, b, and c are collinear (belong to the same straight line) if, and only if

$$\begin{vmatrix} 1 & a & \bar{a} \\ 1 & b & \bar{b} \\ 1 & c & \bar{c} \end{vmatrix} = 0$$

1.24 Show that the four points a, b, c, and d are concyclic (belong to the same circle) if, and only if

$$\begin{vmatrix} 1 & a & \bar{a} & a\bar{a} \\ 1 & b & \bar{b} & b\bar{b} \\ 1 & c & \bar{c} & c\bar{c} \\ 1 & d & \bar{d} & d\bar{d} \end{vmatrix} = 0$$

1.25   The three complex numbers z, z_1, z_2, determine the three vertices of a triangle, and w, w_1, w_2 the corresponding vertices of another triangle. The two triangles are similar if, and only if

$$\begin{vmatrix} 1 & 1 & 1 \\ z & z_1 & z_2 \\ w & w_1 & w_2 \end{vmatrix} = 0$$

The case of degenerate triangles must be appropriately interpreted.

1.26   In the equation

$$c_{11}z_1 + c_{12}z_2 + \cdots + c_{1n}z_n = 0$$
$$c_{21}z_1 + c_{22}z_2 + \cdots + c_{2n}z_n = 0$$
$$\cdot \qquad \cdot \qquad \cdot$$
$$c_{n1}z_1 + c_{n2}z_2 + \cdots + c_{nn}z_n = 0$$

the coefficients and the unknowns are complex numbers:

$$c_{\lambda\mu} = a_{\lambda\mu} + ib_{\lambda\mu} \qquad z_\mu = x_\mu + iy_\mu$$

In order that these equations have not only the trivial solution $z_1 = z_2 = \cdots = z_n = 0$, that is, $x_1 = x_2 = \cdots x_n = y_1 = y_2 = \cdots = y_n = 0$, it is necessary and sufficient that the determinant $|c_{\lambda\mu}|_1^n = 0$. This determinantal equation gives two equations between the $2n^2$ real numbers $a_{\lambda\mu}, b_{\lambda\mu}$. On the other hand, we can write the original system as $2n$ linear homogeneous equations for the $2n$ unknowns. The necessary and sufficient condition for the existence of nontrivial solutions in this cases is the vanishing of a single determinant, that is, one equation connecting the coefficients $a_{\lambda\mu}, b_{\lambda\mu}$. In what way are these results consistent?

1.27   Show that z_1, z_2, z_3 form an equilateral triangle if, and only if,

$$z_1^2 + z_2^2 + z_3^2 = z_2z_3 + z_3z_1 + z_1z_2$$

1.28   On each side of a given (arbitrary) triangle describe an exterior equilateral triangle. Show, by complex numbers, that the centers of these three equilateral triangles are the vertices of a fourth equilateral triangle. (Let $2a$, $2b$, and $2c$ be the vertices of the given triangle, counterclockwise ordered. Express the sides of the triangle whose vertices are the three centers as vectors in terms of

$$u = b - c, \qquad v = c - a, \qquad w = a - b$$

and of ρ which has one of the values considered in Example 1.9.9. Similar to Example 1.7.4.)

1.29   If a quadrilateral inscribed in a circle is such that the centroid of the vertices lies at the center of the circle, then the quadrilateral is a rectangle.

(Choose an advantageous location for the centroid and consider the biquadratic equation of which the four vertices are the roots.)

1.30 PROPOSITION DUE TO GAUSS

Let OA, OB, and OC be the three edges of a cube at its vertex O. Let O coincide with the origin of the complex plane, and let the complex numbers a, b, and c mark the projections onto the complex plane of the three vertices A, B, and C respectively. For this situation, it is necessary and sufficient that the three complex numbers satisfy the equation.

$$a^2 + b^2 + c^2 = 0$$

(Some knowledge of the algebra of orthogonal transformations may help.)

1.31   Assuming that there are three noncollinear points a, b, and c such that the three limits

$$\lim_{n \to \infty} |z_n - a| \qquad \lim_{n \to \infty} |z_n - b| \qquad \lim_{n \to \infty} |z_n - c|$$

exist, show that $\lim_{n \to \infty} z_n$ exists too.

(Could you draw the same conclusion *without* assuming that a, b, and c are noncollinear?)

1.32   Show that

$$|ac - \bar{b}d|^2 + |\bar{a}d + bc|^2 = (|a|^2 + |b|^2)(|c|^2 + |d|^2)$$

1.33   *Quaternions* are obtained by a further extension of the notion of number. We may pass from complex numbers to quaternions just so as we have passed from real numbers to complex numbers: the analogy is strong, although not complete. In spelling out this analogy, the following examples may also shed some light on the nature of complex numbers. The odd numbered examples 1.33 (the present one), 1.35, 1.37, 1.39, and 1.41 form a sequence, and refer to each other; the examples 1.34, 1.36, 1.38, 1.40, and 1.42 are similarly connected, and form a parallel sequence. Examples 1.43 to 1.46 deal with another aspect of the theory of quaternions.

Letters denote real numbers. A *complex number* is defined as an ordered pair of real numbers (x, y). Equality, addition, and multiplication of complex numbers are defined as follows:

$$(a, b) = (c, d) \qquad \text{iff} \qquad a = c, \ b = d$$
$$(a, b) + (c, d) = (a + c, b + d)$$
$$(a, b)(c, d) = (ac - bd, ad + bc)$$

Show that

(a) $(a, b)(c, 0) = (ac, bc)$
(b) $(a, b) = (1, 0)(a, 0) + (0, 1)(b, 0)$

1.34   Letters denote complex numbers. A *quaternion* is defined as an ordered pair of complex numbers (z, w). Equality, addition, and multiplication are defined as follows:

$$(a, b) = (c, d) \quad \text{iff} \quad a = c, \ b = d$$
$$(a, b) + (c, d) = (a + c, b + d)$$
$$(a, b)(c, d) = (ac - \bar{b}d, \bar{a}d + bc)$$

Show that

(a) $(a, b)(c, 0) = (ac, bc)$
(b) $(a, b) = (1, 0)(a, 0) + (0, 1)(b, 0)$

1.35   We make a single exception; just the letter i will be used to denote a complex number. We set
$$(0, 1) = i$$

Show that, with this notation,

$$ii = (-1, 0)$$

$$(a, 0)i = i(a, 0)$$

$$(a, b) = (a, 0) + i(b, 0)$$

1.36   We make a single exception: just the letter j will be used to denote a quaternion. We set
$$(0, 1) = j$$

Show that, with this notation,

$$jj = (-1, 0)$$

$$(a, 0)j = j(\bar{a}, 0)$$

$$(a, b) = (a, 0) + j(b, 0)$$

1.37   We define $\overline{(a, b)}$, the conjugate to the complex number (a, b) by

$$\overline{(a, b)} = (a, -b)$$

Then we define $|(a, b)|$, the absolute value of the complex number (a, b), as the nonnegative square root of the nonnegative real number $|(a, b)|^2$ given by

$$(a, b)\overline{(a, b)} = (|a, b|^2, 0)$$

Express $|(a, b)|^2$ in terms of a and b.

1.38   We define $\overline{(a, b)}$, the conjugate to the quaternion (a, b) by

$$\overline{(a, b)} = (\bar{a}, -b)$$

Then, we define $|(a, b)|$, the absolute value of the quaternion (a, b), as the nonnegative square root of the nonnegative real number $|(a, b)|^2$ given by

$$(a, b)\overline{(a, b)} = (|(a, b)|^2, 0)$$

Express $|(a, b)|^2$ in terms of a and b.

1.39   We introduce a new operation: multiplication of a complex number (a, b) by a real number c, defined by

$$(a, b)c = (ac, bc)$$

[Observe that $(a, b)c = (a, b)(c, 0)$.] Then the complex numbers appear as "vectors" which are "linear combinations" of two "fundamental" vectors, $(1, 0)$ and $(0, 1)$

$$(a, b) = (1, 0)a + (0, 1)b$$

See Example 1.33. We shall write this last equation in the conventionally abbreviated form (see Example 1.35)

$$(a, b) = a + ib$$

Convince yourself that all the rules (I) to (XII) stated in Sections 1.1 and 1.8 remain valid for the complex numbers so introduced.

1.40   We introduce a new operation: multiplication of a quaternion (a, b) by a complex number c, defined by

$$(a, b)c = (ac, bc)$$

[Observe that $(a, b)c = (a, b)(c, 0)$.] Then the quaternions appear as "vectors," which are "linear combinations" of two "fundamental" vectors, $(1, 0)$ and $(0, 1)$

$$(a, b) = (1, 0)a + (0, 1)b$$

See Example 1.34. We shall write this equation in the conventionally abbreviated form (see Example 1.36)

$$(a, b) = a + jb$$

Convince yourself that the rules (I) to (XII) stated in Sections 1.1 and 1.8 remain valid, with two exceptions, (II) and (XI): the multiplication of quaternions is *noncommutative*. As a sort of compensation, we have the rule, which you should prove:

$$\overline{(a + jb)(c + jd)} = \overline{(c + jd)} \cdot \overline{(a + jb)}$$

1.41   The task of verification imposed by Example 1.39 is considerably alleviated if you are acquainted with matrix algebra. Then you can define a complex number as a matrix with real elements:

$$(a, b) = \begin{pmatrix} a & -b \\ b & a \end{pmatrix}$$

By this definition, the rules of Example 1.33 and the rules (I) to (XII) follow from well-known rules on matrices. What is the determinant of the above matrix?

1.42   The task of verification imposed by Example 1.40 is considerably alleviated if you are acquainted with matrix algebra. Then you can define a quaternion as a matrix with complex elements:

$$(a, b) = \begin{pmatrix} a & -\bar{b} \\ b & \bar{a} \end{pmatrix}$$

By this definition, the rules of Example 1.34, and the rules (I) to (X) with the exception of (II), and also (XII), and the rule stated in Example 1.40 follow from well-known rules on matrices. What is the determinant of the above matrix?

1.43   Define the quaternion k by the equation

$$k = j(-i)$$

and show that

$$jk = -kj = i \qquad ki = -ik = j \qquad ij = -ji = k$$
$$ii = jj = kk = -1$$

1.44   Let s, x, y, and z denote real numbers, and put

$$a = s + ix, \qquad b = y - iz$$

so that, by Example 1.43,

$$a + jb = s + ix + jy + kz$$

Of this quaternion $a + jb$

s is termed the *scalar* part, and

$ix + jy + kz$ the *vectorial* part

Show that

$$|a + jb|^2 = s^2 + x^2 + y^2 + z^2$$

1.45   Find the scalar part and the vectorial part of the product

$$(ix + jy + kz)(ix' + jy' + kz')$$

where x', y', z', just as x, y, z, are real. (Do you recognize the result?)

1.46   Prove the identity in real numbers

$$(s^2 + x^2 + y^2 + z^2)(s'^2 + x'^2 + y'^2 + z'^2)$$

$$= (ss' - xx' - yy' - zz')^2 + (sx' + xs' + yz' - zy')^2$$

$$+ (sy' + ys' + zx' - xz')^2 + (sz' + zs' + xy' - yx')^2$$

which is due to Euler and analogous to

$$(x^2 + y^2)(x'^2 + y'^2) = (xx' - yy')^2 + (xy' + yx')^2$$

HINTS AND SOLUTIONS FOR CHAPTER ONE

1.3.1 $|2 + 3i| = \sqrt{13}$ $\arg(2 + 3i) = \arctan \frac{3}{2}$

1.3.4 $0 \leq (x - y)^2$ $2xy \leq x^2 + y^2$

$$(x + y)^2 \leq 2x^2 + 2y^2$$

$$\frac{|x| + |y|}{\sqrt{2}} \leq \sqrt{x^2 + y^2} = |z|$$

1.5.3 $z = 0$, z_1, z_2 are collinear

1.6.1   100

1.6.7 $\cos 2\theta + i \sin 2\theta = (\cos \theta + i \sin \theta)^2 = \cos^2 \theta - \sin^2 \theta + 2i \cos \theta \sin \theta$

1.6.10 $z_1 + \frac{1}{3}(z_2 - z_1)$

1.7.1   (a) The interior of a circle, center at $z = -\frac{2}{3}$, radius $\frac{1}{3}$.
(h) The interior of an ellipse, foci at ± 1, axes 4(major) and $2\sqrt{3}$ (minor).

1.7.6   The midpoint of side b, c is $(b + c)/2$, and the point $\frac{1}{3}$ the distance toward a on the line joining $(b + c)/2$ to a is $(a + b + c)/3$; use symmetry.

1.8.5   From Example 1.8.4

$$|z_1 + z_2|^2 + (1 - |z_1|^2)(1 - |z_2|^2) = (1 + \bar{z}_1 z_2)(1 + z_1 \bar{z}_2)$$

Thus $|z_1 + z_2| < |1 + \bar{z}_1 z_2|$, $|z_1| < 1$, $|z_2| < 1$

1.9.1 $\mathbf{I}\bar{a}b = |a|\,|b|\,\sin(\beta - \alpha)$

The area of the oriented parallelogram (positive if the shortest rotation from a to b is in the positive sense.)

Chapter TWO | COMPLEX FUNCTIONS

Complex functions of a complex variable can be obtained by extension of the usual analytic expressions to complex numbers, and can be used in problems dealing with maps or two dimensional vector fields. Accordingly we have to survey the subject from three different viewpoints in succession.

2.1 EXTENSION TO THE COMPLEX DOMAIN

The most useful functions of a real variable are those that can be represented by some simple formula, such as

$$y = 1 - x^2 \qquad y = \frac{1}{x} \qquad y = \arctan x$$

From our point of view, there is a great difference between the first two examples and the last one; it is much easier to extend the first two functions to the complex domain than the last one.

Dealing with real functions of a real variable, we usually denote the independent variable by x and the dependent variable by y. It is convenient to emphasize that we deal with complex functions of a complex variable by using other letters. Let z stand for the independent complex variable and w for the dependent complex variable.* With this notation, we may consider the expressions

$$(1) \qquad\qquad w = 1 - z^2 \qquad w = \frac{1}{z}$$

for arbitrary complex z (excepting, in the second case, the value $z = 0$ which makes the denominator vanish); to each value of z considered, there

---

* We may easily remember these letters with a little knowledge of German: z means "Zahl" (*number*) and w means "Wert" (the corresponding *value*).

35

corresponds a uniquely determined value w that we can compute, since we know how to add, subtract, multiply, and divide complex numbers. Therefore, each formula (1) represents a well-defined function of the complex variable.

Our examples (1) belong to the simplest type of functions that can be extended to the complex domain. Let $a_0, a_1, \ldots, a_m$ denote any complex (or real) constants. The expression

$$(2) \qquad w = a_0 + a_1 z + a_2 z^2 + \cdots + a_m z^m$$

is a *polynomial* in z of degree m provided that $a_m \neq 0$. It yields a well-defined value w, which we can compute by addition, subtraction, and multiplication, for any complex value z, and so it represents a function of z defined for all complex values, or in the *whole plane*. The last phrase suggests that we should represent the variable complex number z as a "moving" point, ranging over the whole plane.

Let also $b_0, b_1, \ldots, b_n$ be given complex or real constants, $b_n \neq 0$. The *rational fraction*

$$(3) \qquad w = \frac{a_0 + a_1 z + \cdots + a_m z^m}{b_0 + b_1 z + \cdots + b_n z^n}$$

is a quotient of two polynomials. It yields a well-defined value, w, which we can compute by the four fundamental arithmetical operations except when z is a root of the denominator. Except for the points corresponding to these roots, the formula (3) defines w as a rational function of z in the whole plane. A polynomial such as (2) may be considered as an extreme special case of a rational function in which the degree, n, of the denominator ((3)) is equal to 0.

The first example (1) is a polynomial, the second one a rational fraction. It is more difficult to extend to the complex domain functions that are not rational. After due preparation, we shall discuss arctan z in Section 2.8.

2.2 EXPONENTIAL FUNCTION

The function e^x, where $e = 2.71 \cdots$ is the base of natural logarithms, is called the exponential function; it is sometimes convenient to write

$$(1) \qquad e^x = \exp(x)$$

Wishing to extend the definition of the exponential function to complex numbers we cannot start from the expression (1) because we have not yet considered powers with complex exponents. The best we can do is start from the well-known expansion of e^x into powers of x and replace x by z. Thus we define the exponential function for complex values of the variable by the series

(2)
$$\exp(z) = 1 + \frac{z}{1!} + \frac{z^2}{2!} + \frac{z^3}{3!} + \cdots$$

(We postpone considerations of convergence; see Examples 2.1 and 2.2.)

If x and x' are real numbers, then

(3)
$$e^x \cdot e^{x'} = e^{x+x'}$$

by the ordinary rule for the multiplication of powers with the same base. This formula leads us to examine the product $\exp(z) \cdot \exp(z')$ where z and z', as the choice of the letters suggests, denote arbitrary complex numbers. By the definition (2)

$$\exp(z)\exp(z') = \left(1 + \frac{z}{1!} + \frac{z^2}{2!} + \cdots\right)\left(1 + \frac{z'}{1!} + \frac{z'^2}{2!} + \cdots\right)$$

$$= 1 + \frac{z}{1!} + \frac{z^2}{2!} + \frac{z^3}{3!} + \frac{z^4}{4!} + \cdots$$

$$+ \frac{z'}{1!} + \frac{zz'}{1!\,1!} + \frac{z^2z'}{2!\,1!} + \frac{z^3z'}{3!\,1!} + \cdots$$

$$+ \frac{z'^2}{2!} + \frac{zz'^2}{1!\,2!} + \frac{z^2z'^2}{2!\,2!} + \cdots + \frac{z'^3}{3!} + \frac{zz'^3}{1!\,3!} + \cdots$$

$$+ \frac{z'^4}{4!} + \cdots + \cdots$$

$$= 1 + \frac{1}{1!}(z + z') + \frac{1}{2!}(z^2 + 2zz' + z'^2)$$

$$+ \frac{1}{3!}(z^3 + 3z^2z' + 3zz'^2 + z'^3)$$

$$+ \frac{1}{4!}(z^4 + 4z^3z' + 6z^2z'^2 + 4zz'^3 + z'^4) + \cdots$$

$$= 1 + \frac{z + z'}{1!} + \frac{(z + z')^2}{2!} + \frac{(z + z')^3}{3!} + \frac{(z + z')^4}{4!} + \cdots$$

The last line represents $\exp(z + z')$. In order to obtain this result, we have collected in the product of the two infinite series the terms appropriately and we have made use of the binomial theorem. (For discussion of the convergence, refer to 3.10.) We have obtained

$$(4) \qquad \exp(z)\exp(z') = \exp(z+z')$$

for arbitrary complex numbers z and z'.

After this result we should not shrink from defining e^z for arbitrary complex z. We identify e^z with $\exp(z)$ and write henceforward

$$(5) \qquad 1 + \frac{z}{1!} + \frac{z^2}{2!} + \frac{z^3}{3!} + \cdots = e^z$$

This definition is reasonable because it implies the generalization of (2), that is, see (4),

$$(6) \qquad \boxed{e^z e^{z'} = e^{z+z'}}$$

We put this formula in a box to emphasize its importance.

2.2.1   Bearing in mind the definition of e^z for complex values of z, prove:

(a) $e^0 = 1$

(b) $e^{-z} = \dfrac{1}{e^z}$ $\qquad\qquad\qquad\qquad\qquad\qquad\qquad\qquad$ (S)

2.3   TRIGONOMETRIC FUNCTIONS

We now apply the method that has succeeded so well in the case of the exponential function to the trigonometric functions sine and cosine. We define these functions for arbitrary complex z by the expansions

$$(1) \qquad \cos z = 1 - \frac{z^2}{2!} + \frac{z^4}{4!} - \frac{z^6}{6!} + \cdots$$

$$(2) \qquad \sin z = z - \frac{z^3}{3!} + \frac{z^5}{5!} + \frac{z^7}{7!} + \cdots$$

It follows immediately that

$$(3) \qquad \cos(-z) = \cos z \qquad \sin(-z) = -\sin z$$

since the series (1) contains only even powers of z and (2) only odd powers.

The series (1) and (2) appear to be related to the exponential series, Section 2.2 (5). We cannot help observing a certain regularity if we write the three series as follows:

$$e^z = 1 + \frac{z}{1!} + \frac{z^2}{2!} + \frac{z^3}{3!} + \frac{z^4}{4!} + \frac{z^5}{5!} + \cdots$$

$$\cos z = 1 - \frac{z^2}{2!} + \frac{z^4}{4!} - \cdots$$

$$\sin z = \frac{z}{1!} - \frac{z^3}{3!} + \frac{z^5}{5!} + \cdots$$

We may observe that the signs in the last two series recur periodically with the period 4. If we observe also that the powers of i have the same sort of periodicity:

$$i = i, \quad i^2 = -1, \quad i^3 = -i, \quad i^4 = 1, \quad i^5 = i, \quad i^6 = -1, \ldots,$$

then we come very near to the discovery of Euler, who, introducing the number i, found a remarkable relation between the functions $\cos z$, $\sin z$, and e^z.

The foregoing remarks suggest the idea of substituting iz for z in the exponential series. Disposing the terms appropriately, we write

$$e^{iz} = 1 - \frac{z^2}{2!} + \frac{z^4}{4!} - \frac{z^6}{6!} + \cdots + i\left(\frac{z}{1!} - \frac{z^3}{3!} + \frac{z^5}{5!} - \cdots\right)$$

Using (1) and (2), we obtain the very important formula

(4) $$\boxed{e^{iz} = \cos z + i \sin z}$$

This is Euler's theorem. It has many and various applications; its discovery was, in fact, one of the principal incentives to further investigation of complex numbers.

Substituting $-z$ for z in (4) and taking (3) in account, we obtain

(5) $$e^{-iz} = \cos z - i \sin z$$

Combining (4) and (5) by addition and subtraction, we obtain

(6) $$\cos z = \frac{e^{iz} + e^{-iz}}{2} \qquad \sin z = \frac{e^{iz} - e^{-iz}}{2i}$$

Thus we have obtained expressions for the trigonometric functions cosine and sine in terms of the exponential function. If we do not consider complex values of the variable, we have no such expressions.

By (4) and (5), or by (6), the study of trigonometric functions is reduced to that of the exponential function. For instance, we can deduce (3) from (6).

Trigonometric Functions   **39**

Again, by formula 2.2 (6) we have

$$e^{i(z+z')} = e^{iz} \cdot e^{iz'}$$

and hence, applying (4), we obtain

(7) $\cos(z+z') + i(\sin z + z')$

$$= (\cos z + i \sin z)(\cos z' + i \sin z')$$

$$= \cos z \cos z' - \sin z \sin z' + i(\sin z \cos z' + \cos z \sin z')$$

Similarly, starting from

$$e^{-i(z+z')} = e^{-iz} \cdot e^{-iz'}$$

and using (5), we obtain

(8) $\cos(z+z') - i \sin(z+z')$

$$= \cos z \cos z' - \sin z \sin z' - i(\sin z \cos z' + \cos z \sin z')$$

Combining (7) and (8) by addition and subtraction we obtain

(9) $\qquad\qquad \cos(z+z') = \cos z \cos z' - \sin z \sin z'$

(10) $\qquad\qquad \sin(z+z') = \sin z \cos z' + \cos z \sin z'$

These formulas have thus been shown to be true for arbitrary *complex* values of z and z'. (For real values of the variables see Section 2.5, Example 1.) It is worthwhile observing that we have derived them from the simpler analogous equation, Section 2.2 (6).

After the preceding results, we have no difficulty in extending the definition of the other trigonometric functions in a similar way. Thus, we have, for all complex values of z which do not make the denominator vanish,

(11) $\qquad\qquad \tan z = \dfrac{\sin z}{\cos z} = \dfrac{1}{i}\dfrac{e^{iz} - e^{-iz}}{e^{iz} + e^{-iz}}$

2.3.1 Compute e^z for $z = 0,\ i\pi/2,\ i\pi,\ 3i\pi/2$. (S)

2.3.2 Using Section 2.3 (6), verify the following trigonometric identities, valid for all complex numbers:

(a) $\sin\left(\dfrac{\pi}{2} - z\right) = \cos z$ $\qquad\qquad\qquad\qquad\qquad$ (S)

(b) $\sin^2 z + \cos^2 z = 1$

(c) $\tan 2z = \dfrac{2 \tan z}{1 - \tan^2 z}$

2.3.3   Use Section 2.3 (4) to show that

(a) $\overline{e^z} = e^{\bar{z}}$
(b) $|e^z| = e^x$

2.3.4   Show that $e^z \neq 0$ for every complex value of z.   (S)

2.3.5   Show that, for $z = x + iy$, x and y real,

$$|\sin z|^2 = \sin^2 x + \left(\frac{e^y - e^{-y}}{2}\right)^2.$$

2.3.6   Find all zeros of $\sin z$. (Do not neglect the possibility of complex zeros.)   (S)

2.4 CONSEQUENCES OF EULER'S THEOREM

We consider here some general consequences that we shall have to use continually and postpone until the next section certain exercises that are desirable but not indispensable.

(a) Supposing as usual

$$z = x + iy$$

where x and y are real, we have, by Sections 2.2 (6) and 2.3 (4)

(1) $$e^z = e^{x+iy} = e^x \cdot e^{iy}$$

$$= e^x \cos y + ie^x \sin y$$

Using this formula and the usual tables of logarithms and trigonometric functions, we can compute numerically the real and imaginary parts of the exponential function for any given complex value of z.

(b) We know that the trigonometric functions cosine and sine, considered for real values of the variable, have the common period 2π, that is

$$\cos (y + 2\pi) = \cos y, \qquad \sin (y + 2\pi) = \sin y.$$

But, if we change z into $z + 2\pi i$, the real part x remains unchanged, and only the imaginary part y is changed into $y + 2\pi$. Therefore, by (1)

(2) $$e^{z+2\pi i} = e^z$$

Thus the exponential function is periodic; it has the purely imaginary period $2\pi i$.

It is worthwhile to show the same fact in a slightly different manner. If n is an integer.

$$\cos 2n\pi = 1 \qquad \sin 2n\pi = 0$$

and therefore

$$(3) \qquad e^{i2n\pi} = \cos 2n\pi + i \sin 2n\pi = 1$$

Hence

$$(4) \qquad e^{z+2n\pi i} = e^z \cdot e^{2n\pi i} = e^z$$

This last equation is apparently more general than (2) which is contained in it as a special case for $n = 1$. But we may easily derive (4) from (2) by repeated application. For instance,

$$e^{z+4n\pi i} = e^{(z+2n\pi i)+2n\pi i} = e^{z+2n\pi i} = e^z$$

We obtain similarly that

$$e^{z-2\pi i} = e^{(z-2\pi i)+2\pi i} = e^z$$

and so on.

(c) Since

$$\cos \pi = -1 \qquad \sin \pi = 0$$

we have

$$e^{i\pi} = \cos \pi + i \sin \pi = -1$$

We may remember this result in the form

$$(5) \qquad e^{i\pi} + 1 = 0$$

Somebody called this the "most beautiful" formula since, he said, it combines the five "most important" numbers, 0, 1, i, π, and e.

(d) If we write as usual

$$z = x + iy \qquad x = r \cos \theta \qquad y = r \sin \theta$$

then, by 2.3 (4),

$$(6) \qquad z = re^{i\theta}$$

We shall make extensive use of this concise expression of a complex number in polar coordinates, which we may also write in the form

$$(7) \qquad z = |z| \, e^{i \arg z}$$

(e) If, in formula (6), we take $r = 1$, we obtain the complex number $e^{i\theta}$, with real θ. The vector $e^{i\theta}$ is a *unit vector*, that is, its length is 1; unit vectors are useful in specifying directions. The point $e^{i\theta}$ is a point on the circle with center 0 and radius 1; this circle is called the *unit circle*.

2.4.1 Compute numerically (a) e^{13i} (S), (b) e^{2+13i}, (c) e^{-1+2i}.

2.4.2 Show that $\sin (z + 2\pi) = \sin z$ also for complex values of z. (S)

2.4.3   Show that tan z, as a function of the complex variable z, is periodic with the period π.

2.4.4   Write $3 - 2i$ in the form Section 2.4 (6).

2.4.5   Find all solutions of the equation

$$e^z = 1 + i \qquad\qquad\qquad (S)$$

2.4.6   Find all solutions of the equation

$$\sin z = i$$

2.4.7   Find $\mathbf{R}(e^{iz^2})$.   (S)

2.4.8   Find $\mathbf{I}(e^{\sin z})$.

2.4.9   For which complex numbers does the following inequality hold?

$$|e^{-iz}| < 1$$

2.5 FURTHER APPLICATIONS OF EULER'S THEOREM

The formulas 2.2 (6) and 2.3 (4) (extension of the ordinary rule for multiplying powers to the case of complex exponents, and Euler's theorem) are extremely useful in dealing with trigonometric functions, even if we consider such functions only for real values of the variable. We shall discuss a few examples. If the reader has time, he should also solve some of the similar exercises 2.5.1 to 2.5.10 at the end of this section.

Example 1.   We wish to derive the well-known formulas for $\cos(\alpha + \beta)$ and $\sin(\alpha + \beta)$ from Sections 2.2 (6) and 2.3 (4) but, in contrast with Section 2.3, we *restrict ourselves to real values*. This is suggested by the notation; we write α and β instead of z and z', which we used in Section 2.3. We proceed as follows:

$$\cos(\alpha + \beta) + i\sin(\alpha + \beta)$$

$$= e^{i(\alpha+\beta)}$$

$$= e^{i\alpha} \cdot e^{i\beta}$$

$$= (\cos\alpha + i\sin\alpha)(\cos\beta + i\sin\beta)$$

$$= \cos\alpha\cos\beta - \sin\alpha\sin\beta + i(\sin\alpha\cos\beta + \cos\alpha\sin\beta)$$

Comparing the first line with the last line, and separating the real part from the imaginary part, we obtain the desired formulas.

This derivation supposes that the values involved are real but, nevertheless, the result remains valid even if the values involved are complex as shown in Section 2.3. The following examples are of a similar nature but we shall not insist on repeating our remark. Later, we shall be in a better position to understand the true reason of such happenings.

Example 2.   Express $\sin 3\theta$ in terms of $\sin \theta$. We appeal again to our basic formulas 2.2 (6) and 2.3 (4).

$$\cos 3\theta + i \sin 3\theta = e^{i3\theta} = e^{i\theta} \cdot e^{i\theta} \cdot e^{i\theta}$$

$$= (e^{i\theta})^3 = (\cos \theta + i \sin \theta)^3$$

$$= \cos^3 \theta + 3i \cos^2 \theta \sin \theta - 3 \cos \theta \sin^2 \theta - i \sin^3 \theta$$

Equating the imaginary parts of the first and last terms, we have

$$\sin 3\theta = 3 \cos^2 \theta \sin \theta - \sin^3 \theta$$

$$= 3 \sin \theta - 4 \sin^3 \theta$$

which is the desired result; we used the relation

$$\cos^2 \theta = 1 - \sin^2 \theta$$

Example 3.   Express $\cos^4 \theta$ as a linear combination of the cosines of the multiples of θ, that is, of 1, $\cos \theta$, $\cos 2\theta$, ...

Using Section 2.3 (6), we have

$$\cos^4 \theta = \left(\frac{e^{i\theta} + e^{-i\theta}}{2}\right)^4$$

$$= \tfrac{1}{16}(e^{4i\theta} + 4e^{2i\theta} + 6 + 4e^{-2i\theta} + e^{-4i\theta})$$

$$= \tfrac{3}{8} + \tfrac{1}{2} \cos 2\theta + \tfrac{1}{8} \cos 4\theta$$

we used again Section 2.3 (6). The expression we have obtained for $\cos^4 \theta$ is useful when we wish to calculate the integral of this function.

Example 4.   Sum the series

$$1 + 2 \cos \theta + 2 \cos 2\theta + \cdots + 2 \cos n\theta$$

To solve this problem, which is important in the theory of Fourier series, we appeal to Section 2.3 (6). Changing the order of the terms, we may write the proposed expression in the form:

$$e^{-in\theta} + \cdots + e^{-i\theta} + 1 + e^{i\theta} + \cdots + e^{in\theta}$$

This is a geometric series with initial term $e^{-in\theta}$, quotient $e^{i\theta}$, and $2n+1$ terms, whose sum is equal to

$$\frac{e^{-in\theta} - e^{i(n+1)\theta}}{1 - e^{i\theta}} = \frac{e^{i[n+(1/2)]\theta} - e^{-i[n+(1/2)]\theta}}{e^{(1/2)i\theta} - e^{-(1/2)i\theta}}$$

$$= \frac{\sin(2n+1)\dfrac{\theta}{2}}{\sin\dfrac{\theta}{2}}$$

The point is to transform the fraction so that it appears as a quotient of real quantities. We attained this end by multiplying the numerator and denominator by $e^{-i\theta/2}$ and using Section 2.3 (6). We can, however, proceed differently; the fraction in question is also equal to

$$\frac{(e^{-in\theta} - e^{i(n+1)\theta})(1 - e^{-i\theta})}{(1 - e^{i\theta})(1 - e^{-i\theta})} = \frac{\cos n\theta - \cos(n+1)\theta}{1 - \cos\theta}$$

Thus, we have obtained

$$1 + 2\cos\theta + 2\cos 2\theta + \cdots + 2\cos n\theta = \frac{\sin(2n+1)\dfrac{\theta}{2}}{\sin\dfrac{\theta}{2}}$$

(1)

$$= \frac{\cos n\theta - \cos(n+1)\theta}{1 - \cos\theta}$$

2.5.1   Express $\cos nx$ in terms of $\cos x$ for $n = 2, 3, 4$.   (S)

$2\cos^2 x - 1,\qquad 4\cos^3 x - 3\cos x,\ 8\cos^4 x - 8\cos^2 x + 1$

2.5.2   Express $\sin nx$ in terms of $\sin x$ for $n = 5, 7$.

2.5.3   Express $\sin nx/\sin x$ in terms of $\cos x$ for $n = 2, 3, 4$.

2.5.4   Express $(\cos x)^n$ as a linear combination of the multiples of 1, $\cos x$, $\cos 2x, \ldots,$ $\cos nx$ for $n = 2, 3, 5$.

2.5.5   Find $\int \cos^4 x\, dx$.

2.5.6   Express $(\sin x)^n$ as a linear combination of 1, $\cos x$, $\cos 2x, \ldots,$ $\cos nx$ for $n = 2, 4, 6$.   (S)

2.5.7　Express $(\sin x)^n$ as a linear combination of $\sin x$, $\sin 2x, \ldots$, $\sin nx$ for $n = 3, 5, 7$.

2.5.8　Find $\int \sin^5 x \, dx$.

2.5.9　Show that

$$\cos(\alpha + \beta + \gamma) = \cos \alpha \cos \beta \cos \gamma$$
$$- \cos \alpha \sin \beta \sin \gamma$$
$$- \cos \beta \sin \gamma \sin \alpha$$
$$- \cos \gamma \sin \alpha \sin \beta$$

and find an analogous expression for $\sin(\alpha + \beta + \gamma)$.

2.5.10　Show that

$$\int_0^{\pi/2} \cos^{2n} x \, dx = \frac{\pi}{2} \cdot \frac{1}{2} \cdot \frac{3}{4} \cdots \frac{2n-1}{2n} \tag{S}$$

2.6 LOGARITHMS

We now extend to the complex domain the concept of natural logarithms, which we call shortly logarithms. We say that w is the logarithm of z, and write

$$(1) \qquad\qquad\qquad w = \log z$$

if and only if

$$(2) \qquad\qquad\qquad z = e^w$$

Thus, we have defined the logarithmic function for complex values of the variable. By our definition, the two relations (1) and (2) are fully equivalent; both have exactly the same meaning. We have discussed the exponential function before in Section 2.2, defining it by a series, see Section 2.2 (5), and so we know the full import of Equation 2. Knowing how to pass from w to z by (2), we are going to study how to pass in the inverse direction, from z to w, by (1). In mathematical terminology we are going to study the logarithmic function as the inverse function of the exponential function.

We write, see Section 2.4 (6),

$$(3) \qquad\qquad z = re^{i\theta} \qquad w = u + iv$$

with real θ, u, v and positive r. [We exclude the case $r = 0$; see subsection (e).] Then we may write (2) in the form

$$re^{i\theta} = e^{u+iv} = e^u \cdot e^{iv}$$

The modulus of the left-hand side is r, and its argument θ. Now, if two complex numbers are equal, their absolute values are equal, and the difference of their arguments is an integral multiple of 2π (Section 1.3). Therefore

$$e^u = r \qquad v = \theta + 2n\pi$$

where n is an integer. But r is a positive number, and u a real number. Therefore, the connection between r and u is well known; it is the connection between a positive number and its natural logarithm, considered in the theory of real functions. With this meaning of the term "logarithm," we rewrite the last relations in the form

$$(4) \qquad\qquad u = \log r \qquad v = \theta + 2n\pi$$

By (1), (3), and (4)

$$(5) \qquad\qquad \log z = \log r + i(\theta + 2n\pi)$$

which we may also write in the form

$$(6) \qquad\qquad \log z = \log |z| + i \arg z$$

In order to be able to use these formulas correctly we have to emphasize a few points that were taken into account in the derivation but that are not sufficiently expressed by the final form.

(a) The sign "log" is used with two different meanings on the two sides of (5), and the same applies to (6). On the left-hand side we take logarithms of arbitrary complex numbers, using the new, unrestricted meaning of the term. On the right-hand side we take the logarithm of the positive number $r = |z|$, using the old meaning of the term, which is restricted to real and uniquely determined logarithms of positive numbers.

It would be natural to expect that the new definition of logarithm, when applied to positive numbers, should give the same result as the old definition; then the above distinction would be unnecessary. This point will be cleared up in what follows.

(b) The real part of a complex logarithm is uniquely determined, as we have just said, but the imaginary part has infinitely many values, the difference between any two values being an integral multiple of 2π. This indetermination of $\mathbf{I} \log z$ is shown explicitly by (5) and implicitly by (6): $\arg z$ has infinitely many values as we have discussed in Section 1.3.

(c) Out of the infinitely many values of $\log z$ that are originally equally permissible, we select the value whose imaginary part is between 0 and 2π and call it the *principal value* of the logarithm. In other words, in order to obtain the principal value of the logarithm we have to choose, in (6), the principal value of $\arg z$. One of the limits 0 and 2π must be included and the other excluded; the usual convention is that

$$0 \leq \text{I} \log z < 2\pi$$

if $\log z$ represents the principal value. Compare Section 1.3.

(d) The principal value of the logarithm of a positive number is real; it is *the* logarithm in the old acception of the term. For instance

$$\log 10 = 2.302585 \cdots + 2n\pi i$$

to obtain the principal value we have to choose $n = 0$. Thus, as a direct consequence of its definition as the inverse of the exponential function, the logarithm has many values, is a *multivalued* function for all values of the variable, also for positive real values. In this last case, only the principal value is real, and so only this value was encountered in the theory of real numbers. The difference between the usages of the term "log" on the two sides of (6) is that on the right-hand side logarithm is restricted to mean the principal value, while on the left-hand side it is not so restricted.

Observe that no value of the logarithm of -1 is real, since

$$\log(-1) = \pi i + 2n\pi i = (2n + 1)\pi i$$

and that the same is true of any negative number.

(e) In the foregoing we have considered the logarithm of any complex number, except 0. If $z = 0$, our formulas (5) and (6) fail completely. First, $r = |z| = 0$ and no finite real number can be considered as $\log 0$. Second, $\arg 0$ is completely indeterminate, as we have discussed in Section 1.3. It is reasonable to regard the logarithmic function as undefined for $z = 0$.

We cannot be too careful in discussing the several values of the logarithm. They are encountered again and again and have various geometrical as well as physical interpretations, as we shall see later.

2.6.1   Find the principal value of the following logarithms

(a) $\log(1 - i)$ (S)

(b) $\log(1 + e^{i\theta})$ when $\theta = \dfrac{\pi}{4}$

(c) $\log(1 + e^{i\theta})$ $0 < \theta < 2\pi$ $\theta \neq \pi$

2.6.2   Solve the equations

(a) $e^z = 1$

(b) $\cos z = 2$

2.6.3   Show that $\log z_1 z_2 = \log z_1 + \log z_2$ is not necessarily correct if the logarithms are restricted to principal values, but is always correct in the broader sense.

2.6.4   Verify that the usual laws of logarithms remain valid for arbitrary (nonzero) complex numbers.

(a) $\log z^n = n \log z$ n a positive integer

(b) $\log \dfrac{z_1}{z_2} = \log z_1 - \log z_2$

(c) $\log z^{-n} = -n \log z$

2.6.5   If $z = re^{i\theta}$ $z \neq 1$ show that

$$\mathbf{R} \log (z - 1) = \tfrac{1}{2} \log (1 - 2r \cos \theta + r^2)$$

2.7 POWERS

We use the concept of logarithm that we have just acquired to extend the concept of power to complex numbers. Being given a complex constant a, we define z^a by the equation

(1) $z^a = e^{a \log z}$

Having represented the exponential function by a series [2.2 (5)] and discussed the values of the logarithmic function [2.6 (5)] we understand the exact meaning of the right-hand side of (1). If we use polar coordinates r and θ in expressing z, and also Section 2.6 (5), we may rewrite (1) in the form

(2) $z^a = \exp \{a[\log r + i(\theta + 2n\pi)]\}$

The last factor contains the indeterminate integer n, and so the function z^a of z may be multivalued. Whether it actually is, depends on the nature of the number a; we illustrate this point by examples.

Example 1.   Take $a = 2$. Then, by (2),

$$z^2 = \exp [2(\log r + i\theta) + 4n\pi i]$$
$$= e^{2(\log r + i\theta)} = r^2 e^{2\theta i}$$

we used Section 2.4 (4). Thus the function z^2 turns out to be single-valued, which stands to reason, since we have here a rational function.

Example 2. Take $a = \frac{1}{2}$. Then, by (2),

$$z^{1/2} = \exp\left[\tfrac{1}{2}(\log r + i\theta) + n\pi i\right]$$
$$= (-1)^n r^{1/2} e^{i(\theta/2)}$$
$$= \pm r^{1/2} e^{i(\theta/2)}$$

the plus sign being valid when n is even and minus when n is odd. $r^{(1/2)}$ denotes the positive square root of the positive number r. Thus, the square root is a two-valued function in the complex domain, just as it is in the real domain.

This observation can be generalized. If n denotes a positive integer, the nth root is an n-valued function. See Example 2.7.12.

Example 3.　Take $a = i$. Then, by (2),

$$z^i = \exp\left[i(\log r + i\theta + 2n\pi i)\right]$$

this function has an infinity of different values. For instance,

$$(-1)^i = e^{-(2n+1)\pi}$$

the simplest value being obtained for $n = -1$. (Not only e and π, but also e^π is irrational—this fact, however, is very difficult to prove.)

2.7.1　　Find all values of

(a) i^i　　　　　　　　　　　　　　　　　　　　　　　　　　(S)
(b) $(1 + i)^{-i}$
(c) $\log(i^{1/2})$
(d) $\log[\log(\cos\theta + i\sin\theta)]$

2.7.2　　<u>Extraction of roots.</u> From the definition of the general power given in this section, and by the use of de Moivre's (or Euler's) theorem, we can easily evaluate a power of a complex number with any rational exponent. Let p and q be integers, $q > 1$. We define $w = z^{p/q}$ as the solution of the equation

$$w^q = z^p$$

of degree q in w. If we introduce polar coordinates r and θ for z, and ρ and ϕ for w, then

$$\rho^q(\cos q\phi + i\sin q\phi) = r^p(\cos p\theta + i\sin p\theta)$$

Solving for ρ, ϕ, we have

$$\rho = r^{p/q}, \text{ the real positive (usual) root, and}$$

$$q\phi = p\theta + 2n\pi, \ n = 0, \pm 1, \pm 2, \ldots$$

$$\phi = \frac{p}{q}\theta + \frac{2n\pi}{q}$$

Thus,

$$w = \rho e^{i\phi} = r^{p/q}e^{ip\theta/q}e^{2n\pi i/q}, \ n = 0, \pm 1, \pm 2, \ldots.$$

Because of the periodicity of the exponential function, there are only q distinct values of $e^{2n\pi i/q}$, that is, for $n = 0, 1, 2, \ldots, q-1$. The required roots are then given by

$$w = z^{p/q} = r^{p/q}e^{ip\theta/q}e^{2\pi in/q}, \ n = 0, 1, 2, \ldots, q-1$$

Take $z = 1$, $p = 1$, and plot the q points for $q = 2, 3, 4,$ and 6.

2.7.3   Find the indicated roots:

(a) $(-1)^{1/6}$ (S)

(b) $\left(\cos\dfrac{2\pi}{3} + i\sin\dfrac{2\pi}{3}\right)^{3/2}$

(c) $(1 + i)^{1/3}$

2.7.4   Prove that the square roots of i are $\pm\,(1 + i)/\sqrt{2}$.

2.7.5   Show that the cube roots of unity are

$$1 \qquad \frac{-1 + i\sqrt{3}}{2} \qquad \frac{-1 - i\sqrt{3}}{2}$$

2.7.6   Find (a) The cube roots of -1.
   (b) The 6th roots of i.
   (c) The cube roots of $1 - i$.

Illustrate graphically.

2.7.7   Solve $x^4 + 4 = 0$, and factor $x^4 + 4$ into two real quadratic factors.   (S)

2.7.8   Express the square roots of $z = x + iy$ in terms of x and y.

2.7.9   Verify that the ordinary formula for the solution of the quadratic equation

$$az^2 + bz + c = 0$$

holds for complex coefficients a, b, c.

2.7.10   Find all values of $\sqrt{3 + 4i} + \sqrt{3 - 4i}$

2.7.11   Find the nth roots of unity and show that they lie at the vertices of a regular polygon of n sides inscribed in the unit circle.   (S)

2.7.12   Let ρ be a third root of unity and let m and n take all integral values $0, \pm 1, \pm 2, \ldots$ independently of each other. The set of points $m + n\rho$ consists of the vertices of congruent equilateral triangles that cover the whole plane without gaps or overlapping. Discard from this set of points those for which $m + n$ is divisible by 3 and characterize geometrically the remaining set.   (S)

2.8 INVERSE TRIGONOMETRIC FUNCTIONS

Summarizing the foregoing, we may say that, in the complex domain, the exponential function is the common source of trigonometric functions, logarithms, and powers. Also the inverse trigonometric functions can be reduced to logarithms and hence, ultimately, to the exponential function. However, we are not going to treat this subject exhaustively in the text; one example will be enough to show the general trend.

Example.   We define arctan z as the inverse function of the tangent function. We write

(1) $w = \arctan z$

iff

(2) $z = \tan w$

Thus, these two relations, (1) and (2), are fully equivalent, both having exactly the same meaning. Now, by Section 2.3 (11), Equation 2 is equivalent to

$$z = \frac{1}{i} \frac{e^{2iw} - 1}{e^{2iw} + 1}$$

Solving this equation for e^{2iw}, we obtain

$$e^{2iw} = \frac{1 + iz}{1 - iz}$$

and hence, taking logarithms of both sides and using (1), we get

$$\arctan z = \frac{1}{2i} \log \frac{1 + iz}{1 - iz}$$

We know, from Section 2.6, that the logarithmic function has an infinity of values the difference between any two of its values being an integral multiple of $2\pi i$. Therefore, (3) shows that the function arctan z has an infinity of values, the difference between any two being an integral multiple of π. Thus the connection between the various values of the infinitely many-valued function arctan z is exactly the same in the complex domain as in the real domain.

2.8.1   Show that (a) $\arcsin z = -i \log (iz \pm \sqrt{1 - z^2})$
   (b) $\arccos z = -i \log (z \pm \sqrt{z^2 - 1})$

2.8.2   Find (a) arctan $2i$ (S)
   (b) arctan $(1 - i)$

2.8.3   If $\log z = a + ib$, $z \log z = p + iq$ show that

$$\arctan \frac{p}{q} = \arctan \frac{a}{b} - b$$

2.9   GENERAL REMARKS

After having discussed some examples, we consider now the concept of a complex function of a complex variable in general. If, to every complex number z belonging to a certain domain, there corresponds a complex value w, we say that w is a function of z

(1) $w = f(z)$

In the foregoing sections we have discussed a few important cases in which we can compute the value for w corresponding to the number z by some definite simple process. Now we consider the connection between z and w abstractly, apart from any specified, concrete computational operation.

The equation (1) between complex numbers must be equivalent to two equations between real numbers. We write

(2) $z = x + iy$ $w = u + iv$

x, y, u, v being real variables. If z is given, x and y are given, and the converse is also true. But if z is given, the law expressed by (1) permits us to find the complex value w and, therefore, also the real part u of w. Thus u is determined by x and y, that is, the *real variable u is a function of two real variables* x and y. Similarly, v must be a function of x and y, and we express both facts by writing

(3) $$u = \phi(x, y) \qquad v = \psi(x, y)$$

the letters ϕ and ψ standing here for real functions. We may combine (1), (2), and (3) and write

(4) $$f(x + iy) = \phi(x, y) + i\psi(x, y)$$

Summarizing, we see that *a complex function of a complex variable* is equivalent to two coordinated real functions of two real variables; (1) is equivalent to (3).

If this is so, why should we consider complex functions at all? Real functions are more familiar; if a complex function is equivalent to a pair of real functions, what is the point in introducing the less familiar complex functions?

If the two real functions u and v are chosen at random and there is no particular connection between them, then there is really not much point in combining them into a complex function. There are, however, cases in which the two real functions are so closely correlated that it is advantageous to condense the two relations (3) into the one relation (1). We are going to consider, in Sections 2.12 and 2.13, geometrical and physical situations that are more concisely and more intuitively described by one relation of the form (1) than by two relations of the form (3). Before discussing the case in which both variables—the dependent and the independent—are complex, we shall consider, in Sections 2.10 and 2.11, the simpler cases in which one of these two variables is complex and the other real.

2.9.1 SINGLE-VALUED, MULTIVALUED

A function is essentially a rule of correspondence. For instance, by virtue of the equation

$$w = 1 - z^2$$

there corresponds to every complex value assigned to z a complex value of w. Generally, we consider complex functions of a complex variable z. Such a function $f(z)$ determines exactly one complex number w corresponding to each complex number z that belongs to a certain set D; we express such a rule of correspondence by writing

$$w = f(z)$$

and we call D the *domain of definition* of the function $f(z)$.

For instance, all four functions $w = 1 - z^2$, $w = e^z$, $w = \cos z$, $w = \sin z$ have the same domain of definition: the whole z-plane, the set of all complex numbers z. The domain of definition of the function

$$w = \frac{1}{z}$$

can be termed the *punctured plane*: it is "punctured" at the origin, it comprises all complex numbers with the single exception of the value $z = 0$.

A function in the sense here circumscribed should be more precisely termed a *single-valued function*. Some of the functions we have considered in this chapter are not single valued: the logarithm, the inverse trigonometric functions, and a power with a nonintegral exponent are *multivalued* functions.

Certain difficulties inherent in the subject matter of our study oblige us to treat these two kinds of functions differently: unless the contrary is explicitly stated, any *general* explanation about functions refers to *single-valued functions*. We cannot avoid considering *examples* of *multivalued* functions, but each example of this kind should be considered on its own merits until later when we shall be prepared for examining multivalued functions generally. Logarithm, powers, and inverse trigonometric functions are our principal examples; since the logarithm is used for defining the others, the most important example of a multivalued complex function of a complex variable is log z.

The definition of the principal value of log z, given in Section 2.6(c), may appear as "artificial" (Why cut out just this "chunk" of the logarithm?) and this definition certainly introduces an "unnatural" discontinuity (along the positive half of the real axis.) The aim of this definition is, however, to introduce a single-valued function: the principal value of log z *is* single-valued; its domain of definition is the punctured plane (punctured at the origin.)

There is a general hint behind this example: by cutting out some appropriate single-valued "chunk" from any multivalued function (by restricting "artificially" its domain of definition) we could make available for it our general remarks on single valued functions.

2.10 COMPLEX FUNCTION OF A REAL VARIABLE: KINEMATIC REPRESENTATION

We consider the complex variable z as a function of the real variable t.

(1) $$z = f(t)$$

We write, as usual, $z = x + iy$ with real x and y. The law expressed by (1) determines, to each real number t, a corresponding complex value z and, therefore, also the real part x of z. Thus x is a function of t, and so is y.

We express this by writing

$$(2) \qquad\qquad x = \phi(t) \qquad y = \psi(t)$$

where ϕ and ψ stand for real functions.

In choosing the letters x, y, and t we aimed at a simple and natural inter-pretation of the relations (2). We regard t as the time, and x and y as the rectangular coordinates of a point whose motion is described by the two equations (2). Now the position of a point in a plane can be given not only by its rectangular coordinates, x and y, but also by the complex number $z = x + iy$. Equation 1 expresses the fact that the position of the point z depends on the time, and so it characterizes, even more concisely than the two equivalent equations (2), the motion of the point.

Example. If c and p are constants, c complex and p positive, then the equation

$$z = c + pe^{it}$$

expresses that z moves along a circle with center c and radius p. The point z proceeds counterclockwise with angular velocity 1.

If t is measured in minutes and we wish to represent the motion of the point of the minute hand of a watch, we should change the equation into

$$z = c + re^{-2\pi it/60}$$

2.10.1 If a and h denote positive constants and t the time, the three equations

(a) $z_1 = ia + at$
(b) $z_2 = -ihe^{-it}$
(c) $z = ia + at - ihe^{-it}$

represent three moving points. What kind of motion is produced and along what curve?

2.10.2 If a, b, and h denote positive constants and t the time, the three equations

(a) $z_1 = (a + b)e^{it}$
(b) $z_2 = -h \exp [i(a + b)t/b]$
(c) $z = (a + b)e^{it} - h \exp [i(a + b)t/b]$

represent three moving points. What kind of motion is produced and along what curve?

2.10.3   If a, b, and h denote positive constants and t the time, the three equations

(a) $z_1 = (a - b)e^{it}$
(b) $z_2 = h \exp\left[-i(a - b)t/b\right]$
(c) $z = (a - b)e^{it} + h \exp\left[-i(a - b)t/b\right]$

represent three moving points. What kind of motion is produced and along what curve?

2.11   REAL FUNCTIONS OF A COMPLEX VARIABLE: GRAPHICAL REPRESENTATION

We consider the real variable u as a function of the complex variable $z = x + iy$. Then u is determined by x and y, and so u is, in fact, a function of the two real variables x and y

$$(1) \qquad\qquad u = \phi(x, y)$$

In order to represent (1) graphically, we interpret x, y, and u as rectangular coordinates of a point in space. We can choose z arbitrarily (in a certain domain), and then u is determined by (1). Since x and y vary, the point (x, y, u) will move in space and describe a surface. A surface is a two-dimensional locus in space representing a function of two independent real variables just as a curve is a one-dimensional locus in a plane representing a function of one independent real variable.

Surfaces in space and curves in a plane are analogous in theory but very different in practice. A curve in a plane is easy to draw, even fairly accurately, but surfaces exist usually only in our imagination; it is too troublesome to construct them in plaster or other material. Therefore, we often prefer to proceed differently. We consider the numbers x and y for which ϕ takes some preassigned value a, that is, we consider the equation

$$(2) \qquad\qquad \phi(x, y) = a$$

The points (x, y) satisfying Equation 2 form a curve that we may appropriately call a *level curve* of the function $\phi(x, y)$ corresponding to the level a. We plot in a plane a series of level curves, corresponding to a series of equidistant levels

$$a = \ldots, -2h, -h, 0, h, 2h, \ldots$$

where h is a suitably chosen positive number. These curves form a chart of level curves of the function ϕ. Such a chart gives us detailed information about the function $\phi(x, y)$; using it, we can visualize the distribution of the values of the function. If we wish to make our chart accurate we choose a small value of h and, theoretically, nothing prevents us from choosing h arbitrarily small and making our chart arbitrarily precise.

By the way, the two graphical representations are closely connected. Let us imagine, in three-dimensional space, the planes whose equations are

$$u = nh$$

with $n = 0, 1, 2, \ldots$. The plane $u = 0$ is the x, y-coordinate plane, and the other planes are parallel to it and equidistant. Let all these planes intersect the surface representing the function (1), and project all the lines of intersection orthogonally onto the x, y-coordinate plane. What we obtain is the chart of level curves.

(If the surface given by (1) is a part of the earth's surface, a landscape of hills and dales, the chart of level curves obtained by projection onto the horizontal z-plane and reduced on an appropriate scale is called a "contour map.")

In dealing with a complex function of a complex variable we can often use with good effect either of the two representations discussed. Let z and w be complex, x and y real, $z = x + iy$ and $w = f(z)$.

Now, $|w|$, the modulus of $f(z)$, can be considered as a real function of the two real variables x and y. In fact, $|w|$ is determined by x and y since

$$(3) \qquad\qquad |w| = |f(x + iy)|$$

We may represent $|w|$ as a function of x and y either by a surface, called by some authors the "modular surface" of $f(z)$, or by a chart of level lines. The coordinates of a variable point of the modular surface of $f(z)$ are $x, y, |w|$, the last coordinate $|w|$ being given by (3). The chart contains the level lines of $|w|$ whose equations are, with $n = 0, 1, 2, \ldots$,

$$|f(x + iy)| = nh$$

Example.   Describe the modular surface and construct a chart of level lines of the exponential function, $w = e^z$.

We have

$$|w| = |e^{x+iy}| = |e^x \cdot e^{iy}| = e^x$$

Therefore, $|w|$ is constant along straight lines parallel to the imaginary axis. The modular surface is described by a moving straight line parallel to the imaginary axis; it is a cylindrical surface whose cross-section is the well-known exponential curve. To construct a chart we put

$$w = h, 2h, 3h, \ldots$$

and obtain

$$x = \log h, \log 2h, \log 3h, \ldots$$

Any slide rule shows lengths proportional to these numbers. Choosing $h = 1$ we obtain Figure 2.1 on which the real axis is a dotted line and the level lines are marked with the corresponding values of $|w|$ in the z-plane; x and $|w|$ are the coordinates in the plane of the cross-section.

Figure 2.1

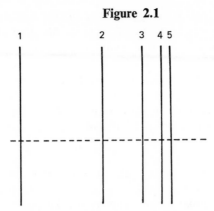

2.12 COMPLEX FUNCTIONS OF A COMPLEX VARIABLE: GRAPHICAL REPRESENTATION ON TWO PLANES

Many methods of constructing maps are known, and many are in actual use, but no one map is perfect. The surface of the earth is spherical, or nearly so. Any map of this curved surface constructed on a flat sheet of paper is bound to be distorted in some respect. (This can be proved but anybody can be led to suspecting it just by peeling potatoes.) Let us now consider two different maps of the globe, on two different planes. Let x and y be the rectangular coordinates of a variable point on the first plane, u and v on the second plane. We write, as usual,

$$x + iy = z \qquad u + iv = w$$

The two maps may look fairly different if they are constructed by different methods, but there is a correspondence between their points. An arbitrarily given point z on the first map represents a certain point on the surface of the globe that is also represented by a certain point w on the second map; this point w corresponds to the given point z. The position of the point w depends on the position of the point z to which it corresponds, that is, w is a function of z:

(1) $$w = f(z)$$

When the point z describes a curve, for instance, the image of a certain river, or the image of the boundary line of two countries, on the first map, the point w also moves along a curve that is the image of the same river, or boundary line, on the second map.

We mentioned the globe in order to connect our ideas with something useful and familiar, but, in fact, we may disregard the globe. We consider two planes: the z-plane, in which the point z varies, and the w-plane, in which the point w varies. If a relation of the form (1) holds, we can assign to any point z of the z-plane, within a certain domain, a corresponding point w

in the w-plane. We call a correspondence of this sort a *mapping* of the z-plane onto the w-plane. Mapping of a plane onto another plane represents graphically a complex function of a complex variable.

It is instructive to realize that the representations discussed in Sections 2.10 and 2.11 are closely connected with the representation we are discussing now.

Let us assign some constant value to y, for example, b. Then $z = x + ib$ moves along a horizontal straight line and

$$(2) \qquad\qquad w = f(x + ib)$$

becomes a complex function of the real variable x. Let x move with unit speed. Then the corresponding point w describes in its plane a certain curve with a certain (variable) speed. This is the kinematic representation we have discussed in Section 2.10.

Let us choose a positive number h, and consider in the w-plane the sequence of concentric circles

$$|w| = h, 2h, 3h, \ldots$$

The points in the z-plane that are mapped onto these circles satisfy the conditions

$$(3) \qquad\qquad |f(x + iy)| = h, 2h, 3h, \ldots$$

respectively. Each equation represents a certain line in the z-plane, a level line of the function's absolute value as we said in Section 2.11. The sequence of these lines (3), or the whole chart of the level lines in the z-plane, is mapped onto the sequence of the circles (2) in the w-plane.

Generalizing this idea, we may consider any family of curves in one plane and investigate onto which family of curves it is mapped in the other plane.

Example.   The z-plane is mapped onto the w-plane by the function

$$w = z^2$$

Which lines of the z-plane are mapped onto the straight lines parallel to the coordinate axes of the w-plane?

We have

$$u + iv = (x + iy)^2$$

and, therefore,

$$u = x^2 - y^2 \qquad v = 2xy$$

A line parallel to the imaginary axis of the w-plane is characterized by the equation $u = a$, where a is a constant and its image in the z-plane has the equation

$$x^2 - y^2 = a$$

Figure 2.2

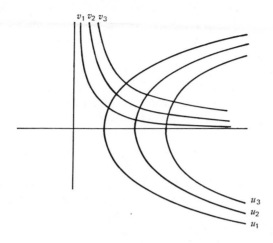

This equation, in which a is regarded as a parameter, represents a family of hyperbolas whose asymptotes bisect the angles between the axes of the z-plane. A line $v = b$ of the w-plane is the image of

$$2xy = b$$

we have here a family of hyperbolas in the z-plane whose asymptotes are the axes (Figure 2.2).

2.13 COMPLEX FUNCTIONS OF A COMPLEX VARIABLE: PHYSICAL REPRESENTATION IN ONE PLANE

Let us focus our attention on a point in a large, steadily flowing river. Many particles of water pass this point, always different particles as time goes by, but they all pass at the same speed if the flow is really steady. Thus a certain speed belongs to this point, represented by a vector having a definite magnitude and direction. In general, to each point of the stream a well-defined vector is attached, representing the velocity of the flow at that point. The steady flow of a river suggests to us the important concept of a *vector-field*.

For the purpose in view, we must specialize the general concept of a vector-field. Let us imagine that the flow of the water is exactly alike in all horizontal layers. This may be approximately so in the middle portion of a large river, sufficiently far from both banks, the bottom, and the surface. We should, however, idealize the case and imagine that the physical state is exactly the same in all horizontal planes, and that, therefore, the vectors attached to the various points of a vertical straight line agree both in magnitude and in direction, all being horizontal. We have sufficient knowledge of such a vector-field if we know the physical state and especially the velocities in any arbitrarily chosen horizontal plane. This sort of field, being adequately represented by the vectors in one plane, in two dimensions, is called a *two-dimensional vector-field*.

Now we may return to the study of a complex function of a complex variable. We consider as before the function

(1)
$$w = f(z)$$

We represent as before the independent variable z as a point in a plane, the z-plane. We represent, however (and here we depart from our former standpoint) the complex value $\bar{w}$, conjugate to w, as a vector issuing from the point z. The complex value w, as function of z, is determined by z, and so is $\bar{w}$. From any point z, within a certain domain, we can draw a vector $\bar{w}$, which we can construct or compute using (1). These vectors attached to the points represent a two dimensional vector-field. The vector varies in general from point to point, both in magnitude and in direction. The function (1) expresses the law of variation and describes adequately the vector-field. We may also say that the vector-field represents concretely the function. As the vector-field may be interpreted as a field of flow or as a field of force of any kind, we see here an opening for various applications.

It may seem strange that we have insisted on regarding $\bar{w}$, and not w, as the vector issuing from the point z but, in fact, this choice is more advantageous, as we shall see in the next chapter.

Example. Find the magnitude and the direction of the vector $\bar{w}$, being given that

$$w = \frac{1}{z}$$

Let r denote as usual the absolute value, and θ the phase, of z [as in Section 2.4 (6)]. Then

$$\bar{w} = \frac{1}{\bar{z}} = \frac{1}{re^{-i\theta}} = \frac{e^{i\theta}}{r}$$

The direction of the vector $\bar{w}$ is the same as the direction of the vector drawn from the origin to the point z where $\bar{w}$ is applied, and the absolute value of $\bar{w}$ is the reciprocal of the value of r. In other words, $\bar{w}$ can be considered as a force radially outward from the origin and inversely proportional to the distance from the origin, that is, a source of repulsive force.

Additional Examples and Comments on Chapter Two

2.1 SOME TESTS FOR CONVERGENCE

The reader is expected to have some facility in treating infinite series with real terms. We shall try, however, to get along without more delicate considerations of convergence, especially in the beginning. We list here some simple tests for convergence that, once established for series with real terms, can be easily extended to series with complex terms.

(I) *If* $|a_1| + |a_2| + |a_3| + \cdots$ *is convergent, then* $a_1 + a_2 + a_3 + \cdots$ *is convergent* (and is called "absolutely convergent").

This fact was established in Example 1.10.3.

(II) *If* $a_1 + a_2 + a_3 + \cdots$ *is convergent, then* $\lim_{n\to\infty} a_n = 0$.

(III) *If* $|a_n|^{1/n}$ *or* $|a_{n+1}/a_n|$ *tends to a limit less than unity when* $n \to \infty$, *then the series* $a_1 + a_2 + a_3 + \cdots$ *is convergent; if, however, one or the other expression exhibited tends to a limit greater than unity or to* ∞, *the series is divergent.* (If the limit is exactly 1, or if there is no limit, the test yields no immediate information.)

The application of (II) to the geometric series

$$c + cz + cz^2 + \cdots + cz^n + \cdots$$

with initial term $c \neq 0$ and ratio z shows that this series is divergent when $|z| > 1$ (it is convergent when $|z| < 1$ as we have seen in Section 1.10). The application of (III) to the same series shows only that it is divergent when $|z| > 1$: here is a case in which (II) is (in one respect) more informative than (III).

If the test (III) is applicable and shows convergence, it shows absolute convergence: in fact, if it is applicable to Σa_n, it is just as applicable to $\Sigma |a_n|$. This simple remark is often useful.

Show that the series introduced in Sections 2.2 (5), 2.3 (1), and 2.3 (2) are absolutely convergent for all complex values of z. (Which test is the most convenient?)

2.2 POWER SERIES

The examples considered in Example 2.1 (the geometric series and the series defining e^z, $\cos z$, and $\sin z$) are particular cases of the general series

$$c_0 + c_1 z + c_2 z^2 + \cdots + c_n z^n + \cdots$$

which is called a *power series*: its successive terms contain the successive powers

$$z^0, z^1, z^2, \ldots z^n, \ldots$$

The constants

$$c_0, c_1, c_2, \ldots c_n, \ldots$$

are called the *coefficients* of the power series. A power series is given if the sequence of its coefficients is given; z stands for a complex variable.

We begin the study of power series by considering a subclass, a more accessible particular case: we assume the existence of the limit

$$\lim_{n \to \infty} |c_n|^{1/n} = 1/R$$

in an extended sense, including the following two extreme cases:
if $|c_n|^{1/n}$ tends to ∞, we set $R = 0$, and if $\lim |c_n|^{1/n} = 0$, we set $R = \infty$.
Show, under this assumption, that

(a) If R is a positive number, the power series is convergent for $|z| < R$ and divergent for $|z| > R$.
(b) If $R = 0$, the power series is convergent for $z = 0$ and for no other value of z.
(c) If $R = \infty$, the power series is convergent for all complex values of z.

Thus, in the case (a) the power series is convergent in each point inside the circle with center 0 and radius R, and in no point outside this circle, which is called the *circle of convergence* of the power series; R is termed the *radius of convergence*. (Observe that nothing was said about convergence or divergence in points on the periphery of the circle of convergence where $|z| = R$.)

In the case (b) the circle of convergence reduces to one point and the radius of convergence is 0.

In the case (c) the circle of convergence encompasses the whole plane and the radius of convergence is infinite.

The power series for e^z, $\cos z$, and $\sin z$ illustrate case (c), the geometric series exemplifies case (a) with $R = 1$.

2.3 (continued) The radius of convergence R can also be computed by the relation

$$\lim_{n \to \infty} |c_{n+1}/c_n| = 1/R$$

provided that the limit on the left-hand side exists.

2.4   Find the radius of convergence of the following power series

(a) $\displaystyle\sum_{0}^{\infty} \frac{z^n}{(n+1)^2}$

(b) $\displaystyle\sum_{1}^{\infty} n^n z^n$

(c) $\displaystyle\sum_{1}^{\infty} n^{-n} z^n$

(d) $\displaystyle\sum_{1}^{\infty} \frac{n^n}{n!} z^n$ (S)

(e) $\displaystyle\sum \frac{(n!)^2}{(2n)!} z^{2n}$

(f) $\displaystyle\sum_{1}^{\infty} \left(1 + \frac{1}{n}\right)^{n^2} z^n$

2.5   If the power series $\Sigma c_n z^n$ belongs to the subclass considered in Example 2.2, it has the same radius of convergence as the power series $\Sigma n c_n z^n$.

2.6 POWER SERIES (continued; see Example 2.2.)

We wish to determine the region of convergence of the power series:

$$1 + 3z + 5^2 z^2 + 2^3 z^3 + 3^4 z^4 + 5^5 z^5 + 2^6 z^6 + \cdots$$

To state more clearly the rule of formation of this series, we exhibit three consecutive terms in a general situation:

$$\cdots + 2^{3k} z^{3k} + 3^{3k+1} z^{3k+1} + 5^{3k+2} z^{3k+2} + \cdots$$

This power series does not belong to the subclass considered in Example 2.2, and the tests of Example 2.1 (III) do not apply: neither $|a_{n+1}/a_n|$, nor $|a_n|^{1/n}$ tends to a limit when $n \to \infty$.

Yet the proposed series is compounded of three geometric series:

$$1 + (2z)^3 + (2z)^6 + \cdots + (2z)^{3k} + \cdots$$
$$3z + (3z)^4 + (3z)^7 + \cdots + (3z)^{3k+1} + \cdots$$
$$(5z)^2 + (5z)^5 + (5z)^8 + \cdots + (5z)^{3k+2} + \cdots$$

These are convergent when the ratio is less than 1 in absolute value, and so they converge for

$$|2z|^3 < 1 \qquad |3z|^3 < 1 \qquad |5z|^3 < 1$$

respectively. When the most restrictive of these three conditions is satisfied, that is, when the point z is contained in the circle

$$|z| < \tfrac{1}{5}$$

all three series converge and so their sum, the proposed series, converges too. When, however, $|z| \geq \tfrac{1}{5}$, the general term of the proposed series does not tend to 0, and so the series cannot converge, by virtue of Example 2.1 (II).

In the case considered in Example 2.2 the limit of $|c_n|^{1/n}$ determines the radius of convergence. In the case just discussed this limit does not exist: the sequence considered has three "points of accumulation", the numbers 2, 3, and 5; and the radius of convergence turns out to be $\tfrac{1}{5}$, the reciprocal of the largest of these three numbers.

The case here discussed suggests the general case. It is generally true that

$$\overline{\lim_{n \to \infty}} |c_n|^{1/n} = 1/R$$

where the left hand side represents the largest value of accumulation, or *upper limit of indetermination*, of $|c_n|^{1/n}$, and R is the radius of convergence of the power series $\Sigma c_n z^n$. We omit the proof that requires, of course, precise knowledge of the concept of $\overline{\lim}$.

2.7

$$(a_0 + a_1 + a_2 + \cdots)(b_0 + b_1 + b_2 + \cdots) = c_0 + c_1 + c_2 + \cdots$$

provided that both factors on the left-hand side are absolutely convergent series, and

$$c_n = a_n b_0 + a_{n-1} b_1 + a_{n-2} b_2 + \cdots + a_0 b_n$$

The usual proof of this proposition given in the theory of infinite series with real terms immediately extends to the case of complex terms.

Revise the proof of Section 2.2 (6) in the light of this proposition.

2.8   Show that

$$\lim_{z \to 0} \frac{e^z - 1}{z} = 1$$

2.9   Show that a power series solution of the functional equation

$$f(z) = z + f(z^2)$$

must be of the form

$$f(z) = c + z + z^2 + z^4 + z^8 + z^{16} + \cdots$$

where c is an (arbitrary) constant.

2.10   Let p denote a positive number, x a positive variable. Show that

(a) $x^p e^{-x} \to 0$ when $x \to \infty$
(b) $x^{-p} \log x \to 0$ when $x \to \infty$
(c) $x^p \log x \to 0$ when $x \to 0$

[(a) is more interesting when p is large, (b) and (c) are more interesting when p is small.]

2.11   Let $a_1, a_2, \ldots a_n$ denote complex numbers, and $b_1, b_2, \ldots b_n$ real numbers,

$$b_1 < b_2 < \cdots < b_n$$

Prove: If

$$a_1 e^{b_1 z} + a_2 e^{b_2 z} + \cdots + a_n e^{b_n z} \equiv 0$$

($\equiv 0$ means identically zero) then

$$a_1 = a_2 = \cdots = a_n = 0$$

2.12   Starting from the origin, go one unit eastward, then the same length northward, then $\frac{1}{2}$ of the length just described westward, the $\frac{1}{3}$ of the last length southward, and so on, going in turn east, north, west, south, and describing $1/n$ of the length described just before at the $(n + 1)$-st step. The path forms a sort of "angular spiral" that winds an infinity of times around a certain point; which point is it?

2.13   Extend the result of Example 1.19, with necessary modifications, to power series.
    Assume that the radius of convergence of the power series

$$f(z) = a_0 + a_1 z + a_2 z^2 + \cdots$$

is R, $R > 0$, and that $a_0 \neq 0$. Prove:
(a) The power series

$$|a_0| - |a_1| x - |a_2| x^2 - |a_3| x^3 - \cdots$$

vanishes either in no point, or in just one point $x = r$, of the interval $0 < x < R$; in the former case set, by definition, $r = R$.
(b) $f(z)$ does not vanish in the circle $|z| < r$

2.14   Extend the result of Example 1.1 to power series.
Prove:
    If $f(z) = \sum_0^\infty a_n z^n$, with a_n positive and nonincreasing, then $f(z) \neq 0$ in the circle $|z| < 1$.

2.15   Prove directly from the power series that $\cos z \neq 0$ for $|z| < \sqrt{2}$.

2.16

$$J_0(z) = \sum_0^\infty (-1)^n \frac{\left(\frac{z}{2}\right)^{2n}}{(n!)^2} \qquad \text{(Bessel function of order zero)}$$

by definition. Show that $J_0(z) \neq 0$ for $|z| < 2$.

2.17   The hyperbolic functions are defined analogously to the trigonometric functions:

$$\cosh z = 1 + \frac{z^2}{2!} + \frac{z^4}{4!} + \frac{z^6}{6!} + \cdots$$

$$\sinh z = z + \frac{z^3}{3!} + \frac{z^5}{5!} + \frac{z^7}{7!} + \cdots$$

$$\tanh z = \frac{\sinh z}{\cosh z} \qquad \coth z = \frac{1}{\tanh z}$$

$$\operatorname{sech} z = \frac{1}{\cosh z} \qquad \operatorname{csch} z = \frac{1}{\sinh z}$$

(A more condensed notation is sometimes used for $\cosh z$ and $\sinh z$, that is, ch z and sh z respectively.)

Verify the following analogues of Euler's formula for the hyperbolic functions:

(a) $\cosh z = \dfrac{e^z + e^{-z}}{2}$

(b) $\sinh z = \dfrac{e^z - e^{-z}}{2}$

(c) $e^z = \cosh z + \sinh z$

2.18   Whereas

$$\cos^2 z + \sin^2 z = 1$$

show that

$$\cosh^2 z - \sinh^2 z = 1$$

2.19   Let t denote a real variable and compare the two complex valued functions of t:

$$z = \cos t + i \sin t$$

$$z = \cosh t + i \sinh t$$

Figure 2.3

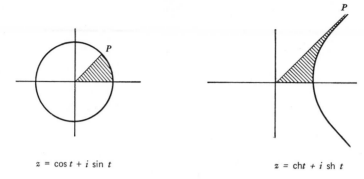

$z = \cos t + i \sin t$ $\qquad\qquad\qquad$ $z = \mathrm{ch}\, t + i\, \mathrm{sh}\, t$

Show (as an illustration to Section 2.10) that the first equation represents motion along the (unit) circle while the second represents motion along an (equilateral) hyperbola; see parts I and II of Figure 2.3.

2.20 (continued)   Compare the two shaded areas I and II in Figure 2.3. Both areas are included in the same way: by an arc of the curve whose end-point corresponds to the general value t, and initial point to the particular value $t = 0$, of the parameter, and by two straight lines drawn to these two points from the origin. Compute both areas.

[The general expression for such a "sectorial" area is

$$\frac{1}{2} \int_0^t \left(x \frac{dy}{dt} - y \frac{dx}{dt} \right) dt \Big]$$

2.21   For real values of x, there is a remarkable analogy between the pair $\cos x$, $\sin x$, and the pair $\cosh x$, $\sinh x$, but there is no "direct" connection. For complex values of the variable, however, there is such a connection. Prove the following identities:

(a) $\cosh z = \cos iz$
(b) $\sinh z = -i \sin iz$
(c) $\cos z = \cos x \cosh y - i \sin x \sinh y$
(d) $|\sin z|^2 = \sin^2 x + \sinh^2 y$
(e) $|\cosh z|^2 = \sinh^2 x + \cos^2 y$

2.22   Verify the following rule concerning identities among the hyperbolic functions:

Every trigonometric identity remains valid when each trigonometric function is replaced by the corresponding hyperbolic function provided we change the SIGN of each term that contains a product of *two* SINES.

For example,

$$\cos (\alpha + \beta) = \cos \alpha \cos \beta - \sin \alpha \sin \beta$$

gives

$$\cosh (\alpha + \beta) = \cosh \alpha \cosh \beta + \sinh \alpha \sinh \beta$$

while

$$\sin (\alpha + \beta) = \sin \alpha \cos \beta + \cos \alpha \sin \beta$$

gives

$$\sinh (\alpha + \beta) = \sinh \alpha \cosh \beta + \cosh \alpha \sinh \beta$$

(Note that a product such as $\tan \alpha \tan \beta$ involves the product of two sines.)

2.23   Consider the equation

$$\frac{1}{z - z_1} + \frac{1}{z - z_2} + \cdots + \frac{1}{z - z_n} = 0$$

where $z_1, z_2, \ldots z_n$ are given points and z is the unknown, and explore its algebraical and mechanical significance.

(a) How is the equation connected with the polynomial $P(z)$ of degree n whose roots are $z_1, z_2, \ldots z_n$?

(b) Each of the terms on the left-hand side is connected with a force that originates in one of the given points and acts on the point z whose situation has to be determined from the equation. How? (Section 2.13.)

2.24. A THEOREM OF GAUSS

The least convex polygon containing the roots of a polynomial contains also the roots of its derivative.

[If the points $z_1, z_2, \ldots z_n$ are marked with nails, a lightly stretched rubber band encompassing all these nails indicates the least-containing convex polygon that is sometimes called the "convex hull" of the set $(z_1, z_2, \ldots z_n)$.]

2.25   If a polynomial has only real roots, then also its derivative has only real roots, and the least interval containing the roots of the polynomial contains also the roots of its derivative.

2.26   (More precise than Example 2.24.) The least convex polygon P containing the roots of a polynomial contains any root of the polynomial's derivative in its *interior*, except in the following two cases:

(a) P reduces to a line segment and so it has no interior (Example 2.25).

(b) A root of the derivative coincides with a (multiple) root of the polynomial on the boundary of P.

2.27   If your proof for Example 2.24 is based on mechanical considerations, find another proof that is free of mechanical considerations. (Such a proof may eventually help you to clarify your mechanical ideas.)

2.28   Consider the equation

$$(z - z_1) + (z - z_2) + \cdots + (z - z_n) = 0$$

where $z_1, z_2, \ldots z_n$ are given points. The solution

$$z = (z_1 + z_2 + \cdots + z_n)/n$$

is called the *centroid* of the points $z_1, z_2, \ldots z_n$. (More exactly, z is the centroid of a system of n equal masses placed in the n given points.)

Show that the relation between the system of n points and their centroid is unchanged by translations and rotations of the coordinate system.

2.29   The centroid of the n vertices of a regular polygon with n sides is the center of the circle circumscribed about the regular polygon.

(The centroid of the n points representing the nth roots of unity is the origin.)

2.30   The least convex polygon containing n given points contains their centroid in its interior, provided it has an interior (that is, provided it does not reduce to a line segment.)

This assertion may appear obvious on mechanical grounds. Give a proof that is free of mechanical considerations.

2.31   A straight line passing through the centroid of n given points divides the plane into two half-planes.
Prove that there are only two possible cases:

(a) Either all n given points lie on the line,
(b) Or each of the two half-planes contains at least one of the given points in its interior (there are "points on both sides.")

2.32   There are $n + 2$ particles: n movable particles constrained to remain in the interval $-1 < x < 1$, each having the same positive unit charge, and two particles fixed at the points $x = -1$ and $x = 1$, respectively, each having the charge $\frac{1}{2}$. The n movable particles are in a state of equilibrium under the forces produced; the force between two particles is repulsive, proportional to the product of the charges, and inversely proportional to the distance between them. Show that the n particles lie at $x_1, x_2, \ldots, x_n$, the zeros of the Legendre polynomial of degree n, $P_n(x)$. [The Legendre polynomial $P_n(x)$ satisfies the differential equation

$$(1 - x^2)P_n''(x) - 2xP_n'(x) + n(n + 1)P_n(x) = 0]$$

2.33   Let

$z_1, z_2, \ldots, z_m$ be complex numbers

$p_1, p_2, \ldots, p_m$ positive numbers

$A(z) = (z - z_1)(z - z_2) \cdots (z - z_m)$

$$B(z) = \left(\frac{p_1}{z - z_1} + \frac{p_2}{z - z_2} + \cdots + \frac{p_m}{z - z_m} \right) A(z)$$

$C(z)$ a polynomial

$P(z)$ a polynomial satisfying the differential equation

$$A(z)\, P''(z) + 2B(z)\, P'(z) + C(z)\, P(z) + 0$$

Show that the least convex polygon containing the roots of $A(z)$ contains also the roots of $P(z)$.

2.34 CRITICISM

Reconsider the contents of the chapter and the problems solved and notice such points that seem to need a more rigid treatment supplementing the more intuitive and heuristic considerations offered.

HINTS AND SOLUTIONS FOR CHAPTER TWO

2.2.1 $e^{-z} \cdot e^z = e^0 = 1$

2.3.1 $1, i, -1, -i$

2.3.2(a) $\dfrac{\exp i\left(\dfrac{\pi}{2} - z \right) - \exp -i\left(\dfrac{\pi}{2} - z \right)}{2i} = \dfrac{i e^{-iz} + i e^{iz}}{2i} = \cos z$

2.3.4 $|e^z| = e^x \neq 0$ all x

2.3.6 $0 = \sin z = \dfrac{e^{iz} - e^{-iz}}{2i}$ or $e^{2iz} = 1$

$$e^{-2y} \cos 2x = 1 \qquad e^{-2y} \sin 2x = 0$$

$$\text{Hence } \sin 2x = 0 \qquad \cos 2x = 1 \qquad y = 0$$

$$x = n\pi, \, n = 0, \pm 1, \pm 2, \ldots$$

$$z = n\pi \qquad n = \text{integer}$$

2.4(d)
$$\left|\frac{u_{n+1}}{u_n}\right| = \frac{(n+1)^{n+1}}{(n+1)!}\frac{n!}{n^n}|z|$$

$$= \left(1 + \frac{1}{n}\right)^n |z| \to e\,|z|$$

$$e\,|z| < 1 \Rightarrow |z| < \frac{1}{e}$$

2.4.1(a) $e^{13i} = \cos(13) + i\sin(13)$

2.4.2 $\sin(z + 2\pi) = \dfrac{e^{i(z+2\pi)} - e^{-i(z+2\pi)}}{2i} = \dfrac{e^{iz} - e^{-iz}}{2i} = \sin z$

2.4.5 $e^z = 1 + i = e^x(\cos y + i\sin y)$

$e^x \cos y = 1 \qquad e^x \sin y = 1$

$e^{2x}(\cos^2 y + \sin^2 y) = e^{2x} = 2$

$\tan y = 1$

$x = \frac{1}{2}\log 2 \qquad y = \dfrac{\pi}{4} + 2n\pi \qquad n = \text{integer}$

2.4.7 $\mathbf{R}(e^{iz^2}) = \mathbf{R}(e^{i(x^2-y^2+2ixy)})$

$\qquad = e^{-2xy}\cos(x^2 - y^2)$

2.5.1 $\cos 2x = 2\cos^2 x - 1$

$\cos 3x = 4\cos^3 x - 3\cos x$

$\cos 4x = \mathbf{R}(\cos x + i\sin x)^4 = 8\cos^4 x - 8\cos^2 x + 1$

2.5.6 $\sin^6 x = (1 - \cos^2 x)^3 = 1 - 3\cos^2 x + 3\cos^4 x - \cos^6 x$ (see Section 2.5.1)

2.5.10 $I_n = \displaystyle\int_0^{\pi/2} \cos^{2n} x \, dx = \int_0^{\pi/2} \cos^{2n-2} x(1 - \sin^2 x)\, dx$

$\qquad = I_{n-1} - \dfrac{1}{2n-1} I_n$ on integration by parts

$I_n = \dfrac{2n-1}{2n} I_{n-1} \qquad n \geqq 1$

$I_0 = \dfrac{\pi}{2}$

$$I_n = \frac{I_n}{I_{n-1}} \cdot \frac{I_{n-1}}{I_{n-2}} \cdots \frac{I_1}{I_0} I_0$$

$$= \frac{2n-1}{2n} \cdot \frac{2n-3}{2n-2} \cdots \frac{1}{2} \cdot \frac{\pi}{2}$$

2.6.1 $\quad \log(1-i) = \tfrac{1}{2}\log 2 - \dfrac{\pi i}{4} + 2\pi i$

2.7.1 $\quad i^i = e^{i\log i} = \exp i\left[\dfrac{\pi i}{2} + 2n\pi i\right] = \exp\left(-\dfrac{\pi}{2} - 2n\pi\right)$

2.7.3 $\quad (-1)^{1/6} = (e^{\pi i})^{1/6} = e^{\pi i/6} \cdot e^{2k\pi i/6} \qquad k = 0, 1, 2, \ldots, 5$

2.7.7 $\quad (x^2 + 2x + 2)(x^2 - 2x + 2)$

2.7.11 $\quad z_k = e^{2\pi i k/n}, \; k = 0, 1, \ldots, n-1$

$$|z_k| = 1 \qquad |z_{k+1} - z_k| = |e^{2\pi i/n} - 1|$$

so that all sides are equal.

2.7.12 A quilted diagram formed by the vertices of regular hexagons.

2.8.2 $\quad \arctan 2i = \dfrac{1}{2i} \log \dfrac{1-2}{1+2}$

$$= \frac{1}{2i}[-\log 3 + i\pi] + n\pi$$

Chapter | **DIFFERENTIATION:**
THREE | **ANALYTIC FUNCTIONS**

Differentiation is the most important analytic operation that we can perform on real functions. We are going to extend this operation to the complex domain. A complex function of a complex variable is equivalent to a pair of real functions of two real variables; only if the pair is exceptionally well assorted can we differentiate the complex function. Such well-assorted pairs of real functions have remarkable properties and are represented by mappings and vector fields of exceptional geometrical and physical importance.

3.1 DERIVATIVES

Not much knowledge of differential calculus is needed to convince us of the usefulness of derivatives in the solution of geometrical and physical problems. Thus we naturally wish to extend the concept of the derivative to the complex domain. We consider the complex variable w as a function of the independent complex variable z

$$w = f(z)$$

and we say, repeating word for word the familiar definition, that the *derivative of w with respect to z is defined to be the limit*

(1) $$\lim_{\Delta z \to 0} \frac{f(z + \Delta z) - f(z)}{\Delta z}$$

The denominator of the fraction considered, Δz, is the increment of the independent variable z, the numerator is the corresponding increment of the function $f(z)$, and the derivative itself is defined as the limit of the quotient of the increments when Δz, the increment of the independent variable, approaches 0. The derivative may or may not exist; the function $f(z)$ may, or may not, be differentiable. If the limit (1) exists, we use for it the familiar

75

notation; we write

$$(2) \qquad \lim_{\Delta z \to 0} \frac{f(z + \Delta z) - f(z)}{\Delta z} = f'(z) = \frac{df(z)}{dz} = w' = \frac{dw}{dz}$$

and we speak of the derivative of w with respect to z, or the derivative of the function $f(z)$.

For the simplest complex functions, which can be obtained by the extension of familiar operations to the complex domain (see Chapter 2), the derivatives can often be found by the methods used in the real domain.

Example. Find the derivative of the function $w = z^2$.

Let z receive the increment Δz. The corresponding increment of w is

$$\Delta w = (z + \Delta z)^2 - z^2 = 2z\Delta z + \Delta z^2$$

and the quotient of increments

$$(3) \qquad \frac{\Delta w}{\Delta z} = 2z + \Delta z$$

When Δz approaches 0, the right-hand side of (3) tends obviously to $2z$. Therefore,

$$(4) \qquad (z^2)' = \frac{dz^2}{dz} = 2z$$

3.1.1 Using the definition of the derivative, obtain the derivatives of

(a) $f(z) = \dfrac{z}{1 + z}$

(b) $f(z) = z^{1/2}$ (S)

(c) $f(z) = e^z$

3.1.2 Show that the following functions do not have a derivative with respect to the complex variable z.

(a) $f(z) = \bar{z}$

(b) $f(z) = x$ (S)

3.1.3 $f(z) = \dfrac{x^3 y(y - ix)}{x^6 + y^2}$ $z \neq 0$

$\qquad f(0) = 0$

Prove that $(f(z) - f(0))/(z - 0) \to 0$ as $z \to 0$ along any radius vector, but not as $z \to 0$ *in any manner.* (S)

3.2 RULES FOR DIFFERENTIATION

We can extend the familiar rules for differentiation to the complex domain without adding any essentially new ideas to the proofs. All we have to do is to convince ourselves, as in the example of the foregoing section, that the intervening rules and concepts apply to complex numbers and complex functions just as they apply to real numbers and real functions.

Let c denote a constant, and w and W functions of the complex variable z. We have the following rules:

(1) $$(w + W)' = w' + W'$$

(2) $$(wW)' = w'W + wW'$$

(3) $$\left(\frac{w}{W}\right)' = \frac{w'W - wW'}{W^2}$$

(4) $$c' = 0$$

(5) $$(cw)' = cw'$$

(6) $$\frac{dw}{dz} = \frac{1}{\dfrac{dz}{dw}}$$

(7) $$\frac{dW}{dz} = \frac{dW}{dw}\frac{dw}{dz}$$

Some of these rules require an explanation that, however, is scarcely different from the familiar explanations required in the real domain. Thus, (3) supposes that W does not vanish at the point z considered. Rule 6 supposes that the two relations

$$w = f(z) \quad \text{and} \quad z = g(w)$$

are exactly equivalent (so that the two functions considered are inverse to each other) and that

$$\frac{dz}{dw} = g'(w)$$

exists and does not vanish. Rule 7 supposes that

$$W = f(w) \qquad w = g(z)$$

and that the derivatives on the right-hand side of (7) exist. This rule allows us to calculate the derivative of the composite function

$$W = f[g(z)]$$

It can also be written in the more explicit but somewhat less intuitive form

$$\frac{df[g(z)]}{dz} = f'[g(z)]g'(z)$$

It is obvious that

$$z' = \frac{dz}{dz} = 1$$

Applying this and Rule 2 to $w = z$ and $W = z$, we find that

$$(z^2)' = 1 \cdot z + z \cdot 1 = 2z$$

the result we have derived directly in Section 3.1 (4). Using what we have just obtained, and applying Rule 2 again, we obtain

$$(z^3)' = (z^2 \cdot z)' = 2z \cdot z + z^2 \cdot 1 = 3z^2$$

Repeating the process we see that, in general,

(8) $$(z^n)' = nz^{n-1}$$

for any positive integer n. (We may use mathematical induction if we insist on formal rigor.)

Using Rule 8 together with the foregoing rules we are able to differentiate the simplest functions (in fact, any rational function) in the usual way. For instance, if $c_0, c_1, \ldots, c_n$ are constants, we find by combined application of (1), (4), (5), and (8) that

$$(c_0 + c_1 z + c_2 z^2 + \cdots + c_n z^n)' = c_1 + 2c_2 z + \cdots + nc_n z^{n-1}$$

It is natural to suspect that this formula extends to an infinite series of the form

$$c_0 + c_1 z + c_2 z^2 + \cdots + c_n z^n + \cdots$$

that we call a *power series* in z; the constants $c_0, c_1, c_2, \ldots, c_n, \ldots$ are called the *coefficients* of the power series. Thus, we are led to the formula

(9) $$(c_0 + c_1 z + c_2 z^2 + c_3 z^3 + \cdots)' = c_1 + 2c_2 z + 3c_3 z^2 + \cdots$$

Of course, we have not proved (9) rigorously. Its truth, however, is strongly suggested by the preceding and a rigorous proof may be given (Section 3.10).

Applying the foregoing rules to a few well-chosen examples, we may convince ourselves that all the familiar *formal rules of the differential calculus remain valid* in the complex.

Example 1. For complex values of z we defined the exponential function by the power series

$$e^z = 1 + \frac{z}{1!} + \frac{z^2}{2!} + \frac{z^3}{3!} + \cdots + \frac{z^n}{n!} + \cdots$$

see Section 2.2 (1). Applying (9), we find

$$(e^z)' = \frac{1}{1!} + \frac{2z}{2!} + \frac{3z^2}{3!} + \cdots + \frac{nz^{n-1}}{n!} + \cdots$$

$$= 1 + \frac{z}{1!} + \frac{z^2}{2!} + \cdots + \frac{z^{n-1}}{(n-1)!} + \cdots$$

That is, we have the same rule as in the real domain:

(10) $$(e^z)' = e^z$$

Example 2. For complex values of z, we have defined the trigonometric functions $\cos z$ and $\sin z$ by the power series

$$\cos z = 1 - \frac{z^2}{2!} + \frac{z^4}{4!} - \cdots$$

$$\sin z = z - \frac{z^3}{3!} + \frac{z^5}{5!} - \cdots$$

see Section 2.3. Applying (9), we find

$$(\cos z)' = -\frac{2z}{2!} + \frac{4z^3}{4!} - \frac{6z^5}{6!} + \frac{8z^7}{8!} - \cdots$$

$$(\sin z)' = 1 - \frac{3z^2}{3!} + \frac{5z^4}{5!} - \frac{7z^6}{7!} + \cdots$$

After obvious simplification, we find that just as in the real domain:

(11) $$(\cos z)' = -\sin z$$

(12) $$(\sin z)' = \cos z$$

Example 3. Find $(\tan z)'$.
From (3), (11), and (12),

$$(\tan z)' = \left(\frac{\sin z}{\cos z}\right)' = \frac{\cos z \cdot \cos z - \sin z(-\sin z)}{\cos^2 z} = \frac{1}{\cos^2 z}$$

3.2.1 Show that (a) $(\arctan z)' = \dfrac{1}{1 + z^2}$

(b) $(\arcsin z)' = \dfrac{1}{\sqrt{1 - z^2}}$

3.2.2   Verify that (a) $(\cosh z)' = \sinh z$

(b) $(\tanh z)' = \text{sech}^2 z$
$$= (\cosh z)^{-2}$$

3.2.3   Show that $(\log z)' = \dfrac{1}{z}$

3.2.4   Find (a) $(e^{\sin z})'$ (b) $(\log \cos z)'$

3.2.5   Find $(\cos \sqrt{z})'$ and $(e^{z^2})'$ and verify that the answers are consistent with the answers obtained by differentiating the appropriate infinite series term by term

Solution:
$$(\cos \sqrt{z})' = -\frac{\sin \sqrt{z}}{2\sqrt{z}}$$

$$= \left(1 - \frac{z}{2!} + \frac{z^2}{4!} - \frac{z^3}{6!} + \cdots\right)'$$

$$= -\frac{1}{2!} + \frac{2z}{4!} - \frac{3z^2}{6!} + \frac{4z^3}{8!} - \cdots$$

$$= -\frac{1}{2}\left(1 - \frac{z}{3!} + \frac{z^2}{5!} - \frac{z^3}{7!} + \cdots\right)$$

$$= -\frac{1}{2}\frac{\sin \sqrt{z}}{\sqrt{z}}$$

3.2.6   Show that $\lim\limits_{z \to 0} \dfrac{\sin z}{z} = (\sin z)'\Big|_{z=0}$

$$\lim\limits_{z \to 0} \frac{1 - \cos z}{z} = -(\cos z)'\Big|_{z=0}$$

3.3   ANALYTIC CONDITION FOR DIFFERENTIABILITY: THE CAUCHY-RIEMANN EQUATIONS

Formally, that is, from the stand point of formal rules, the differentiation of complex functions appears to be scarcely different from the differentiation of real functions. Materially, there is a great difference. In the real domain, differentiability is the normal case; practically all real functions of a real variable that intervene naturally in geometrical or physical problems possess a derivative. In the complex domain, differentiability is exceptional; that is, only in exceptional cases does the expression in Section 3.1 (1) have a definite limiting value.

Proceeding with our study, we shall have ample opportunity to understand this fundamental point. We can, however, already understand it to a certain extent. Let us remember that a complex function of a complex variable is equivalent to a pair of real functions of two real variables (2.9). If w is a function of z

$$w = f(z)$$

and

$$w = u + iv, \qquad z = x + iy$$

with real u, v, x, y, then both u and v are functions of x and y,

$$u = \phi(x, y), \qquad v = \psi(x, y)$$

Conversely, we may take any two real functions u and v of the variables x and y and compound them into a complex function $u + iv$. Now, if we take two real functions u and v at random, there are chances that they will turn out to be an ill-assorted pair, and then the resulting complex function will not be differentiable. Thus, the odds appear to be against differentiability.

If we wish to see the point more clearly we have to investigate the conditions that u and v must satisfy in order that $w = u + iv$ should be a differentiable function of $z = x + iy$. If w is differentiable, the quotient of the increments $\Delta w/\Delta z$ must tend to a limit when Δz approaches 0 in any manner whatever. Now Δz can approach 0 in many ways,* especially (a) through real values and (b) through purely imaginary values.

(a) We change z into $z + \Delta z$ by changing x alone and leaving y unchanged. Then the increment of z is real,

(1a) $$\Delta z = \Delta x$$

and

(2a) $$\Delta w = \Delta_x u + i\, \Delta_x v$$

where

$$\Delta_x u = \phi(x + \Delta x, y) - \phi(x, y), \qquad \Delta_x v = \psi(x + \Delta x, y) - \psi(x, y)$$

The subscript x in $\Delta_x u$ and $\Delta_x v$ emphasizes that these increments are due to a partial change of z; only the real part x is changed, the imaginary part y does not vary. From (1a) and (2a) it follows that

$$\frac{\Delta w}{\Delta z} = \frac{\Delta_x u}{\Delta x} + i\,\frac{\Delta_x v}{\Delta x}$$

and hence

(3a) $$\lim_{\Delta z \to 0} \frac{\Delta w}{\Delta z} = \lim_{\Delta x \to 0} \frac{\Delta_x u}{\Delta x} + i \lim_{\Delta x \to 0} \frac{\Delta_x v}{\Delta x} = \frac{\partial u}{\partial x} + i\,\frac{\partial v}{\partial x}$$

* This remark, the meaning of which is not entirely obvious, is illustrated by Example 3.1.3.

The derivatives on the right-hand side are partial derivatives with respect to x since the increments $\Delta_x u$ and $\Delta_x v$ resulted from a partial change of z, y being constant.

(b) Now we change z into $z + \Delta z$ by changing y alone and leaving x unchanged. Then the increment of z is purely imaginary.

$$\text{(1b)} \qquad \Delta z = i \, \Delta y$$

$$\text{(2b)} \qquad \Delta w = \Delta_y u + i \, \Delta_y v$$

$$\Delta_y u = \phi(x, y + \Delta y) - \phi(x, y) \qquad \Delta_y v = \psi(x, y + \Delta y) - \psi(x, y)$$

$$\frac{\Delta w}{\Delta z} = \frac{1}{i} \frac{\Delta_y u}{\Delta y} + \frac{\Delta_y v}{\Delta y}$$

and hence

$$\text{(3b)} \qquad \lim_{\Delta z \to 0} \frac{\Delta w}{\Delta z} = \frac{1}{i} \lim_{\Delta y \to 0} \frac{\Delta_y u}{\Delta y} + \lim_{\Delta y \to 0} \frac{\Delta_y v}{\Delta y} = -i \frac{\partial u}{\partial y} + \frac{\partial v}{\partial y}$$

In (3a) and (3b), we arrived at different expressions, but this is understandable; the difference of expressions originates from the difference in the choice of the increment Δz, see (1a) and (1b). Yet in whatever manner Δz approaches 0, the quotient $\Delta w/\Delta z$ is supposed to tend to the same limit, according to the definition of the derivative. Therefore, (3a) and (3b), although different in form, must be equal in value and so we obtain

$$\text{(4)} \qquad \frac{\partial u}{\partial x} + i \frac{\partial v}{\partial x} = -i \frac{\partial u}{\partial y} + \frac{\partial v}{\partial y}$$

This is a *necessary condition for the existence of the derivative*. We can present condition (4) in various ways. We can separate the real and imaginary parts and write

$$\text{(5)} \qquad \frac{\partial u}{\partial x} = \frac{\partial v}{\partial y} \qquad \frac{\partial u}{\partial y} = -\frac{\partial v}{\partial x}$$

Or we may prefer the (not quite orthodox) form

$$\text{(6)} \qquad \frac{\partial w}{\partial x} = \frac{\partial w}{i \, \partial y}$$

that exhibits strikingly the main point of the foregoing argument [(compare (3a) and (3b)]: differentiating horizontally or vertically, in the real or in the purely imaginary direction, we obtain the same value for the derivative.

It can be shown that the conditions (5) are not only necessary but also (essentially) sufficient for the existence of the derivative. We do not insist on the proof here (see Example 3.10) but we state the result. *The validity of the equations (5) is a necessary and sufficient condition for the existence of the derivative of the complex function* $u + iv$, provided that the partial derivatives $\partial u/\partial x$, $\partial u/\partial y$, $\partial v/\partial x$, $\partial v/\partial y$ exist and are continuous.

The equations (5) are called the *Cauchy-Riemann equations*: they form a system of two linear homogeneous partial differential equations of the first order. The main point is that they clear up essentially the question raised at the beginning of this section. We know now under which conditions a pair of real functions u and v is well- assorted and forms a differentiable complex function $u + iv$: the functions u and v must satisfy the Cauchy-Riemann equations (5). A pair of functions so assorted is called a pair of *conjugate functions*.

Example 1. Are the real and imaginary parts of the exponential function e^z conjugate functions?
From

$$u + iv = e^{x+iy} = e^x(\cos y + i \sin y)$$

follows

$$\frac{\partial u}{\partial x} = e^x \cos y \qquad \frac{\partial v}{\partial x} = e^x \sin y$$

$$\frac{\partial u}{\partial y} = -e^x \sin y \qquad \frac{\partial v}{\partial y} = e^x \cos y$$

The Cauchy-Riemann equations are satisfied, and so our u and v *are* conjugate functions; therefore the exponential function must possess a derivative. This we know already, see Section 3.2 (10). Using (6), we can compute the derivative in two different ways:

$$\frac{dw}{dz} = \frac{\partial w}{\partial x} = e^x(\cos y + i \sin y)$$

$$= \frac{1}{i}\frac{\partial w}{\partial y} = -ie^x(-\sin y + i \cos y) = e^z$$

Example 2. Are the real and imaginary parts of $\bar{z}$ conjugate functions?
From

$$u + iv = \overline{x + iy} = x - iy$$

follows

$$\frac{\partial u}{\partial x} = 1 \qquad \frac{\partial v}{\partial x} = 0$$

$$\frac{\partial u}{\partial y} = 0 \qquad \frac{\partial v}{\partial y} = -1$$

The first Cauchy-Riemann equation is not satisfied, our u and v are not conjugate functions, and so $\bar{z}$ considered as function of z is not a differentiable function.

3.3.1   Examine the Cauchy-Riemann conditions for the following functions, and state which of them are differentiable:

(a) $w = z + \bar{z} = 2x$ (S)

(b) $w = x + iy^2$

(c) $w = e^{-y}(\cos x + i \sin x)$

(d) $w = \dfrac{x}{x^2 + y^2} + \dfrac{iy}{x^2 + y^2}$

3.3.2   Show that for the single-valued analytic function

$$w = \sqrt{z} = \sqrt{r}\left(\cos\frac{\theta}{2} + i \sin\frac{\theta}{2}\right) \qquad 0 \leqq \theta < 2\pi$$

u and v satisfy the Cauchy-Riemann equations

3.3.3   In polar coordinates

$$w = f(z) = f(re^{i\theta})$$

Show that the Cauchy-Riemann equations become

$$u_r = \frac{1}{r} v_\theta \qquad v_r = -\frac{1}{r} u_\theta \qquad \text{(S)}$$

Solution: $\dfrac{\Delta f}{\Delta z} = \dfrac{\Delta(u + iv)}{\Delta(re^{i\theta})} = \dfrac{\Delta(u + iv)}{e^{i\theta}\,\Delta r}$ if θ constant

$$= \frac{\Delta(u + iv)}{r\,\Delta(e^{i\theta})} \text{ if } r \text{ constant}$$

Passing to the limit ($\Delta r \to 0$ or $\Delta\theta \to 0$) we have

$$f'(z) = e^{-i\theta}(u_r + iv_r)$$

$$= \frac{e^{-i\theta}}{ir}(u_\theta + iv_\theta)$$

3.3.4   Find a conjugate to

$$u(x, y) = x^3 - 3xy^2 \qquad \text{(S)}$$

3.3.5   Find a conjugate to

$$u(x, y) = \cos x \cosh y$$

3.3.6   Assume that $f(z)$ is analytic, and $f'(z) \equiv 0$. Show that $f(z) \equiv$ constant.   (S)

3.4 GRAPHICAL INTERPRETATION OF DIFFERENTIABILITY: CONFORMAL MAPPING

Any complex function of a complex variable can be represented graphically by the mapping of one plane onto another plane, but not every such function is differentiable. In order that w should be a differentiable function of z, a certain condition must be satisfied. Analytically, this condition is expressed by the Cauchy-Riemann equations. But what does this condition signify graphically? Which property of the mapping distinguishes a differentiable complex function from a nondifferentiable complex function?

The derivative is the limit of the quotient of the increments. If w as function of z is differentiable, this limit must exist in whatever manner the increment of z approaches 0. Let us compare two different increments of z; let $\Delta_1 z$ and $\Delta_2 z$ denote these increments, and let $\Delta_1 w$ and $\Delta_2 w$ denote the corresponding increments of w. Then the three points in the z-plane

$$z, \qquad z + \Delta_1 z, \qquad z + \Delta_2 z$$

are mapped onto the three points in the w-plane

$$w, \qquad w + \Delta_1 w, \qquad w + \Delta_2 w$$

respectively, that is,

$$w = f(z) \qquad w + \Delta_1 w = f(z + \Delta_1 z) \qquad w + \Delta_2 w = f(z + \Delta_2 z)$$

See Figure 3.1. We let $\Delta_1 z$ and $\Delta_2 z$ approach 0. We suppose that the derivative exists; since its value is independent of the manner in which the increment of Δz approaches 0, we have

(1)
$$\lim_{\Delta_1 z \to 0} \frac{\Delta_1 w}{\Delta_1 z} = \lim_{\Delta_2 z \to 0} \frac{\Delta_2 w}{\Delta_2 z} = f'(z)$$

Figure 3.1

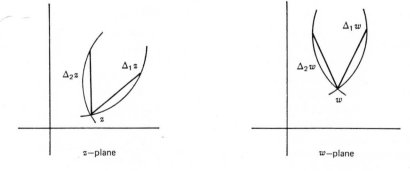

z—plane   w—plane

Graphical Interpretation of Differentiability: Conformal Mapping   **85**

Let us restrict ourselves to the consideration of the case in which the derivative at the point z does not vanish,

$$f'(z) \neq 0$$

If this is so, we can infer from (1) that

(2)
$$\lim_{\substack{\Delta_1 z \to 0 \\ \Delta_2 z \to 0}} \frac{\Delta_1 w \, \Delta_2 z}{\Delta_1 z \, \Delta_2 w} = \frac{f'(z)}{f'(z)} = 1$$

The meaning of this equation is that *approximately* (symbolized by $\doteq$ instead of $=$)

(3)
$$\frac{\Delta_1 w \, \Delta_2 z}{\Delta_1 z \, \Delta_2 w} \doteq 1$$

provided that $\Delta_1 z$ and $\Delta_2 z$ are both small. With the same degree of approximation, we have

(4)
$$\frac{\Delta_2 z}{\Delta_1 z} \doteq \frac{\Delta_2 w}{\Delta_1 w}$$

and the smaller $\Delta_1 z$ and $\Delta_2 z$ are, the better is the approximation.

Now, the complex equation (4) is equivalent to the following two real equations:

(5)
$$\left| \frac{\Delta_2 z}{\Delta_1 z} \right| \doteq \left| \frac{\Delta_2 w}{\Delta_1 w} \right|$$

(6)
$$\arg \frac{\Delta_2 z}{\Delta_1 z} \doteq \arg \frac{\Delta_2 w}{\Delta_1 w}$$

In order to see the geometric significance of these equations, let us fix our attention on the triangle in the z-plane with vertices z, $z + \Delta_1 z$, $z + \Delta_2 z$, and on the corresponding triangle in the w-plane with vertices w, $w + \Delta_1 w$, $w + \Delta_2 w$. Equation 6 expresses the fact that the angle at z in the first triangle is equal to the angle at w in the second triangle (equal in magnitude and sense). Equation 5 expresses the fact that the corresponding sides of the two triangles, which include the equal angles at z and w, are proportional. Therefore, by virtue of the two equations (5) and (6), or of the single equation (4), the triangles are similar.

The equation (4) on which the whole argument is based is only approximate, not exact; it tends to become exact as $\Delta_1 z$ and $\Delta_2 z$ approach 0. Therefore, we have to state our result carefully: *If the derivative exists and is different from 0, corresponding triangles tend to become similar when their dimensions tend to 0. Or, in other words, if the function represented by the mapping is differentiable, infinitesimal parts of the planes that are mapped onto each other are similar,* provided that $dw/dz \neq 0$. A mapping of this sort in which the infinitesimal parts are of the same form is termed *conformal*. Thus, we can say quite shortly that a *differentiable function yields a conformal mapping*. We may add (see Example 3.10.12) that only differentiable functions yield conformal mappings. Thus we know now which property of the mapping distinguishes the differentiable complex functions from the nondifferentiable complex functions.

The conformality of the mapping can be expressed in another manner that is, in fact, more distinct. We may be able to visualize the situation more clearly if we take it more concretely. Let us imagine (as in Section 2.12) that we have two geographic maps, one in the z-plane and one in the w-plane, both representing the same part of the surface of the globe. The corresponding points z and w represent the same geographic point, for example, a point where a river crosses the boundary line of two countries (Figure 3.2). Let $z + \Delta_1 z$ be a point on the boundary line in the first map. Then, of course, $w + \Delta_1 w$ is the corresponding point on the boundary line in the second map. Similarly, let the points $z + \Delta_2 z$ and $w + \Delta_2 w$ represent the same point of the river, each in its plane. When $\Delta_1 z$ is small it almost coincides in direction with the tangent to the boundary line at the point z. In fact, all four increments $\Delta_1 z$, $\Delta_2 z$, $\Delta_1 w$, $\Delta_2 w$ tend to coincide with the direction of certain tangents, of which two are issued from the point z in the z-plane and two are issued from the point w in the w-plane; on the other hand, two are tangent to a boundary line and two to a river line. But, as we have said before, the angle between $\Delta_1 z$ and $\Delta_2 z$ tends to be equal, in magnitude and sense, to the angle between $\Delta_1 w$ and $\Delta_2 w$. Therefore, the angle between the boundary line and the river must be the same on both maps. *If the derivative exists, the angles between corresponding lines are equal*, in magnitude and sense, unless the derivative vanishes at the point of intersection.

Figure 3.2

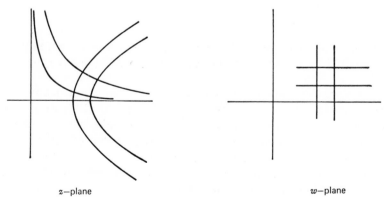

z—plane   w—plane

Graphical Interpretation of Differentiability: Conformal Mapping   **87**

We can summarize our argument by saying that a conformal transformation is necessarily angle preserving. We take this term to mean that corresponding angles are equal both in magnitude and sense.

Example. Let us reconsider the example of Section 2.12. The function $w = z^2$ is obviously differentiable and so the mapping is angle preserving (except at $z = 0$ where the derivative $2z$ vanishes). Now, the straight lines parallel to the imaginary axis of the w-plane intersect, at right angles, the straight lines parallel to the real axis. Therefore, the two sets of hyperbolas, which we have found (Section 2.12) to be the mapping of the above two sets of straight lines onto the z-plane, also intersect at right angles (Figure 3.2).

3.4.1 At what angle do the curves

$$\mathbf{R}z^3 = 1, \qquad \mathbf{R}z^3 = \mathbf{I}z^3 \quad \text{meet?}$$

3.4.2 Plot a few of the curves $r = \text{const}, \theta = \text{const}$ for $w = 1/z$, and verify that they intersect orthogonally.

3.4.3 Plot a few of the curves $u = \text{const}, v = \text{const}$ for $w = 1/z$, and verify that they intersect orthogonally.

3.4.4 Plot a few of the curves $u = \text{const}, v = \text{const}$ for $w = (z - 1)/(z + 1)$ and verify that they intersect orthogonally.

3.4.5 $w = f(z)$ is analytic. Prove that the curves $u(x, y) = \text{const}$, $v(x, y) = \text{const}$ intersect orthogonally provided $f'(z) \neq 0$ at the point of intersection.

3.4.6 Show that the slopes of the level curves $u(x, y) = \text{const}, v(x, y) = \text{const}$ for the analytic function $f(z) = u + iv$ are, respectively, $\cot [\arg f'(z)]$, $-\tan [\arg f'(z)]$.

3.5 PHYSICAL INTERPRETATION OF DIFFERENTIABILITY: SOURCELESS AND IRROTATIONAL VECTOR-FIELDS

Any complex function of a complex variable can be represented by a two-dimensional vector-field (as we have discussed in Section 2.13) but not every such function is differentiable. In order that w should be a differentiable function of z, a certain condition must be satisfied. Analytically this condition is expressed by the Cauchy-Riemann equations. Graphically, the condition is expressed by a particular property of the mapping representing the complex function: the mapping is conformal. But what does the condition signify physically? Which particular property of the vector-field distinguishes a differentiable complex function from a nondifferentiable complex function?

The clarity of the answer to this question depends very much on a good notation. Let z denote a variable point of our two-dimensional vector-field, and let $\bar{w}$ be the vector attached to the point z, as in Section 2.13. Let x and y denote the coordinates of the point z, as usual, and u and v the components of the vector $\bar{w}$. Then we have

$$z = x + iy$$

$$\bar{w} = u + iv$$

and, passing to the conjugate quantity,

(1) $$w = u - iv$$

We consider w as function of z

$$w = u - iv = f(z) = f(x + iy)$$

If this function is differentiable, we can obtain the value of the derivative by differentiating horizontally or vertically, that is, giving to z either a real or a purely imaginary increment, and therefore, according Section 3.3 (6)

(2) $$\frac{\partial w}{\partial x} = \frac{1}{i} \frac{\partial w}{\partial y}$$

or [we insist on notation (1)]

$$\frac{\partial u}{\partial x} - i \frac{\partial v}{\partial x} = \frac{1}{i}\left(\frac{\partial u}{\partial y} - i \frac{\partial v}{\partial y}\right)$$

Decomposing the foregoing complex equation into two real equations we obtain

(3) $$\frac{\partial u}{\partial x} + \frac{\partial v}{\partial y} = 0$$

(4) $$\frac{\partial v}{\partial x} - \frac{\partial u}{\partial y} = 0$$

These equations express that the function represented by our vector-field is differentiable. They are susceptible of an important *physical interpretation* that we shall explain in the present section. In explaining it, we consider alternately fields of flow and fields of force, and we take for granted certain physical concepts that will be discussed more thoroughly in the next section.

(a) We can express the physical significance of Equation 3 more intuitively if we regard our two-dimensional vector-field as a field of flow, and the vector $\bar{w}$ as a velocity, the intensity of the current at the point z. The expression

(5)
$$\frac{\partial u}{\partial x} + \frac{\partial v}{\partial y}$$

is termed the *divergence* of the vector $\bar{w}$ and is usually represented by the symbol div $\bar{w}$. The divergence measures the *outflow per unit volume* in a closed neighborhood of the point z. [More details follow in Section 3.6 (d).] If the divergence is positive, the point z acts as a "source," and if the divergence is negative, the point acts as a "sink." Equation 3 states that the outflow vanishes, that is, that the point is neither a source nor a sink. If the divergence vanishes at every point, the field is called *sourceless*. For instance, the field of steady flow of an incompressible fluid is sourceless; across any closed surface just as much matter goes in as out, so that the net outflow is zero. Equation 3 characterizes a sourceless vector-field.

(b) We can express the physical significance of Equation 4 more intuitively if we regard our two-dimensional vector-field as a field of force, and the vector $\bar{w}$ as a force, the intensity of the field at the point z. The expression

(6)
$$\frac{\partial v}{\partial x} - \frac{\partial u}{\partial y}$$

is termed the *curl* of the vector $\bar{w}$. The curl measures *work per unit area*. We mean the work done by the field when a small particle, under the influence of the field, describes a closed curve surrounding the point z. This work divided by the area enclosed by the curve, tends to the expression (6), the curl of $\bar{w}$, when the dimensions of the curve approach 0. More details follow in Section 3.6 (b). In a conservative field of force, the work along any closed curve is zero, and therefore the curl vanishes at every point. If the curl vanishes at each point, the field is called *irrotational*. (This term is connected with still another interpretation of the curl that we can just hint at here: the curl measures the rotation of an element of volume when $\bar{w}$ is the small displacement of a point in an elastic or plastic body.) Equation 4 characterizes an irrotational vector-field.

(c) Let us summarize the foregoing discussion. The differentiability of the function

$$w = u - iv = f(z) = f(x + iy)$$

is linked to the complex equation (2), which is equivalent to the two real equations (3) and (4), expressing that the divergence and the curl of $\bar{w}$ vanish; and if these two quantities vanish throughout a field, the field is both sourceless and irrotational. Therefore *a differentiable complex function of a complex variable is represented by a sourceless and irrotational field* but a nondifferentiable function is not represented by such a field.

Example. The vector-field considered in the example of Section 2.13 is sourceless and irrotational, since the function is differentiable

$$\frac{d}{dz}\left(\frac{1}{z}\right) = -\frac{1}{z^2}$$

We must, however, be careful to exclude from the field the origin where the function is not differentiable, not even continuous.

In fact, if we visualize the magnitude and direction of the vector $\bar{w} = 1/\bar{z}$, which we have discussed in Section 2.13, we may regard the field as representing a horizontal sheet of water, flowing uniformly in all directions away from the origin where, we may imagine, it is being pumped up continually from some underground source. The existence of the derivative in all points different from 0 shows that the origin is the only source.

3.6 DIVERGENCE AND CURL

The aim of the present section is to discuss more thoroughly the two concepts on which the physical interpretation of the differentiability of a complex function is based: divergence and curl. We consider a two-dimensional vector-field. The magnitude and direction of the vector with components u and v depend on the position of the variable point with coordinates x and y to which the vector is attached; that is

$$u = u(x, y) \qquad v = v(x, y)$$

We should not forget that a so-called two-dimensional field is, in fact, extended throughout three-dimensional space but has a particularly simple structure: it is stratified, its physical state being the same in all strata parallel to the x, y-plane (as we have discussed in Section 2.13). Various questions concerning the special case of two-dimensional fields can be better understood after a short consideration of the general case. When we have to discuss general, three-dimensional fields we call the vector of the field **w**, using heavy print.

(a) We begin with the consideration of a general, three-dimensional field. We regard it as a field of force and we let **w** denote the force in a (variable) point of the field. We wish to express the work performed by the field when a particle, under the influence of the field, describes a given curve C.

Consider a point of C, and let ds denote the element of the arc and $\mathbf{w}_t$ the tangential component of **w**; that is, the projection of **w** on the tangent of C. Then the work is obviously

$$\int \mathbf{w}_t \, ds$$

the integral being extended over the whole curve C.

Figure 3.3

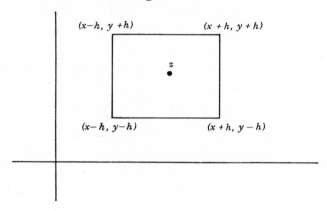

(b) Now, we descend to the consideration of a special two dimensional field. We regard $\bar{w}$ as a force. We wish to compute W, the work performed by the field when a particle, moving counterclockwise, describes the perimeter of a square of center z and side $2h$; the sides are parallel to the axes (Figure 3.3)

Of the four sides of the square, let us consider the east side first. This side is parallel to the y-axis and, along it, the particle moves northward. The position of the moving particle is given by the complex number

$$z + h + is$$

where s varies from $-h$ to h. The tangential component of $\bar{w}$, that is, the component in the direction of the path, is v or, more explicitly

$$v(x + h, y + s)$$

and so the work along the east side of the square is

(1)
$$\int_{-h}^{h} v(x + h, y + s) \, ds$$

This is a simple special case of the expression mentioned under (a). The work along the west side is

(2)
$$-\int_{-h}^{h} v(x - h, y + s) \, ds$$

the minus sign before the integral is due to the southward movement of the particle. The work along the remaining two sides of the square is computed similarly, and the expression for the work around the whole perimeter of the square may be written as

(3)
$$W = \int_{-h}^{h} [v(x + h, y + s) - v(x - h, y + s)$$
$$- u(x + s, y + h) + u(x + s, y - h)] \, ds$$

The two integrals (1) and (2) contribute the first two terms to the integrand in (3).

We compute now the work per unit area, dividing (3) by the area of the square:

(4)
$$\frac{W}{4h^2} = \frac{1}{2h} \int_{-h}^{h} \left[\frac{v(x+h, y+s) - v(x-h, y+s)}{2h} \right. $$
$$\left. - \frac{u(x+s, y+h) - u(x+s, y-h)}{2h} \right] ds$$

Finally, we pass to the limit, letting the side of the square approach zero. Observe that for any (continuous and differentiable) function $\phi(x)$

(5)
$$\lim_{h \to 0} \frac{\phi(x+h) - \phi(x-h)}{2h} = \phi'(x)$$

(6)
$$\lim_{h \to 0} \frac{1}{2h} \int_{-h}^{h} \phi(x+s) \, ds = \phi(x)$$

[The left-hand side of (6) is the mean value of the function $\phi(x)$ over an interval that shrinks into the point x as h approaches 0.] Dealing with (4), we face a sort of superposition of the two processes (5) and (6); we obtain*

(7)
$$\lim_{h \to 0} \frac{W}{4h^2} = \frac{\partial v}{\partial x} - \frac{\partial u}{\partial y} = \text{curl } \bar{w}$$

Thus, the curl of the vector $\bar{w} = u + iv$ measures the work per unit area in the immediate neighborhood of the point $z = x + iy$, as stated in Section 3.5 (b).

(c) We return to the consideration of a general three-dimensional field. We regard it now as a field of flow, and for the sake of simplicity, we admit that the flow is steady and the streaming fluid incompressible. We wish to express F, the volume of fluid crossing a given surface S in unit time.

Let dS denote an element of S and $\mathbf{w}$ the velocity of the current across the surface-element dS. If all the particles passing dS in a unit span of time were to retain their velocities, they would fill, at the end of that span of time, a prism whose (infinitesimal) base is dS and whose lateral edge is $\mathbf{w}$. This prism is in general oblique; its volume is $\mathbf{w}_n \, dS$ where $\mathbf{w}_n$ denotes the normal component of $\mathbf{w}$, that is, the altitude of the prism. Therefore, the volume of the fluid crossing the whole surface S in unit time is

(8)
$$\iint \mathbf{w}_n \, dS$$

---

* A quite strict proof has to appeal to two so-called "mean value theorems," first to the theorem of the differential calculus, then to the theorem of the integral calculus.

the integral being extended over the whole surface S. (The surface S has two sides. We have to designate one of the sides as the "inner" side and the other as the "outer" side. The component $\mathbf{w}_n$ is taken as positive when it is directed from the inner to the outer side, or parallel to the so-called *exterior normal*.)

The integral (8) is called the *flux* of the vector $\mathbf{w}$ across the surface S. The flux may have a physical significance even if $\mathbf{w}$ is not the intensity of a current, but some other sort of vector.

(d) Now, we redescend to the consideration of a special two-dimensional field. We regard the complex number $\bar{w}$ as the intensity of a two-dimensional stationary current. We wish to compute the flux F across the perimeter of a square of center z and side $2h$.

The term "flux" stands in this statement as an abbreviation for "flux per unit thickness" and even the latter, more detailed expression needs explanation. Therefore, let us visualize the situation. Between the z-plane, which we regard as horizontal and another horizontal plane at unit distance above the z-plane, there is a layer of the field of unit thickness. The particles of this layer passing above the perimeter of the square mentioned in our problem pass, in fact, across the lateral surface of a right prism whose base is the given square and whose altitude is 1. "Flux across the perimeter of the square" means indeed the flux across the lateral surface of the prism described. We should add that we consider the outgoing flow as positive.

Having well understood the terms of the problem, we pass to its solution. Of the four sides of the square, let us consider the east side first. The component of $\bar{w}$ normal to this side is u or, more explicitly,

$$u(x + h, y + s)$$

where s varies from $-h$ to h. The flux across the east side is

(9)
$$\int_{-h}^{h} u(x + h, y + s)\, ds$$

In fact, we compute the flux across a lateral face of the prism just described, which is a rectangle perpendicular to the z-plane; its base is the east side of our square and its altitude is 1. The element of the area of this rectangle is $1\, ds$ that, by the definition of the flux (8), has to be multiplied by the normal component of the vector $\bar{w}$, which is u. Integrating, we obtain (9).

The flux across the west side of the square is

(10)
$$-\int_{-h}^{h} u(x - h, y + s)\, ds$$

The minus sign results from the fact that the component of $\bar{w}$ in the direction of the exterior normal is $-u$ (and *not* u). The flux across the remaining two sides of the square is computed similarly, and the expression for the flux across the whole perimeter of the square may be written as

(11)
$$F = \int_{-h}^{h} [u(x + h, y + s) - u(x - h, y + s)$$
$$+ v(x + s, y + h) - v(x + s, y - h)] \, ds$$

The integrals (9) and (10) contribute the first two terms to the integrand in (11).

We obtain the flux per unit volume by dividing (11) by the volume of the prism:

(12)
$$\frac{F}{4h^2} = \frac{1}{2h} \int_{-h}^{h} \left[\frac{u(x + h, y + s) - u(x - h, y + s)}{2h} \right.$$
$$\left. + \frac{v(x + s, y + h) - v(x + s, y - h)}{2h} \right] ds$$

Finally, we let the side of the square approach 0. The mathematical difficulty is certainly not greater than it was in the case of (4); we obtain

(13)
$$\lim_{h \to 0} \frac{F}{4h^2} = \frac{\partial u}{\partial x} + \frac{\partial v}{\partial y}$$

Thus, the divergence of the vector $\bar{w} = u + iv$ measures the outflow per unit volume in the immediate neighborhood of the point $z = x + iy$ as stated in Section 3.5 (a).

(e) It is desirable to complete the foregoing discussion by studying the concepts of divergence and curl in a general vector-field. By seeing clearly how the special case of two-dimensional fields fits into a more general frame, we may improve our understanding of the whole subject. See Example 3.39.

3.6.1 Regarding the vector field $\bar{w} = 1/\bar{z}$ first as a velocity field of a fluid, and then as a field of force, compute the flux across, and the work done by the field along, a circle of radius r centered at the origin.

3.6.2 For the vector field $\bar{w} = 1/\bar{z}$, find the curves (streamlines) whose tangent at each point have the direction of the vector-field (whose slope is v/u).

3.7 LAPLACE'S EQUATION

We return to the analytic condition of differentiability; $w = u + iv$ is a differentiable function of $z = x + iy$ if, and only if, it satisfies the Cauchy-Riemann equations

$$\frac{\partial u}{\partial x} = \frac{\partial v}{\partial y} \qquad \frac{\partial u}{\partial y} = -\frac{\partial v}{\partial x}$$

See Section 3.3 (5). We observe that we can eliminate one of the two conjugate functions u and v from these two equations. In fact, differentiating the first equation with respect to x, the second with respect to y, and adding, we obtain

(1)
$$\frac{\partial^2 u}{\partial x^2} + \frac{\partial^2 u}{\partial y^2} = 0$$

This equation is called *Laplace's equation* or, more precisely, Laplace's equation for two-dimensional fields. Laplace's equation is of great importance in mathematical physics, and we shall return to it in later chapters. At the present stage, however, we restrict ourselves to a short comment.

Equation 1 shows that the real part u of a differentiable complex function $w = u + iv$ of the complex variable $z = x + iy$ cannot be chosen at random. In fact, u must satisfy Laplace's equation (1). Neither can the imaginary part be chosen arbitrarily; in fact, v must also satisfy Laplace's equation,

(2)
$$\frac{\partial^2 v}{\partial x^2} + \frac{\partial^2 v}{\partial y^2} = 0$$

To show (2), we can proceed as we did in deriving (1), by eliminating u from the Cauchy-Riemann equations. Or we may observe that, $w = u + iv$ being differentiable,

$$-iw = v - iu$$

is differentiable too and that, therefore, its real part, which is v, must satisfy Laplace's equation and so we have (2).

Both the real and imaginary parts of a differentiable complex function of a complex variable satisfy Laplace's equation.

3.7.1 $u(x, y)$ is a solution of Laplace's equation (is *harmonic*) in the interior of a rectangular region. Show that there exists an analytic function $u + iv$ in this region, which is uniquely determined, except for an additive constant.

Solution: Define $v_y = u_x$

$$v = \int_{y_0}^{y} u_x(x, s)\, ds + h(x)$$

where $h(x)$ is (with respect to y) an arbitrary "constant of integration." Then

$$v_x = \int_{y_0}^{y} u_{xx}(x, s)\, ds + h'(x)$$

$$= -\int_{y_0}^{y} u_{yy}(x, s)\, ds + h'(x)$$

$$= -u_y(x, y) + u_y(x, y_0) + h'(x)$$

The C-R equations are satisfied if $h'(x) = u_y(x, y_0)$, which determines $h(x)$ except for a constant of integration.

3.7.2 Show that the product of two harmonic functions is itself harmonic if one is the conjugate of the other.

3.8 ANALYTIC FUNCTIONS

We call a complex function of a complex variable *analytic* if it is differentiable. The term is appropriate. Differentiation is the most important operation of mathematical analysis, and therefore it is natural to call a function admitting this operation analytic.*

A function $f(z)$ is said to be analytic in a region of the z-plane if it has a derivative at each point of that region. For instance, the function e^z is analytic in the whole plane, and $\bar{z}$ is nowhere analytic; this is shown by the solution of Examples 1 and 2 of Section 3.3.

Turning to a general example, let us suppose that both w and W are analytic functions of z. That is, we suppose that the derivatives w' and W' exist. But then also the derivative of wW exists; in fact, it can be computed according to the familiar rule:

$$(wW)' = w'W + wW'$$

Consequently, wW is also analytic; *the product of two analytic functions in an analytic function.*

We have used the familiar rule for the differentiation of a product. Using, in the same way, the other simple rules of the differential calculus, we find analogous results, especially the following: *a function obtained from analytic functions by the four fundamental arithmetical operations—addition, subtraction, multiplication, and division—is an analytic function provided, of course, that points at which a division by 0 occurs are excluded.*

Apply this rule to z (which is an analytic function of z, the derivative being 1) and to any constants (which are also analytic functions, the derivative being 0. From z and constants we can obtain any rational function by the four fundamental arithmetic operations, and so we find the proposition that *any rational function is an analytic function.* We must, of course, exclude the points at which the denominator vanishes; at such points the function is, strictly speaking, not even defined [as we shortly mentioned in connection with Section 2.1 (3)].

Another familiar rule of the differential calculus, Section 3.2 (7), yields the proposition that *an analytic function of an analytic function is an analytic*

---

* Various other terms are used by some authors, instead of analytic, such as holomorphic regular, regular analytic, and synectic, for example.

function. For instance, e^{z^2} is an analytic function. Still another rule, Section 3.2 (6), leads to the proposition that *the inverse function of an analytic function is an analytic function,* provided that the points at which the derivative of the original function vanishes are excluded. For instance, log z is an analytic function since it is the inverse of the exponential function.

We can summarize the preceding by the somewhat vague but suggestive statement that the class of analytic functions is "self-sufficient." If we need the result of simple operations performed on analytic functions, we need not look for it outside the class; we can find the desired result within the class of analytic functions. The same fact is expressed in other terms by saying that the set of analytic functions is *closed* with respect to the operations considered.

3.8.1   Show that

$$\left(\sum_1^\infty (-1)^{n+1} \frac{z^n}{n}\right)' = \sum_0^\infty (-1)^n z^n = \frac{1}{1+z}$$

so that

$$\sum_1^\infty (-1)^{n+1} \frac{z^n}{n} = \log(1+z) \qquad |z| < 1$$

The series is single valued, while the logarithm is infinitely multivalued. Which one of the determinations of log $(1 + z)$ is represented by the series?

Solution: $(\log 1 + z)' = \dfrac{1}{1+z} = \sum_0^\infty (-1)^n z^n \qquad |z| < 1$

$$= \left(\sum_1^\infty (-1)^{n+1} \frac{z^n}{n}\right)',$$

Thus, by Example 3.3.6

$$\log(1+z) = \sum_1^\infty (-1)^{n+1} \frac{z^n}{n} + C$$

Let $z = 0$, then we have log $1 = C$, and here $C = 0$ so that we have the principal value branch of the logarithm.

3.9 SUMMARY AND OUTLOOK

Taking a complex function of a complex variable "at random," we have little chance to hit upon an analytic function. Among complex functions of a complex variable, analytic functions form a small and select class; only functions fulfilling certain strict conditions are admitted. In the foregoing sections, we have investigated the conditions characterizing differentiable functions (i.e., analytic functions) from three different standpoints: analytically, graphically, and physically.

A function is analytic if its real and imaginary parts satisfy the Cauchy-Riemann equations.

A function is analytic if it is represented by a conformal, angle-preserving mapping of one plane onto another.

A function is analytic if it is represented by a sourceless and irrotational two-dimensional vector-field.

Strictly speaking, we should have said "if and only if" instead of the short "if" in all of the three foregoing statements; the conditions given are both necessary and sufficient. Therefore, the three conditions, although strikingly different in form, are equivalent. And we have not yet exhausted the various forms of the conditions that characterizes analyticity. We shall find still other remarkable and surprising forms later, in studying the expansion and integration of analytic functions in Chapters Four, Six, and Seven.

There is a condition for analyticity that is so simple that we can state it here although we are very far yet from proving it. We know that any rational function is an analytic function (Section 3.8). It is plausible (although not so easy to prove) that a function that is a limit of rational functions is also an analytic function (provided that the convergence is uniform). Yet it is surprising that the converse is also true: any analytic function is either a rational function or the limit of a (uniformly convergent) sequence of rational functions (Section 6.24). The class of analytic functions is small as we said before, but it is an extremely interesting class and contains almost all functions useful in physical problems. In the sections that follow, we shall study only analytic complex functions of a complex variable.

Although we have scarcely taken the very first steps in the theory of analytic functions, we may forsee already that this theory has manifold and remarkable connections. In particular (this is the point most important for us), analytic functions constitute a powerful tool for solving problems of mathematical physics and engineering.

Additional Examples and Comments on Chapter Three

COMMENT 1.   In Section 3.3, we obtained the Cauchy-Riemann equations as being *necessary* for the existence of a derivative with respect to the complex variable z. We show here that these conditions are also *sufficient*.

We are given two functions $u(x, y)$ and $v(x, y)$ satisfying

$$u_x = v_y \qquad u_y = -v_x$$

in a neighborhood of a point (x_0, y_0), and know in addition that these derivatives are continuous in this neighborhood. Accordingly, we must show that with $f(z) = u + iv$, $\Delta f/\Delta z = [f(z_0 + \Delta z) - f(z_0)]/\Delta z$ tends to a unique limit as $\Delta z \to 0$ in any manner whatever. Let us set $\Delta z = h + ik$.

$$\frac{\Delta f}{\Delta z} = \frac{u(x_0 + h, y_0 + k) - u(x_0, y_0)}{h + ik} + i\,\frac{v(x_0 + h, y_0 + k) - v(x_0, y_0)}{h + ik}$$

$$= \frac{u(x_0 + h, y_0 + k) - u(x_0, y_0 + k)}{h + ik} + \frac{u(x_0, y_0 + k) - u(x_0, y_0)}{h + ik}$$

+ two similar terms with $v(x, y)$ in place of $u(x, y)$.

Each of the two quotients in $u(x, y)$ now have one of the two variables the same ($y_0 + k$ in the first, x_0 in the second), and we are essentially involved with a function of one variable only.

What is useful to know here is the mean value theorem for the differential calculus, as well as the definition of continuity. Let us recall these.

$f(x)$ is continuous at $x = x_0$ if $|f(x) - f(x_0)|$ can be made less than any preassigned quantity merely by restricting $|x - x_0|$ to a suitable interval. An equivalent statement, more useful for our purposes, is

$$f(x_0 + h) = f(x_0) + \varepsilon(x_0, h) \qquad \text{where} \qquad \lim_{h \to 0} \varepsilon(x_0, h) = 0$$

The mean value theorem states

$$\frac{f(x_0 + h) - f(x_0)}{h} = f'(\xi) \qquad x_0 < \xi < x_0 + h$$

If, in addition (as we shall require), $f'(x)$ is also continuous at $x = x_0$, we can combine the two statements into the single equation

3.02 $\qquad f(x_0 + h) = f(x_0) + hf'(x_0) + h\varepsilon(x_0, h)$

$\qquad$ where $\lim_{h \to 0} \varepsilon(x_0, h) = 0$

Applying the result 3.02 to 3.01, we have

3.03 $\qquad \dfrac{\Delta f}{\Delta z} = \dfrac{hu_x(x_0, y_0)}{h + ik} + \dfrac{ku_y(x_0, y_0)}{h + ik} + i\,\dfrac{hv_x(x_0, y_0)}{h + ik} + i\,\dfrac{kv_y(x_0, y_0)}{h + ik}$

$$+ \frac{h\varepsilon_1}{h + ik} + \frac{k\varepsilon_2}{h + ik} + \frac{ih\varepsilon_3}{h + ik} + \frac{ik\varepsilon_4}{h + ik}$$

where

$$\lim_{h, k \to 0} \varepsilon_\nu = 0 \qquad \nu = 1, 2, 3, 4$$

Using the Cauchy-Riemann conditions, Section 3.10.3 becomes

$$\frac{\Delta f}{\Delta z} = u_x(x_0, y_0) + iv_x(x_0, y_0) + \frac{h(\varepsilon_1 + i\varepsilon_3) + k(\varepsilon_2 + i\varepsilon_4)}{h + ik}$$

Now

$$\left|\frac{h}{h+ik}\right| \leqq 1 \qquad \left|\frac{k}{h+ik}\right| \leqq 1 \qquad \text{so that}$$

$$\left|\frac{\Delta f}{\Delta z} - u_x(x_0, y_0) - iv_x(x_0, y_0)\right| \leqq |\varepsilon_1| + |\varepsilon_2| + |\varepsilon_3| + |\varepsilon_4| \to 0 \text{ as } h, k \to 0$$

That is, $f'(z)$ exists, and equals $u_x + iv_x$.

COMMENT 2.   We have verified, in special cases, that the result of differentiating a power series term by term is consistent with the result obtained from differentiating other forms of the function. For example

$$(\cosh z)' = \left(\sum_0^\infty \frac{z^{2n}}{(2n)!}\right)' = \sum_0^\infty \frac{z^{2n+1}}{(2n+1)!} = \sinh z$$

and

$$(\cosh z)' = \left(\frac{e^z + e^{-z}}{2}\right)' = \left(\frac{e^z - e^{-z}}{2}\right) = \sinh z$$

In addition, we have seen (Example 2.14.4) that a differentiated power series has the same radius of convergence as the original series.

To prove that, in general, we can differentiate power series term by term, two things are required: (1) that the function $f(z) = \sum_0^\infty a_n z^n$ is indeed analytic, and (2) that the formally differentiated series is itself the derivative of $f(z)$.

LEMMA.   If $f(z) = \sum_0^\infty a_n z^n$ converges for $|z| < R$, then $f(z)$ is analytic in $|z| < R$, and $f'(z) = \sum_0^\infty na_n z^{n-1}$.

PROOF.   For any z in $|z| < R$, let h be an arbitrary complex number so that $|z| + |h| < R$. Then

$$\frac{\Delta f}{\Delta z} = \frac{f(z+h) - f(z)}{h} = \sum_0^\infty a_n \frac{(z+h)^n - z^n}{h}$$

Now

$$\frac{(z+h)^n - z^n}{h} = \left[z^n + nz^{n-1}h + \frac{n(n-1)z^{n-2}}{2!}h^2 + \cdots - z^n\right]\frac{1}{h}$$

$$= nz^{n-1} + \frac{n(n-1)z^{n-2}}{2!}h + \frac{n(n-1)(n-2)z^{n-3}}{3!}h^2 + \cdots$$

This gives us

$$\frac{\Delta f}{\Delta z} - \sum_0^\infty n a_n z^{n-1} = \sum_2^\infty a_n \left[\frac{n(n-1)z^{n-2}}{2!} h + \frac{n(n-1)(n-2)z^{n-3}}{3!} h^2 + \cdots\right]$$

$$\left|\frac{\Delta f}{\Delta z} - \sum_0^\infty n a_n z^{n-1}\right| \leqq |h| \sum_2^\infty |a_n| \frac{n(n-1)}{2}\left[|z|^{n-2} + \frac{n-2}{3!} 2 |z|^{n-3} |h|\right.$$

$$\left. + \frac{(n-2)(n-3)}{4!} 2 |z|^{n-4} |h|^2 + \cdots\right]$$

$$\leqq |h| \sum_2^\infty |a_n| \frac{n(n-1)}{2} (|z| + |h|)^{n-2}$$

We can now draw our conclusions: the series on the right converges (and hence is finite) for $|z| < R$, $|z| + |h| < R$, since the radius of convergence of a differentiated power series is the same as for the given series; when $h \to 0$, then, the right-hand side tends to zero, so that $\Delta f/\Delta z$ does indeed tend to a limit, that is, $f(z)$ is analytic. The derivative is clearly $\sum_0^\infty n a_n z^{n-1}$.

3.1 Regarding $x = (z + \bar{z})/2$, $y = (z - \bar{z})/2i$ as a change of independent variables from x, y to $z, \bar{z}$, show (in relying on the usual rules) that

$$\frac{\partial}{\partial x} = \frac{\partial}{\partial z} + \frac{\partial}{\partial \bar{z}}$$

$$\frac{\partial}{\partial y} = i\left(\frac{\partial}{\partial z} - \frac{\partial}{\partial \bar{z}}\right)$$

$$\frac{\partial^2}{\partial x^2} + \frac{\partial^2}{\partial y^2} = 4\frac{\partial^2}{\partial z \, \partial \bar{z}}$$

(*Note.* $\partial/\partial z$ is not to be confused with d/dz: the first is the partial derivative with respect to z, with $\bar{z}$ held fixed; the second is the usual derivative with respect to the complex variable z in the sense of Section 3.1.)

3.2 With the notation of Example 3.1, any function of the two variables $x, y, f(x, y)$ can be written

$$f(x, y) = f\left(\frac{z + \bar{z}}{2}, \frac{z - \bar{z}}{2i}\right) = g(z, \bar{z})$$

Show that the Cauchy-Riemann equations become

$$\frac{\partial f}{\partial \bar{z}} = 0 \quad \text{in which case} \quad g'(z) = \frac{\partial f}{\partial z}$$

3.3 With the notation of Example 3.1 show that Laplace's equation becomes $\partial^2 u/\partial z \partial \bar{z} = 0$, whose general solution (involving two arbitrary functions) is

$$u = f(z) + g(\bar{z})$$

Conversely, if f, g are analytic functions, then show that $u(x, y) = f(z) + g(\bar{z})$ is harmonic.

3.4 A standard abbreviation for the Laplacian operator $\partial^2/\partial x^2 + \partial^2/\partial y^2$ is Δ. Show that, for $f(z)$ analytic

$$\Delta \, |f(z)|^2 = 4 \, |f'(z)|^2$$

3.5 Given $\phi(x, y)$, let $\Delta_{xy}\phi$ be its Laplacian. Let $w = f(z)$ be analytic; make the change of variables x, y, to u, v and consider ϕ as a function of u, v. Show that

$$\Delta_{xy}\phi = \Delta_{uv}\phi \, |f'(z)|^2$$

3.6 Find a particular solution of

$$\Delta\phi = xy$$

Solution: $\quad 4\phi_{z\bar{z}} = \dfrac{z + \bar{z}}{2} \cdot \dfrac{z - \bar{z}}{2i} = \dfrac{z^2 - \bar{z}^2}{4i}$

Formal integration gives $\phi = \dfrac{z^3\bar{z}}{48i} - \dfrac{\bar{z}^3 z}{48i}$

$$= \frac{xy}{12}(x^2 + y^2)$$

A general solution is thus $(xy/12)(x^2 + y^2) + \mathbf{R}f(z)$, where $f(z)$ is any analytic function.

3.7 Find (a) a general solution of $\Delta(\Delta\phi) = 0$ and (b) a particular integral for $\Delta(\Delta\phi) = xy$.

3.8 Let $\partial/\partial\xi$ and $\partial/\partial\eta$ be directional derivatives in two perpendicular directions so that a positive rotation by $90°$ leads from ξ to η. Then show that $w = u + iv$ is analytic if, and only if, $u_\xi = v_\eta$, $u_\eta = -v_\xi$, and the derivatives are continuous.

3.9 $f_1(z)$, $f_2(z)$, $\ldots$, $f_n(z)$ are analytic, and $|f_1(z)|^2 + |f_2(z)|^2 + \cdots + |f_n(z)|^2$ is harmonic. Prove that each $f_k(z)$ is a constant.

3.10   Show that $f(z) = u + iv$, analytic in a region R, is identically a constant in R if any of the following statements hold:

(a) $u = 0$
(b) $v = 0$
(c) $|f(z)| = \text{const}$
(d) $\overline{f(z)}$ is differentiable

3.11 $f(z) = Re^{i\psi}$ is analytic. Show that

$$R_x = R\psi_y \qquad R_y = -R\psi_x$$

and that

$$\frac{f'(z)}{f(z)} = \frac{\partial \log R}{\partial x} + i \frac{\partial \psi}{\partial x}$$

3.12   Suppose $w = f(z)$ gives rise to a conformal map of a region R in the z-plane onto a region R' of the w-plane. That is, any curve in R with a continuously turning tangent is mapped onto a corresponding curve in R', with a continuously turning tangent; moreover, any two such curves in R that intersect at an angle θ are mapped onto curves in R' that intersect at the same angle θ; the two angles agree not only in magnitude, but also in sense. Assume that u and v have continuous partial derivatives of the first order ($w = u + iv$ as usual) and prove that $f(z)$ is analytic in R.

Solution: Let $x = \phi(t)$, $y = \psi(t)$ be the parametric representation of a curve C in R with continuously turning tangent, that is $\dot\phi(t)$ and $\dot\psi(t)$, the derivatives with respect to t, exist and are continuous; C passes through a certain point z_0 of R and its slope at z_0 is $\tan \tau = \dot\psi(t_0)/\dot\phi(t_0)$.

Then $u = u[\phi(t), \psi(t)]$, $v = v[\phi(t), \psi(t)]$ is a parametric representation of the image C' of C in R'; let w_0 be the image of z_0; then the slope of C' at the point w_0 is

$$\frac{v_x\dot\phi + v_y\dot\psi}{u_x\dot\phi + u_y\dot\psi} = \frac{v_x + v_y \tan \tau}{u_x + u_y \tan \tau}$$

Let $\tan \gamma = v_x/u_x$ stand for the slope of the image in R' at w_0 of a parallel to the x-axis through z_0 in R. The conformality the mapping requires that C and the horizontal direction at z_0 include the same angle as C' and the image of this direction at w_0, and so the slope of C' at w_0 must be $\tan(\gamma + \tau)$ which quantity developed stands on the left-hand side of the equation:

$$\frac{v_x + u_x \tan \tau}{u_x - v_x \tan \tau} = \frac{v_x + v_y \tan \tau}{u_x + u_y \tan \tau}$$

This must hold whatever the direction of C at the point z_0 may be, that is, we have here an identity in $\tan \tau$. By straightforward manipulation, the identity yields the two equations

$$u_x(v_y - u_x) - v_x(u_y + v_x) = 0$$

$$v_x(v_y - u_x) + u_x(u_y + v_x) = 0$$

This is a system of two linear homogeneous equations with determinant $u_x^2 + v_x^2$; unless this determinant vanishes, both unknowns must vanish and so we must have

$$v_y - u_x = 0 \qquad u_y + v_x = 0$$

the Cauchy-Riemann equations. If, however, the determinant vanishes, that is $u_x = 0$ and $v_x = 0$, we interchange the roles of the x- and the y-axis. If $u_y^2 + v_y^2 > 0$, we obtain again the same conclusion. If, however, all four partial derivatives u_x, v_x, u_y, and v_y vanish, the Cauchy-Riemann equations trivially hold.

3.13   Given $u(x, y) = x^2/y^2$, $v(x, y) = x^2 + y^2$, show that the level lines $u = $ const, $v = $ const intersect orthogonally, but that the mapping is not conformal.

Solution: See 3.12.

3.14   Under a conformal mapping by $w = f(z)$, show that the *linear magnification* at z is $|f'(z)|$ (that is, an element of arc ds in the z-plane is multiplied by $|f'(z)|$ in the w-plane); also show that the area or superficial magnification is $|f'(z)|^2$. Thus show that a region D of the z-plane is mapped onto a region of the w-plane whose area

$$A = \int_D \int |f'(z)|^2 \, dx \, dy.$$

3.15   The rectangle in the z-plane bounded by the lines $x = 0$, $x = 2$, $y = 0$, $y = \pi/4$ is mapped by $w = \cosh z$ onto a domain in the w-plane. Show that the area of this image domain is $(\pi \sinh 4 - 8)/16$.

Solution: See Example 3.14.

3.16   The numbers a, c are real, $0 < c < \pi$. Find the area in the w-plane of the image of the rectangle

$$a - c < x < a + c \qquad -c < y < c$$

under the mapping $w = e^z$. Find the ratio of the areas of the corresponding domains, and show that this ratio approaches e^{2a} as $c \to 0$.

3.17 $w = f(z)$ is analytic in $|z| < R$, and maps the circle $|z| = r < R$ onto a curve C. Show that the length of C is

$$\int_0^{2\pi} |f'(re^{i\theta})|\, r\, d\theta$$

Solution: See Example 3.14.

3.18   Locate the points of equal linear magnification for the mapping $w = z^2$ (Example 3.14). The change in the direction of a line element under the mapping $w = f(z)$ is $\arg f'(z)$, and we might call it the *phase shift* at z [provided $f'(z) \neq 0$]. Locate the points of equal phase shift.

3.19   Locate the points of equal linear magnification for $w = (az + b)/(cz + d)$, $ad - bc = 1$ and, in particular, for unit magnification.

3.20   Verify that $\int_0^{2\pi} e^{in\theta}\, d\theta = 0$ if n integral, $\neq 0$

$$= 2\pi \text{ if } n = 0.$$

3.21 $f(z) = \sum_0^\infty a_n z^n$ is analytic in $|z| < $ R. Show that

(a) $|f(re^{i\theta})|^2 = \sum_0^\infty a_k \bar{a}_{n-k} e^{i\theta(2k-n)}$

(b) $\int_0^{2\pi} |f(re^{i\theta})|^2\, d\theta = 2\pi \sum_0^\infty r^{2n} |a_n|^2$

Solution: Example 3.20.

3.22 $w = f(z)$ maps the circle $|z| = r$ onto a curve C of the w-plane. Show that the curvature of C is

$$k = \frac{1}{\rho} = \frac{1 + \mathbf{R}z\dfrac{f''(z)}{f'(z)}}{|zf'(z)|}$$

($1/\rho = d\phi/ds$, where ϕ is the angle between the tangent to the curve and the positive real axis and s the arc length.)

3.23   If the radius of curvature of a curve is of constant sign, the curve stays on the same side of its tangent. Such curves are called *convex*. A line segment joining any two points of the convex set bounded by a closed convex curve passes through points of the set only. Show that a sufficient condition for the convexity of the image of $|z| = r$ under $w = f(z)$ is that

$$\mathbf{R}z\frac{f''(z)}{f'(z)} > -1$$

Solution: Example 3.22.

3.24   Prove that

$$f(z) = 1 + \sum_1^\infty \frac{\alpha(\alpha - 1) \cdots (\alpha - n + 1)}{n!} z^n$$

is analytic for $|z| < 1$, and that $f'(z) = \alpha f(z)/(1 + z)$. Thus show that

$$[(1 + z)^{-\alpha} f(z)]' = 0 \qquad \text{and} \qquad f(z) = (1 + z)^\alpha$$

Solution: The coefficients

$$\left| \frac{\alpha(\alpha - 1) \cdots (\alpha - n + 1)}{n!} \right| \leq |\alpha| \frac{(|\alpha| + 1) \cdots (|\alpha| + 1 + n)}{n!}$$

$$\leq \frac{(n + k)!}{n!}$$

The nth root test shows that

$$\sum_0^\infty \frac{(n + k)!}{n!} r^n$$

converges for $r < 1$, hence so does our series. Differentiating,

$$f'(z) = \sum_1^\infty \frac{\alpha(\alpha - 1) \cdots (\alpha - n + 1)}{(n - 1)!} z^{n-1}$$

$$(1 + z)f'(z) = \sum_0^\infty \frac{\alpha(\alpha - 1) \cdots (\alpha - n)}{n!} z^n + \sum_1^\infty \frac{\alpha(\alpha - 1) \cdots (\alpha - n + 1)}{(n - 1)!} z^n$$

$$= \alpha + \sum_1^\infty \frac{\alpha(\alpha - 1) \cdots (\alpha - n + 1)}{(n - 1)!} z^n \left(\frac{\alpha - n}{n} + 1 \right)$$

$$= \alpha f(z)$$

This expresses the fact that $[(1 + z)^{-\alpha} f(z)]' = 0$ so that

$$f(z) = (1 + z)^\alpha \cdot \text{const}$$

Let $z = 0$; $f(0) = 1$ so that the constant is 1.

3.25   As well as infinite series, there are other limiting operations of considerable importance in analysis. One of these is infinite products. For example, see Example 2.5.10

$$\int_0^{\pi/2} \cos^{2n} x \, dx = \frac{\pi}{2} \cdot \frac{1}{2} \cdot \frac{3}{4} \cdot \frac{5}{6} \cdots \frac{2n - 1}{2n}$$

The integral tends to zero as $n \to \infty$ and, more precisely, it can be shown (Example 10.4.1) that

$$\sqrt{\pi} = \lim_{n \to \infty} \frac{2 \cdot 4 \cdot 6 \cdots 2n}{1 \cdot 3 \cdot 5 \cdots (2n-1)} \frac{1}{\sqrt{n}}$$

or that (this is Wallis' product)

$$\frac{\pi}{2} = \lim_{n \to \infty} \frac{2 \cdot 2}{1 \cdot 3} \frac{4 \cdot 4}{3 \cdot 5} \frac{6 \cdot 6}{5 \cdot 7} \cdots \frac{2n \cdot 2n}{(2n-1)(2n+1)}$$

$$P_n = u_1 u_2 \cdots u_n = \prod_1^n u_k$$

where the symbol $\prod$ is to products as $\sum$ is to sums. If P_n is to tend to a limit as $n \to \infty$, then P_{n+1} tends to the same limit, so that $(P_{n+1})/P_n = u_{n+1} \to 1$ provided, of course that $P_n \neq 0$ and $P_n \nrightarrow 0$. We exclude, for the moment, the possibility that $P_n = 0$ for any n, that is, that any $u_k = 0$, and regard the case $P_n \to 0$ as divergence. Then a necessary condition for convergence, that is, that $\lim_{n \to \infty} P_n$ exists and differs from 0, is that $u_n \to 1$. We emphasize this fact by writing $u_n = 1 + a_n$

$$P_n = \prod_1^n (1 + a_k)$$

and then the necessary condition of convergence is

$$\lim_{n \to \infty} a_n = 0$$

Show three facts:

(a) For all real values of x, except for $x = 0$,

$$1 + x < e^x$$

(b) If $a_n \geqq 0$,   and

$$a_1 + a_2 + \cdots + a_n = S_n$$

then

$$1 + S_n \leqq P_n \leqq e^{S_n}$$

(c) If $a_n \geqq 0$

$$\sum_1^\infty a_n \quad \text{and} \quad \prod_1^\infty (1 + a_n)$$

are equiconvergent, that is, they converge or diverge together, the convergence or divergence of one implies the same behavior of the other.

Solution: (a) That

$$e^x = 1 + \frac{x}{1!} + \frac{x^2}{2!} + \frac{x^3}{3!} + \frac{x^4}{4!} + \cdots > 1 + x$$

is obvious when $x > 0$, and is seen from the series when $-1 < x < 0$ (the terms alternate in sign and decrease in absolute value), and is again obvious when $x \leqq -1$.

(b) Since $a_1, a_2, \ldots, a_n$ are positive,
$$
\begin{aligned}
P_n &= (1 + a_1)(1 + a_2) \cdots (1 + a_n) \\
&= 1 + a_1 + a_2 + \cdots + a_n + a_1 a_2 + \cdots + a_1 a_2 \cdots a_n \\
&> 1 + S_n
\end{aligned}
$$

On the other hand,
$$
P_n = (1 + a_1)(1 + a_2) \cdots (1 + a_n) < e^{a_1} e^{a_2} \cdots e^{a_n} = e^{S_n}
$$

(c) Both S_n and P_n are monotonically increasing. It has been shown under (b) that the boundedness of one implies the boundedness of the other. Hence they are equiconvergent.

3.26   Assume that $0 < a_n < 1$ and prove that
$$
\sum_1^\infty a_n \qquad \text{and} \qquad \prod_1^\infty (1 - a_n)
$$
are equiconvergent.

3.27   Using the result of Example 3.8.1, that is, that the principal value branch of
$$
\log(1 + z) = \sum_1^\infty (-1)^{n+1} \frac{z^n}{n} = z - \frac{z^2}{2} + \frac{z^3}{3} - \cdots
$$
show that, for $|z| < \frac{1}{2}$,
$$
|\log(1 + z)| < 2\,|z|
$$

3.28   If $\prod_1^\infty (1 + |a_n|)$ converges, the product $\prod_1^\infty (1 + a_n)$ is absolutely convergent. Show that an absolutely convergent product of complex terms a_n is also convergent. (Use Example 3.27.)

3.29   Show that
$$
\prod_1^\infty \left(1 - \frac{z^2}{n^2}\right)
$$
converges for all z. (It can be shown that this product equals $\sin \pi z / \pi z$)

3.30   Show that
$$
P_n = \cos \frac{z}{2} \cos \frac{z}{4} \cos \frac{z}{6} \cdots \cos \frac{z}{2n}
$$

tends to a continuous limit, for all z, as $n \to \infty$.

Solution: With $\cos z = 1 + a$,

$$a = \cos z - 1 = -\frac{z^2}{2} + \frac{z^4}{4!} - \frac{z^6}{6!} + \cdots$$

$$|a| \leqq \frac{|z^2|}{2}\left(1 + \frac{2|z|^2}{4!} + \frac{2|z^4|}{6!} + \cdots\right)$$

$$\leqq \frac{|z|^2}{2}\left(1 + \frac{|z|^2}{2!} + \frac{|z|^4}{4!} + \cdots\right)$$

$$\leqq \frac{|z|^2}{2}\cosh 1 \qquad \text{for} \qquad |z| < 1$$

The nth term in the product is $\cos z/2n = 1 + a_n$ with

$$|a_n| \leqq \frac{|z|^2}{8n^2}\cosh 1$$

provided n is sufficiently large. This series is absolutely convergent (Example 3.28).

3.31   Show that

$$\cos\frac{x}{2}\cos\frac{x}{4}\cos\frac{x}{6}\cdots = \frac{\sin x}{x} \cdot \frac{3\sin\frac{x}{3}}{x} \cdot \frac{5\sin\frac{x}{5}}{x}\cdots$$

3.32   Show that there exists a power series

$$w = f(z) = 1 + \sum_1^\infty a_n z^n$$

satisfying

$$zw'' + w' + zw = 0$$

Furthermore, show that this series converges for all z.

3.33   A conformal mapping, by an analytic function $w = f(z)$, is called *univalent* in a region R if the image point corresponding to each z in R is covered precisely once. That is, $f(z_2) \neq f(z_1)$ for $z_2 \neq z_1$. Show that

(a) z^2 is univalent for $\text{I}z > 0$
(b) $z^2 + 2z + 3$ is univalent for $|z| < 1$

3.34   Show that $w = \sin z$ is univalent for $|z| < \pi$.

3.35   If $w = f(z)$ is univalent in $|z| < 1$, and $f(0) = 0$, show that $z = g(w)$, the inverse function, is single valued (and univalent) for $|w|$ sufficiently small.

3.36 Let $w = f(z) = z + \sum_2^\infty a_n z^n$ be univalent in $|z| < 1$. Show that the area of the image of $|z| \le r$, where $r < 1$, is

$$\pi\left(r^2 + \sum_2^\infty n\,|a_n|^2\,r^{2n}\right)$$

Solution: See Examples 3.14, 3.21, and 3.33.

3.37 $w = f(z)$ is univalent in $|z| < 1$. Show that the area of the image of the circle $|z| \le r$, $r < 1$, is

$$\tfrac{1}{4}r\,\frac{d}{dr}\int_0^{2\pi} |f(re^{i\theta})|^2\,d\theta$$

Solution:

$$\text{Area} = \iint |f'(z)|^2\,dx\,dy = \frac{1}{4}\iint \Delta\,|f(z)|^2\,dx\,dy$$

$$= \frac{1}{4}\int_0^{2\pi}\int_0^r \left[\frac{d}{dr}\left(r\,\frac{d}{dr}\right) + \frac{1}{r}\frac{d^2}{d\theta^2}\right] |f(re^{i\theta})|^2\,dr\,d\theta$$

$$= \frac{1}{4}\int_0^{2\pi} r\,\frac{d}{dr}\,|f(re^{i\theta})|^2 \Big|_0^r\,d\theta$$

$$= \tfrac{1}{4}r\,\frac{d}{dr}\int_0^{2\pi} |f(re^{i\theta})|^2\,d\theta$$

(from Example 3.4).

3.38 If $f(z)$ is an analytic function defined in the region R there is an analytic function $f^*(z)$ defined and analytic in the region $\bar{R}$, symmetric to R with respect to the real axis, such that

$$f(\bar{z}) = \overline{f^*(z)}$$

Solution: $w = f(z)$ maps the region R of the z-plane conformally, with conservation of the sense of the angles, onto a region S of the w-plane; $f(\bar{z})$ maps $\bar{R}$ onto S reversing the angles. By symmetry with respect to the real axis of the w-plane we obtain $f^*(z)$ that maps $\bar{R}$ onto $\bar{S}$ conformally without reversing the angles (Example 3.12).

3.39 Let us visualize a region in three-dimensional space in which we have a general vector-field W, with components $u(x, y, z)$, $v(x, y, z)$, and $w(x, y, z)$. At a point x, y, z of our region, we construct a cube, centered at x, y, z with sides of length $2h$, and faces parallel to the coordinate planes.

(a) Regarding W as the velocity vector of a fluid moving through our region compute the total flux of fluid (volume per unit time) crossing the six faces of the cube. The limiting value, as $h \to 0$, of the flux divided by the volume, is called the *divergence* of the vector-field at the point x, y, z. Show that this is $u_x + v_y + w_z$.

(b) Regarding W as a field of force, compute the work done by the field in moving a particle in the positive sense (i.e., so that a right-hand screw advances in the direction of the outer normal when rotated in this direction) around the periphery of each face of the cube. If we assign the direction of the outer normal to each of these elements of work, we have three pairs of equal vectors, forming the three components of a vector. The limiting form of this vector, per unit area, as $h \to 0$, is called the *curl* of the vector-field. Show that the curl of W has the components

$$w_y - v_z, \qquad u_z - w_x, \qquad v_x - u_y$$

SOLUTIONS FOR CHAPTER THREE

3.1.1(b)   Call $\Delta z = h$

$$\frac{\Delta f}{\Delta z} = \frac{\sqrt{z + h} - \sqrt{z}}{h} = \frac{\sqrt{z + h} - \sqrt{z}}{h} \cdot \frac{\sqrt{z + h} + \sqrt{z}}{\sqrt{z + h} + \sqrt{z}}$$

$$= \frac{1}{\sqrt{z + h} + \sqrt{z}} \to \frac{1}{2\sqrt{z}}$$

3.1.2(b) $\Delta f/\Delta z = \Delta x/(\Delta x + i\Delta y)$ tends to different limits as

$$\Delta x \to 0 \text{ (giving zero)}$$

and as

$$\Delta y \to 0 \text{ (giving 1)}$$

3.1.3   Consider $y = kx^3$.

3.3.1(a) $u = 2x \qquad v = 0$

$u_y + v_x = 0$

$u_x - v_y = 2 \neq 0 \qquad$ not analytic in z

3.3.4 $u = x^3 - 3xy^2 \qquad u + iv$ analytic

(a) $v_y = u_x = 3x^2 - 3y^2$

$v = 3x^2 y - y^3 + h(x)$

(b) $v_x = 6xy + h'(x) = -u_y = 6xy$

$h'(x) = 0, \qquad h(x) = c$

(c) $u + iv = z^3 + ci, \qquad c = \text{any real constant}$

3.3.6 $f'(z) \equiv 0$ says $u_x = v_x = u_y = v_y = 0$

identically in x, y. Hence $u = \text{const}$, $v = \text{const}$.

| **CONFORMAL MAPPING BY
GIVEN FUNCTIONS**

We examine some particular—and particularly simple and useful—conformal mappings and vector-fields generated by the elementary analytic functions studied in Chapters 2 and 3. We investigate, so to speak, the "conformal geometry of elementary functions."

4.1 THE STEREOGRAPHIC OR PTOLEMY PROJECTION

A map is a representation of the whole surface of the earth, or of a part of it, on a plane. Lines on the earth's surface (coastlines, rivers, and boundaries) are represented by corresponding plane curves; angles included by rivers and boundary lines, for example, at their intersection are represented by angles formed by the corresponding intersecting plane curves. The map would be perfect if it preserved all angles and reduced all distances measured along curves on the same scale. Unfortunately, such a perfect map is impossible (see end of p. 142). Yet, it is possible to construct a "semiperfect" map that, although it distorts the distances, preserves all angles; such a map is called *conformal*.

We begin by constructing a conformal map of the sphere; it is usually called the *stereographic projection*. This is an important map in many respects, also historically; it was invented by the great Greek astronomer Ptolemy.

As indicated, we regard the earth as a perfect sphere and scale its radius down to unity. In terms of rectangular coordinates (ξ, η, ζ), the equation of the sphere is

$$\xi^2 + \eta^2 + \zeta^2 = 1$$

We regard the point $(0, 0, 1)$ as the north pole N; then the equator is the intersection of the sphere with the plane $\zeta = 0$, which we shall call the *equatorial plane*. If we pass a straight line through the north pole N, and any other point, P of the earth's surface, this line intersects the equatorial plane $\zeta = 0$ in a unique point P' (Figure 4.1).

Let us denote the rectangular coordinates of P' by $(x, y, 0)$ or, since $\zeta = 0$ for all points in this plane, we merely write (x, y) for P'. Thus, with each point

Figure 4.1

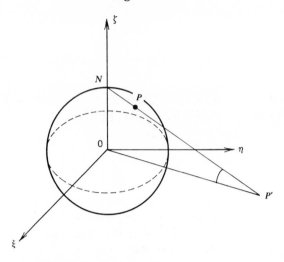

$P(\xi, \eta, \zeta)$ on the sphere, we associate a unique point $P'(x, y)$ in the equatorial plane. Conversely, to each point $P'(x, y)$ in the equatorial plane, we find a unique corresponding point on the sphere by joining P' to the north pole (producing the line beyond P' if necessary), and obtaining the intersection $P(\xi, \eta, \zeta)$ of this line with the sphere. [For instance, if P' is the origin $(0, 0)$, P is the south pole $(0, 0, -1)$.] This mapping, or one-to-one correspondence of points of the sphere with points of the plane, is called the *stereographic projection*.

Actually, we have only obtained a map of those points of the sphere that are distinct from the north pole. It is convenient to include into our mapping the north pole as well. To achieve this, let $P(\xi, \eta, \zeta)$ on the sphere approach the north pole and observe what happens to the image point $P'(x, y)$ in the equatorial plane. As P approaches N, the line through P and N approaches tangency at N, and P' tends to infinity. Since there is only one north pole, we say that the plane has only *one point at infinity*: as $P(\xi, \eta, \zeta)$ approaches N, the point P' approaches the point at infinity. Thus, by adding one "infinite point" to the x-, y-plane, we make the plane *closed* in the sense that now each point of the plane corresponds to precisely one point of the *closed sphere*, and conversely.

This concept of the point at infinity is an extremely useful one, and the stereographic mapping helps us to visualize the behavior of points at large distances from the origin by interpreting them as being near the north pole on the sphere.

The relation between $P(\xi, \eta, \zeta)$ and $P'(x, y)$ is readily obtained in terms of the coordinates of these points. Thus, from similar triangles in Figure 4.2,

Figure 4.2

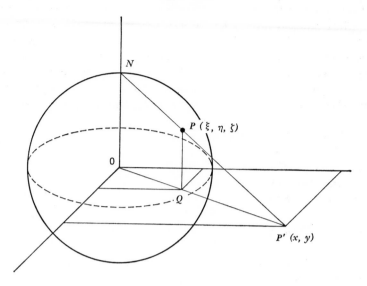

we obtain

$$\frac{x}{\xi} = \frac{y}{\eta} = \frac{OP'}{OQ} = \frac{1}{1 - \zeta}$$

(the right-hand part of Figure 4.2 is the orthogonal projection of the spatial configuration onto the z-plane.) Hence

(1)
$$x = \frac{\xi}{1 - \zeta} \qquad y = \frac{\eta}{1 - \zeta}$$

or, associating $z = x + iy$ with the point (x, y),

(2)
$$z = x + iy = \frac{\xi + i\eta}{1 - \zeta}$$

From now on, we shall often say "the point z" instead of the point $P'(x, y)$ and call the equatorial plane the *z-plane*, or the $x + iy$-plane.

Many of the properties of the stereographic projection become almost transparent when we consider what happens to the parallels of latitude and the meridians of longitude under the mapping. Any parallel of latitude, when its points are joined to N, gives rise to a cone (the axis of which is the diameter through N, the axis of the globe) so that the curve in the z-plane corresponding to a parallel of latitude is the intersection of this cone with the z-plane, that is, a circle, centered at the origin. Similarly, meridians of longitude are intersections of the sphere with planes passing through the axis of the sphere (the axis of the globe, the diameter through N and the south pole). The image of a meridian of longitude is thus the intersection of two planes through the origin, or a straight line, which passes through the origin. Thus, the grid formed by the parallels of latitude and meridians of longitude maps onto a grid of concentric circles and radial straight lines (identical with the coordinate lines $r = $ const and $\theta = $ const in polar coordinates). We observe further that in both grids the two families of curves intersect at right angles. This is a particular case of a more comprehensive fact: we shall show that the *stereographic projection is angle preserving*.

4.1.1   Find the stereographic image of

   (a) The equator of the sphere $\zeta = 0$.
   (b) The northern hemisphere $\zeta > 0$.
   (c) The hemisphere $\eta > 0$.

Inspect the figure *and* use (2).

4.1.2   Show that the coordinates of the point on the sphere of which z is the stereographic projection are

$$\xi = \frac{z + \bar{z}}{|z|^2 + 1} \qquad \eta = -\frac{i(z - \bar{z})}{|z|^2 + 1} \qquad \zeta = \frac{|z|^2 - 1}{|z|^2 + 1}$$

Another form of these relations is sometimes preferable:

$$\xi + i\eta = \frac{2z}{1 + |z|^2} \qquad 1 - \zeta = \frac{2}{1 + |z|^2}$$

[Start from (2).]   (S)

4.1.3   Let z_1 and z_2 be the stereographic images of the points P_1 and P_2 of the sphere, respectively. Show that the distance between P_1 and P_2 (the length of the chord joining them) is

$$\frac{2\,|z_1 - z_2|}{\sqrt{(1 + |z_1|^2)(1 + |z_2|^2)}}$$

(This "chordal distance" may be small even if $|z_1 - z_2|$ is large.)   (S)

4.1.4   Let P_1' and P_2' be the stereographic images of the points P_1 and P_2 of the sphere, respectively. Show that the following triangles are similar:

$$\triangle NP_1'\, P_2' \sim \triangle NP_2\, P_1$$

(Example 4.1.5)   (S)

4.1.5   Express in terms of z the distances NP', NP, and their product (notation of Section 4.1.)   (S)

4.1.6 The two points P_1 and P_2 of the sphere are at opposite ends of a diameter. Show that their stereographic images, z_1 and z_2, satisfy $z_1\bar{z}_2 = -1$. (S)

4.2 PROPERTIES OF THE STEREOGRAPHIC PROJECTION

In the last section we saw that certain circles on the sphere (parallels of latitude) map onto circles while other circles on the sphere (meridians of longitude) map onto straight lines. In addition, there is a more comprehensive fact: every circle of the sphere maps onto a circle or a straight line of the plane.

(a) A circle on the sphere is the intersection of the sphere with a plane. If the plane passes through the north pole, we obtain a circle through the north pole N. The projecting line through N and any point of such a circle lies in the plane determining the circle. Therefore, the image point lies in the intersection of this plane with the equatorial plane, that is, in a *straight line*: this straight line is the image of the circle through the north pole. What we have shown contains the particular case of the meridians. On the other hand, a circle on the sphere not passing through the north pole maps onto some finite closed curve (not passing through the point at infinity), and so its image is certainly not a straight line.

(b) To see all this analytically, we express by equations the fact that a circle on the sphere is the intersection of the sphere with some plane: a point (ξ, η, ζ) of such a circle satisfies both equations

$$\xi^2 + \eta^2 + \zeta^2 = 1, \qquad \lambda\xi + \mu\eta + \nu\zeta = p$$

where we suppose that $\lambda^2 + \mu^2 + \nu^2 = 1$ and $p \geq 0$. We recall that in this notation, λ, μ, ν are direction cosines of the normal to the plane, and p is the distance from the origin to the plane. In our case, $p < 1$ in order that the sphere and the plane should actually intersect. By using Section 4.1 (1), we have

$$\xi = x(1 - \zeta), \qquad \eta = y(1 - \zeta)$$

and we rewrite our equations in the form

$$(\lambda x + \mu y)(1 - \zeta) + \nu\zeta = p$$
$$(1 - \zeta)^2 (x^2 + y^2) + \zeta^2 = 1$$

We can divide the second equation by $1 - \zeta$, by discarding only the north pole, $\zeta = 1$. Elimination of ζ yields the desired image in the $x + iy$-plane of the circle on the sphere:

(1) $$(p - \nu)(x^2 + y^2 + 1) = 2(\lambda x + \mu y - \nu)$$

This is the equation of a circle if $p \neq \nu$, and the equation of a straight line if $p = \nu$; our derivation is completely general and gives the answer in all cases. The significance of the case $p = v$ is seen from the equation of the plane in this case:

$$\lambda\xi + \mu\eta + v(\zeta - 1) = 0$$

Such a plane passes through the north pole $(0, 0, 1)$, determines a circle passing through the north pole, and so our calculation confirms what we have seen above by a simple geometric argument; the image of such a circle is a straight line.

We see that a circle on the sphere always maps onto a circle on the plane, unless the circle on the sphere passes through the north pole. Let us now approach this exceptional case by continuous change; (1) can be transformed into

$$\left(x - \frac{\lambda}{p - v}\right)^2 + \left(y - \frac{\mu}{p - v}\right)^2 = \frac{1 - p^2}{(p - v)^2}$$

and hence we see that as $p - v \to 0$, both the center and the radius of the circle tend to infinity (p should not tend to 1), and the circle becomes a straight line. It is convenient to think of straight lines as being circles of infinite radius, and with this understanding we have the comprehensive statement: *the stereographic mapping sends circles on the sphere onto circles on the plane.*

(c) We are now in a position to show that angles are preserved under the stereographic mapping. The curves C_1 and C_2 on the sphere intersect at an angle α at the point P. This statement tacitly assumes that the curves have tangents at their point of intersection, and that the angle α is the angle between these tangents. Let C_1' and C_2' be the images in the plane of C_1 and C_2 under the stereographic mapping, respectively, and let α' be the angle between these image curves at their point of intersection P', the image of P; see Figure 4.3. Now the angles α and α' are determined solely by the tangents to the curves in question at their points of intersection; moreover, it is intuitively clear that curves tangent to each other are projected onto image-curves tangent to each other. Thus, we may replace each of the curves C_1 and C_2 by any other curve (e.g., a circle) that passes through and has the same tangent at the point of intersection P. In particular, let us replace each curve by a special circle that passes through the north pole. This is a natural choice; in fact, the tangents to C_1' and C_2' at their point of intersection P' (which form the angle α') are precisely the stereographic images of the two circles just introduced, respectively. Next, the two circles through N intersect at the same angle α, both at the point P and at the north pole N, by virtue of the sphere's symmetry. Now, the tangent plane to the sphere at the north pole N is obviously parallel to the equatorial z-plane. Observe that these two parallel planes are intersected in parallel lines by any third (nonparallel) plane. The plane of a circle passing through the north pole N is such a third plane. Therefore, the sides of the angle α with vertex at N are parallel to the sides of the angle α'

Figure 4.3

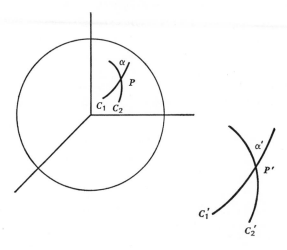

with vertex at the intersection P' of C_1' and C_2'. This shows that $\alpha' = \alpha$; the angles are preserved.

(*Remark on method.* Our argument has reduced *the proof of a general theorem to the proof of a special case*—to the case in which the intersecting curves in the equatorial plane are straight lines.)

(d) Let us apply some of the geometry we have learned. We map the z-plane stereographically onto the sphere, then rotate the sphere $180°$ about the ξ-axis, and finally map the sphere back stereographically onto the z-plane. The point originally at z has moved so eventually to another location of the plane that we call w; thus, w is a function of z. The explicit form of the mapping is obtained as follows: first z is projected into (ξ, η, ζ); then the rotation sends ξ into ξ, η into $-\eta$, and ζ into $-\zeta$; finally $(\xi, -\eta, -\zeta)$ is projected back into w. Thus by a proper application of Section 4.1 (2),

$$w = \frac{\xi - i\eta}{1 + \zeta} \qquad \text{whereas} \qquad z = \frac{\xi + i\eta}{1 - \zeta}$$

Elimination of ξ, η, ζ between these relations and $\xi^2 + \eta^2 + \zeta^2 = 1$ yields $w = w(z)$. A shorter computation gives

$$w \cdot z = \frac{\xi - i\eta}{1 + \zeta} \cdot \frac{\xi + i\eta}{1 - \zeta} = \frac{\xi^2 + \eta^2}{1 - \zeta^2} = \frac{1 - \zeta^2}{1 - \zeta^2} = 1$$

or, $w = 1/z$; it is remarkable that a rotation of the sphere induces such a simple mapping of the plane onto itself.

Now, a conformal map of a conformal map is obviously conformal, and a circle-preserving map of a circle-preserving map is obviously circle-preserving. We already know that the mapping given by the analytic function $w = 1/z$ is conformal, as long as we avoid $z = 0$. From its relation to the rotation of the sphere we have just considered, we see a geometric reason for this fact and, in addition, we see that $w = 1/z$ maps circles onto circles, with straight lines as limiting cases of circles. Thus the special mapping $w = 1/z$ possesses a very elegant and simple geometric property connecting circles and straight lines. In the next section we shall get acquainted with further conformal transformations that leave the family of circles invariant, that is, that map circles onto circles; we shall investigate some of their more important geometrical properties.

4.2.1   Show that

$$w = \frac{\xi - i\eta}{1 + \zeta} = \frac{1}{z}$$

by substituting for ξ, η, ζ their expressions in terms of z found in Example 4.1.2.   (S)

4.2.2 $w = f(z)$ is obtained by mapping z stereographically onto P on the sphere, then rotating the sphere 90° in the clockwise direction about the ξ-axis, sending P to a new position P^*, and then mapping P^* stereographically onto the $\zeta = 0$ plane, calling this latter point w. Find $w = f(z)$ explicitly.   (S)

4.2.3   Verify that repeated application of the mapping function of Example 4.2.2 corresponds to a rotation of 180° about the ξ-axis, that is

$$z_1 = i\,\frac{z - i}{z + i} \quad \text{and} \quad w = i\,\frac{z_1 - i}{z_1 + i} \quad \text{imply} \quad w = \frac{1}{z}$$

4.2.4   Find the center and radius of the stereographic image of the circle on the sphere:

$$\xi^2 + \eta^2 + \zeta^2 = 1, \qquad \xi = p, \qquad 0 < p < 1$$

4.3 THE BILINEAR TRANSFORMATION

(a) Besides the special mapping considered in the previous section, there are other, more obvious mappings that are conformal and send circles into circles. As the first of these, we consider the *parallel translation* that carries the point (x, y) into $(x + a, y + b)$, or $z = x + iy$ into $z + c = x + a + i(y + b)$:

(1) $$w = z + c$$

Not only does the parallel translation leave circles invariant: it moves the plane as a rigid body and leaves the shape and size of any figure unaltered.

In fact, it leaves all directions unaltered; it carries any straight line onto a parallel straight line. It carries all points along parallel paths; hence the name.

Another motion of the plane is *rotation about a fixed point*. It is sufficient to consider only rotations about the origin, so that z goes into

$$(2) \qquad\qquad w = e^{i\alpha}z$$

We recall that in light of our interpretation of complex numbers as vectors, (2) merely states that the vector z is rotated through an angle α in the counter-clockwise direction: $e^{i\alpha}$ is a unit vector.

A third elementary mapping that preserves circles is the *uniform stretching* or *dilatation* that carries x into px, and y into py where $p > 0$, so that z is carried into

$$(3) \qquad\qquad w = pz, \qquad p > 0$$

(If $p < 1$ we have, in fact, a compression, but we may call it "stretching" as we call division by 2 multiplication by $\frac{1}{2}$— such is the mathematical language.)

(b) The mappings given by (1), (2), and (3) change any figure into a similar figure, and any mapping formed by compounding them has the same property and so, especially, it will preserve angles and map circles onto circles. The most general mapping we can achieve by a succession of these three mappings has the form $w = az + b$, where a and b are complex constants; if we denote by α the argument of a, so that $a = |a|\,e^{i\alpha}$, then the transformation of z into w:

$$w = az + b = |a|e^{i\alpha}z + b$$

can be achieved by three successive transformations:

$$z_1 = e^{i\alpha}z \qquad \text{rotation}$$

$$z_2 = |a|\,z_1 \qquad \text{stretching}$$

$$w = z_2 + b \qquad \text{translation}$$

(c) We now introduce the mapping $w = 1/z$ as a fourth basic circle preserving conformal mapping. The most general combination of $w = 1/z$ with (1), (2), and (3) yields an expression of the form*

$$(4) \qquad\qquad w = \frac{az + b}{cz + d}$$

where a, b, c, d are complex constants. Conversely, every function of the form (4) can be conceived as a combination of the four basic mappings

---

* This mapping was studied extensively by the German mathematician Möbius and in many references, transformations of the form (4) are called Möblus transformations.

described above. If $c = 0$, we must have $d \neq 0$ and we fall back on the case discussed in subsection (b). If, however $c \neq 0$, we can represent (4) in the form

(5)
$$w = \frac{a}{c} + \frac{bc - ad}{c} \frac{1}{cz + d}$$

and

$$z_1 = cz + d$$

$$z_2 = \frac{1}{z_1}$$

$$w = \frac{a}{c} + \frac{bc - ad}{c} z_2$$

shows a sequence of three successive transformations achieving the desired result.

(d) From (5) we see that when $ad - bc = 0$, the mapping becomes $w = a/c$ which is not a particularly interesting mapping: it throws the whole z-plane into a single point; we call it a *degenerate* mapping. In the rest of this chapter we exclude this degenerate case and insist on $ad - bc \neq 0$.

(We could go even further. The mapping function $w = (az + b)/(cz + d)$ is unaltered if we multiply numerator and denominator by the same nonzero constant k; such multiplication changes the expression $ad - bc$ into $k^2(ad - bc)$, and we may choose k so that this expression becomes unity. When it is convenient to do so, we may thus consider $ad - bc$ to be equal to 1.)

A more symmetric form of (4) is obtained by multiplying through by $cz + d$, giving

(6)
$$cwz + dw - az - b = 0$$

which is linear in z for each fixed w, and linear in w for each fixed z. Accordingly we call the expression in (6) *bilinear* and the transformation given by $w = (az + b)/(cz + d)$ a *bilinear transformation*. From the symmetry of Equation (6) we see that the inverse of a bilinear mapping is also bilinear; we find, in fact, that

$$z = \frac{-dw + b}{cw - a} \qquad (-a)(-d) - b \cdot c \neq 0$$

4.4 PROPERTIES OF THE BILINEAR TRANSFORMATION

(a) The bilinear transformation is built up by repeated applications of the four basic transformations $w = z + c$, $w = e^{i\alpha}z$, $w = pz$, and $w = 1/z$. For example, if we follow a translation by taking the reciprocal, we have

$$z_1 = z + c$$

$$w = \frac{1}{z_1}$$

or, eliminating z_1, $w = 1/(z + c)$. That is, we obtain $w = 1/(z + c)$ by substituting $z + c$ for z in $w = 1/z$.

The most general transformation we can compound by applying successively our four basic transformations is the bilinear transformation.

If in $w = (az + b)/cz + d)$ we set $z = (\alpha z_1 + \beta)/(\gamma z_1 + \delta)$, the result can be nothing else but another bilinear transformation

$$w = \frac{a_1 z_1 + b_1}{c_1 z_1 + d_1}$$

(Express a_1, b_1, c_1, d_1 in terms of a, b, c, d, α, β, γ, δ.) In this way we see that the collection of *all* bilinear transformation is "self-sufficient," forming a *group* in technical language: the inverse of any such transformation, and the successive application of any two such transformations, yields another bilinear transformation.

(b) We already know that the mapping given by $w = (az + b)/(cz + d)$ is conformal, since it is composed of a sequence of conformal mappings. Moreover, we have here an analytic function; for a verification, we may compute $w' = (ad - bc)/(cz + d)^2$ which exists and is different from zero for each finite $z \neq -d/c$. We know further that each bilinear transformation is circle preserving and maps circles into circles.

Let us ask now whether it is possible to map the circumference of any given circle C_1 onto the circumference of another given circle C_2 by a bilinear transformation. Now, a circle is determined by any three points on its circumference. We could solve our mapping problem by choosing three points z_1, z_2, z_3 on C_1 and requiring them to map into w_1, w_2, w_3 on C_2, respectively; see Figure 4.4. Is this a reasonable approach?

We note that $w = (az + b)/(cz + d)$ contains in fact, four arbitrary constants a, b, c, d, but only three independent ratios. If $ad - bc \neq 0$, at least one of the two coefficients c and d is not zero. Suppose $c \neq 0$; then

$$w = \frac{\dfrac{a}{c} z + \dfrac{b}{c}}{z + \dfrac{d}{c}}$$

and the transformation is completely determined by knowing a/c, b/c, and d/c. Thus we should be able to require three conditions of our mapping, and the mapping of z_1 into w_1, z_2 into w_2 and z_3 into w_3 should determine the mapping. Remember also that circles map into circles; therefore, if z_k maps into

Figure 4.4

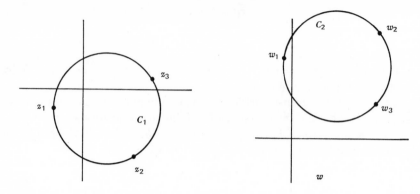

w_k for $k = 1, 2, 3$, the circle through z_1, z_2, z_3 will map into a circle passing through w_1, w_2, w_3, and so will be uniquely determined. If the three points in either plane are collinear then, of course, the circle in that plane becomes a special circle, a straight line.

The mapping of C_1 onto C_2 is thus accomplished if we can find a, b, c, d so that

$$(1) \qquad w_k = \frac{az_k + b}{cz_k + d} \qquad k = 1, 2, 3$$

A glance at (1) shows that the system of equations is equivalent to three homogeneous algebraic equations of first degree for a, b, c, d, so that a solution always exists and can be computed from linear equations.

(c) There is, however, another approach that is often preferable. Let us recall that inversion and successive application of bilinear transformations yield bilinear transformations. Hence the required bilinear relation between w and z could be given in the form

$$\frac{\alpha w + \beta}{\gamma w + \delta} = \frac{az + b}{cz + d}$$

where we have plenty of freedom in choosing the eight constants α, β, δ, γ, a, b, c, and d.

Let us now observe this: when the numerator on one side vanishes, the numerator on the other side must vanish too, and the denominators are analogously connected. This leads us to set up the desired relation in the form

$$(2) \qquad \frac{w - w_1}{w - w_2} = k \frac{z - z_1}{z - z_2}$$

wherein the constant k is not yet determined (we expect, however, that $k \neq 0$). The form (2) renders certain that $w = w_1$ corresponds to $z = z_1$

and $w = w_2$ to $z = z_2$. It remains to express that $z = z_3$ corresponds to $w = w_3$:

$$\frac{w_3 - w_1}{w_3 - w_2} = k \frac{z_3 - z_1}{z_3 - z_2}$$

By division, we get rid of k and obtain the desired bilinear transformation in the form

(3)
$$\frac{w - w_1}{w - w_2} \frac{w_3 - w_2}{w_3 - w_1} = \frac{z - z_1}{z - z_2} \frac{z_3 - z_2}{z_3 - z_1}$$

From (3) we see immediately the correspondence of

$$w = w_1, \ w_2, \ w_3 \qquad \text{to} \qquad z = z_1, \ z_2, \ z_3$$

respectively, since both sides become simultaneously 0, ∞ and 1 in these three cases, respectively. This makes (3) easy to remember.

(d) The expression on the left-hand side of (3) is usually termed the *cross-ratio* of the four points w, w_1, w_2, w_3. Hence, we can express the contents of (3) by saying that a *bilinear transformation leaves the cross-ratio invariant*.

The cross-ratio of the four points w, w_1, w_2, w_3 is often denoted by (w, w_1, w_2, w_3). We shall seldom use this symbol, and we do not enter into explaining the importance of the cross-ratio in other branches of mathematics.

(e) There is much more to say about bilinear transformations in general, and about certain particular bilinear transformations: we refer to the exercises and comments following this section and the ones at the end of this chapter.

4.4.1 Find the image of the circle $|z - i| = 1$ under the bilinear mapping $w = (z - i)/(z + i)$. (S)

Solution: (a) Since the given circle does not pass through $z = -i$ (where w becomes infinite), the image is a finite curve, and hence a circle.

(b) w is real for z on the imaginary axis (i.e., the imaginary axis of the z-plane maps onto the real axis of the w-plane). Since the imaginary axis of the z-plane forms a diameter of the given circle, conformality shows that the real axis forms a diameter of the image circle. Thus, corresponding to $z = 0$, $2i$ we have $w = -1, \frac{1}{3}$, and these are the endpoints of a diameter. The image circle is thus $|w + \frac{1}{3}| = \frac{2}{3}$.

4.4.2 Find the image of $|z - i| = 1$ under $w = 1/z$.

4.4.3 Show that formula 4.4 (3) remains essentially valid when one of the points z_1, z_2, z_3 lies at infinity, or one of w_1, w_2, w_3 lies at infinity, or both.

4.4.4   Set

$$(z_1, z_2, z_3, z_4) = \frac{z_1 - z_2}{z_1 - z_3} \frac{z_4 - z_3}{z_4 - z_2} = \lambda$$

Then, the 24 permutations of z_1, z_2, z_3, z_4 yield 24 expressions for the cross-ratio. Show that there are only six distinct values of these expressions, and that all six are bilinear functions of λ.   (S)

Solution: The value of the cross-ratio remains unchanged when we interchange

   (a) z_1 with z_4 and z_3 with z_2
   (b) z_1 with z_3 and z_2 with z_4
   (c) z_1 with z_2 and z_3 with z_4

as well as (d) when we leave z_1, z_2, z_3, z_4 unchanged (the identical permutation).

  Thus there are 6 blocks each containing four equivalent expressions of the cross-ratio. We find the six different values by leaving z_1 unchanged and permuting z_2, z_3, z_4. Thus we obtain

$$\lambda \qquad \frac{1}{\lambda} \qquad 1 - \lambda \qquad \frac{1}{1-\lambda} \qquad \frac{\lambda}{\lambda - 1} \qquad \frac{\lambda - 1}{\lambda}$$

4.4.5   Find the six distinct values of $(0, 1, \infty, z)$.

Solution: See Example 4.4.4.

4.4.6   Assuming that $w = (az + b)/(cz + d)$ gives the mapping corresponding to a rotation of the sphere, by 90° clockwise about the ξ-axis, deduce the values of a, b, c, d from the fact that a second application of the same mapping gives $w = 1/z$.

4.4.7   Find the bilinear mapping that maps $z = 1$, i, $-i$ into $w = -1$, i, $-i$ respectively. Into what region does the interior of the unit circle of the z-plane map?

4.4.8   The mapping $w = -1/z$ of Example 4.4.7 sends $z = i$ into $w = i$, and $z = -i$ into $w = -i$. These points are called *fixed points* of the mapping. For the general bilinear mapping, the fixed points are obtained by solving the equation

$$w = \frac{az + b}{cz + d} = z$$

Find the fixed points of the mappings

(a) $w = -i\dfrac{(z - i)}{z + i}$ (S)

(b) $w = \dfrac{z + 1}{z - 1}$

(c) $\dfrac{w - i}{w + i} = \dfrac{z + 1}{z - 1}$

4.4.9   Find the most general bilinear transformation $w = (az + b)/(cz + d)$ having $z = \pm 1$ as its fixed points.

4.4.10   In Exercise 4.4.9, show that

$$\frac{w - 1}{w + 1} = k\frac{z - 1}{z + 1}$$

for a suitable k.

4.4.11   Assuming that $w = (az + b)/(cz + d)$ has two fixed points z_1 and z_2, show that the transformation can be written

$$\frac{w - z_1}{w - z_2} = k\frac{z - z_1}{z - z_2}$$

for a suitable constant k.

4.4.12   GEOMETRICAL INVERSION

Two points are inverse points with respect to a given circle if:

(a) They lie on the same "radius" (half line through the center).
(b) The product of their distances from the center equals the square of the radius of the circle.

Show that z_1, z_2 are inverse points with respect to the circle $|z - a| = R$ if, and only if

$$(z_1 - a)\overline{(z_2 - a)} = R^2$$ (S)

4.4.13   The general equation of a circle can be written $(A + \bar{A})z\bar{z} + \bar{B}z + B\bar{z} + C + \bar{C} = 0$ (see Example 1.22).
If z_1, z_2 are related by the equation

$$(A + \bar{A})z_1\bar{z}_2 + \bar{B}z_1 + B\bar{z}_2 + C + \bar{C} = 0$$

show that z_1 and z_2 are inverse points of the above circle. (Assume $A \neq 0$.)

4.4.14 In Example 4.4.13, if $A \to 0$, the circle becomes a straight line as a limit. What is the limiting position of a pair of inverse points in the circle, when $A \to 0$?

4.5 THE TRANSFORMATION $w = z^2$

We now turn to a study of some of the mapping properties of the other elementary functions we have been considering. One of the simplest of these is the general power $w = z^n$ with integral exponent n. We will content ourselves with the case $n = 2$. (For $n = 1$, $w = z$ is merely the identity;

Figure 4.5

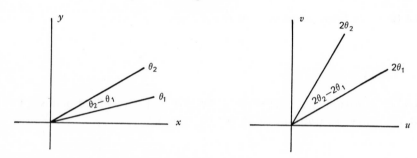

$n = 0$ gives $w = 1$, sending every point z into $w = 1$; and $w = 1/z$, $n = -1$, we have already considered.)

Since

$$\frac{d}{dz}(z^2) = 2z$$

which is continuous for all finite z, and $\neq 0$ for all z except $z = 0$, we observe that $w = z^2$ gives a conformal mapping of any finite region of the z-plane that excludes $z = 0$, into a corresponding region of the w-plane. Of particular interest here is the mapping near $z = 0$. Since the derivative vanishes at $z = 0$, we know nothing as yet about the nature of the mapping there. Let us examine some special cases. Consider a region bounded by two radial lines emanating from the origin (Figure 4.5).

$$w = \rho e^{i\phi} = z^2 = (re^{i\theta})^2 = r^2 e^{2i\theta}$$

Hence $\rho = r^2$, $\phi = 2\theta$. The line $\theta = \theta_1$ maps into $\phi = 2\theta_1$, while $\theta = \theta_2$ maps into $\phi = 2\theta_2$, and the angle $\theta_2 - \theta_1$ goes over into $2(\theta_2 - \theta_1)$. Thus angles are doubled at the origin. (It is equally easy to see that in the general case $w = z^n$ the angles at the origin are multiplied by n if $n > 0$.) We cannot expect a mapping to be conformal at a point, then, unless $w' = f'(z)$ is distinct from zero at this point. Points at which the derivative vanishes are called *critical points* of the mapping function. (We shall see later, that the case $w = z^n$ is typical; see Section 6.5.)

We can take advantage of this scaling of angles; for example, we can map the quarter-plane $x > 0$, $y > 0$ conformally onto the upper half-plane $v > 0$ by employing the mapping $w = z^2$ (Figure 2.2).

4.5.1 Show that, under the transformation $w = z^2$, the upper half-plane $Iz > 0$ is mapped onto the whole of the w-plane with the exception of the points on the positive real axis of w. Similarly show that the region $Iz < 0$ is mapped onto the same region.

Figure 4.6

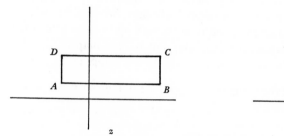
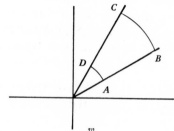

4.5.2 Find the image of the unit semicircle of the z-plane, $0 < \theta < \pi$, $0 < r < 1$, under "inversion" about $z = -1$, that is, under $w = 1/(z + 1)$. Thus find a mapping (or sequence of mappings) that transform the unit semicircle onto the upper half w-plane.

4.6 THE TRANSFORMATION $w = e^z$

Setting $w = \rho e^{i\phi}$ and $z = x + iy$, we obtain $w = \rho e^{i\phi} = e^z = e^x \cdot e^{iy}$ and hence

$$\rho = e^x \qquad \phi = y + 2n\pi$$

where, if we insist on the principal value of the phase, $0 \le \phi < 2\pi$, so that n must be chosen accordingly depending on the value of y. Since

$$\frac{d}{dz} e^z = e^z$$

the mapping is conformal everywhere. Observe that a rectangle in the z-plane bounded by lines $x = \text{const}$ and $y = \text{const}$ maps into a "mixed quadrilateral" bounded by arcs of concentric circles $\rho = \text{const}$ and radial lines $\phi = \text{const}$ (Figure 4.6). (Consider at first only rectangles of which the vertical sides, parallel to the y-axis, are shorter than 2π; afterwards we may increase the

Figure 4.7

length of the vertical sides to, and beyond, 2π and see what happens.) Some limiting cases of the above mapping are of interest. Thus the half strip $0 < y < \pi$, $x < 0$ gives Figure 4.7. while $0 < y < \pi$, $-\infty < x < \infty$ gives Figure 4.8 which shows that $w = e^z$ maps the infinite strip $0 < y < \pi$, $-\infty < x < \infty$ conformally onto the upper half-plane. We could now combine this result with a bilinear map and map the half plane onto some other circular region, the unit circle, for example, and have a map of the strip directly onto the unit circle. What function will do the mapping?

4.6.1 In what region of the w-plane is mapped the infinite strip $-\pi < y < \pi$ by the exponential function $w = e^z$?

Figure 4.8

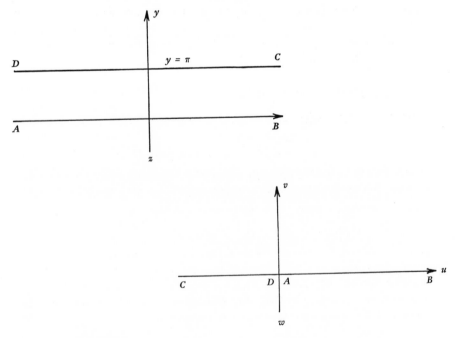

4.7 THE MERCATOR MAP

The stereographic projection yields a conformal map of the sphere on the plane (see Section 4.1), and this is the only map of this kind we have considered so far. Yet, in starting from this one map, we can easily obtain any number of other angle-preserving maps of the sphere, by mapping the plane conformally onto itself by an arbitrary analytic function; in fact, a conformal map of a conformal map is again conformal. The exponential function leads here to a result of practical importance in navigation.

Figure 4.9

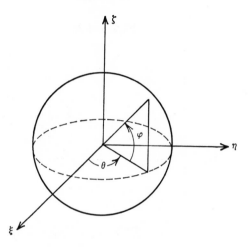

Let the point (ξ, η, ζ) on the sphere go over into the point z of the complex plane, see Section 4.1 (2), and set

$$z = x + iy = \frac{\xi + i\eta}{1 - \zeta} = e^w = e^{u+iv}$$

Thus the (ξ, η, ζ)-sphere is mapped conformally onto the w-plane. We introduce spherical polar coordinates; let ϕ denote the latitude, and θ the longitude. Then, see Figure 4.9,

$$-\frac{\pi}{2} \leqq \phi \leqq \frac{\pi}{2} \qquad -\pi \leqq \theta < \pi \qquad \text{and we have}$$

$$\xi = \cos \phi \cos \theta$$

$$\eta = \cos \phi \sin \theta$$

$$\zeta = \sin \phi$$

The Mercator Map   **131**

and by Section 4.1 (2)

$$z = \frac{\cos\phi}{1 - \sin\phi}\, e^{i\theta} = \tan\left(\frac{\phi}{2} + \frac{\pi}{4}\right) e^{i\theta} = e^{u} \cdot e^{iv}$$

Thus

$$u = \log\tan\left(\frac{\phi}{2} + \frac{\pi}{4}\right) \qquad v = \theta$$

and the spherical surface is represented conformally on the infinite strip $-\pi \leqq v < \pi$, $-\infty < u < \infty$. Lines $u = $ const correspond to parallels of latitude, and $v = $ const to meridians of longitude; especially, $u = 0$ corresponds to the equator, and $v = \pi$, and also $v = -\pi$, to the date line

Figure 4.10

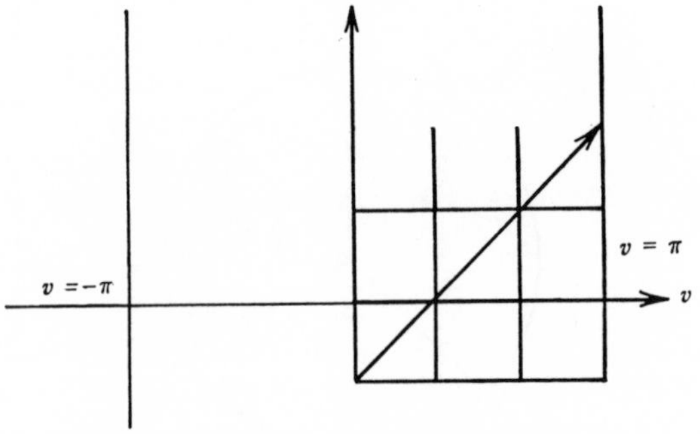

(see Figure 4.10). This is the well-known Mercator map; let us say a word on its use in navigation.

If the helmsman steers the ship so that its motion includes a fixed angle with the due north, the ship sails a "constant compass course," or "rhumb line" on the sphere that cuts each meridian of longitude at the same angle. (Do not neglect, of course, magnetic deviations, or use a gyro-compass—we have neglected already the deviation of the globe from the exactly spherical shape.) Thanks to conformality, the image of a rhumb line in the w-plane cuts each straight line $v = $ const at the same angle—and so it must be a straight line itself. Hence a constant compass course can be plotted on the Mercator map with a straightedge.

Additional Examples and Comments on Chapter Four

4.1 The mapping $w = 1/z$ corresponds to a rotation of the sphere $180°$ about the ξ-axis. What transformation of the sphere is induced by $w = 1/\bar{z}$?

4.2 Show that each circle cutting $|z| = 1$ orthogonally is mapped into itself under $w = 1/z$.

4.3   Find the circle on the sphere that is the stereographic image of the circle

$$|z - a| = R$$

That is, find λ, μ, ν, and p of the equation of the plane

$$\lambda\xi + \mu\eta + \nu\zeta = p$$

that cuts the sphere to form the circle (see Section 4.2).   (S)

4.4   Find the center and radius of the image of the circle $|z - a| = r$ under the mapping $w = 1/z$.

Figure 4.11

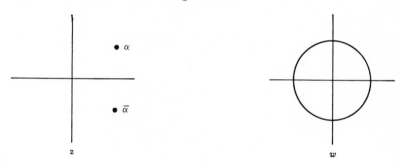

4.5   Prove that the bilinear transformation maps circles onto circles by showing that

$$(A + \bar{A})z\bar{z} + \bar{B}z + B\bar{z} + C + \bar{C} = 0$$

becomes

$$(A' + \bar{A}')w\bar{w} + \bar{B}'w + B'\bar{w} + C' + \bar{C}' = 0$$

when $w = (az + b)/(cz + d)$.

4.6   Given any circle, and a pair of points z_1, z_2 that are inverse points in this circle, show that the bilinear transformation $w = (az + b)/(cz + d)$ maps the circle and inverse points of the z-plane into an image circle, and points w_1, w_2, which are themselves inverse points in the image circle (see Examples 4.4.12 and 4.4.13).

COMMENT 1.   The results of Exercises 4.5 and 4.6 provide us with a rapid and elegant method of constructing particular bilinear mappings. Consider, for example, the problem of finding the most general bilinear transformation that maps the upper half-plane onto the unit circle (Figure 4.11). Let $w = (az + b)/(cz + d)$. Then some point of the upper half-plane is mapped onto $w = 0$; call this point α. The inverse of α in the z-plane "circle" is $\bar{\alpha}$, which maps onto the inverse of $w = 0$ in the circle $|w| = 1$, that is, onto $w = \infty$.

Thus

$$w = \frac{a(z - \alpha)}{c(z - \bar{\alpha})}$$

So far, we have not specified that the circle in the w-plane is the unit circle, just that it is centered at $w = 0$. Setting $z = x$, we have

$$\left| \frac{a}{c} \frac{x - \alpha}{x - \bar{\alpha}} \right| = 1 = \left| \frac{a}{c} \right| \left| \frac{x - \alpha}{x - \bar{\alpha}} \right| = \left| \frac{a}{c} \right|$$

since $x - \bar{\alpha} = \overline{x - \alpha}$. In other words, $a/c = e^{i\beta}$, for some real β, and we have

$$w = e^{i\beta} \frac{z - \alpha}{z - \bar{\alpha}} \qquad I\alpha > 0$$

4.7 Show that the most general bilinear mapping that sends the unit circle $|z| < 1$ onto the unit circle $|w| < 1$ is

$$w = e^{i\beta} \frac{z - \alpha}{\bar{\alpha}z - 1} \qquad |\alpha| < 1$$

4.8 For the elementary functions we have considered so far, the mappings from the z-plane to the w-plane are continuous in the closed domains consisting of the regions and their boundaries. In order to determine the map of a given region, then, we merely determine the image of the boundary of the region, and determine on which side of this boundary the image region lies. Show that the following rule follows from conformality: if we walk along the boundary of a region in the z-plane in that direction for which our left hand points to the interior of the region, then the same orientation holds for the image curve and region; that is, we proceed from A towards C in the z-plane so that our left hand points to the interior of the region, and if the image of the boundary arc ABC is $A'B'C'$, then the interior of the image region lies on the left as we proceed from A' towards C' (Figure 4.12).

Solution: At any point B of the boundary curve (which we assume has a tangent), let BC be one arc, and BD another arc making a small angle α at B, with BD lying in the interior of the given region. From conformality, the images of BC, BD are $B'C'$, $B'D'$, meeting at the same angle α, and in the same sense. In the diagram above, we must rotate in a counterclockwise direction to go from BC to BD, and thus also to go from $B'C'$ to $B'D'$.

Figure 4.12

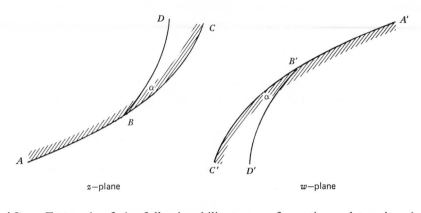

z−plane   w−plane

4.9 For each of the following bilinear transformations, determine the image region corresponding to $|z| < 1$.

(a) $\dfrac{z - i}{z + i}\dfrac{1 + i}{1 - i} = \dfrac{w + 1}{w + i}\dfrac{i + i}{i + 1}$

(b) $w = \dfrac{z - 1}{z + 1}$

(c) $w = \dfrac{2z - 3}{z + i}$

See Example 4.8.

4.10 For an arbitrary bilinear transformation, with two fixed points z_1, z_2,

$$w = \frac{az + b}{cz + d}$$

can be written as

$$\frac{w - z_1}{w - z_2} = k\frac{z - z_1}{z - z_2}$$

See Examples 4.4.8 and 4.4.10.
Set

$$\frac{z - z_1}{z - z_2} = \rho e^{i\phi} \qquad \frac{w - z_1}{w - z_2} = \rho' e^{i\phi'} \qquad k = |k|\, e^{i\gamma}$$

Then the bilinear transformation states that

$$\rho' = |k|\, \rho \qquad \text{and} \qquad \phi' = \phi + \gamma$$

Interpret these statements geometrically.

Solution: $\rho = $ const gives a circle, centered on the line joining z_1, z_2, so that z_1, z_2 are inverse points in this circle. $\phi = $ const is a circle through z_1 and z_2, and the bilinear transformation transforms each of the two systems of circles into itself, while each circle of one system cuts each circle of the second system orthogonally.

4.11 What is the analogue of Example 4.10 for a bilinear transformation having only a single fixed point? (In this case, if z_1 is the fixed point,

$$\frac{1}{w - z_1} = \frac{1}{z - z_1} + k\Big)$$

4.12 Show that the determinant of a product of two bilinear transformations is the product of their determinants.

4.13 Find the mapping $w = (az + b)/(cz + d)$ that sends 1, ρ, ρ^2 into 0, 1, ∞. (ρ is a cube root of unity, $e^{2\pi i/3}$.)

4.14 Let z_1, z_2, z_3, z_4 in the plane project stereographically onto P_1, P_2, P_3, and P_4 on the sphere, and let $P_1 - P_2$ denote the length of the line joining P_1 and P_2. Show that

$$(P_1, P_2, P_3, P_4) = |(z_1, z_2, z_3, z_4)|$$

Solution: See Examples 4.1.3 and 4.4.4.

4.15 Show that the mapping $w = f(z)$ induced by an arbitrary rotation of the sphere about a diameter (preceded and followed by a stereographic projection) is bilinear. What are the fixed points in this case?

4.16 If z_1, z_2 are the stereographic projections of the ends of a diameter of the sphere, and the sphere is rotated through an angle θ about this diameter, show that

$$\frac{w - z_1}{w - z_2} = e^{i\theta}\frac{z - z_1}{z - z_2}$$

Solution: See Example 4.15.

4.17 In order that

$$w = \frac{az + b}{cz + d}$$

corresponds to a rotation of the sphere, it is necessary and sufficient that it can be written in the form

$$w = \frac{az - \bar{c}}{cz + \bar{a}}$$

Solution: See Examples 4.15 and 4.16.

4.18   Given a circle and a curve, the set of points inverses with respect to the circle of the points of the curve is a new curve. Show that the inverse of a circle (not cutting the fixed circle) is itself a circle.

4.19   Verify that the equation of a straight line can be put in the forms

(a) $\lambda x + \mu y = p$ $\lambda^2 + \mu^2 = 1$ $\lambda + i\mu = \zeta$

(b) $\mathbf{R}(z\zeta) = p$ $|\zeta| = 1$

(c) $\mathbf{R}(ze^{-ir}) = p$

(d) $b\bar{z} + \bar{b}z + c + \bar{c} = 0$

4.20   For the circle $z\bar{z} = R^2$, let z_1 be a point outside the circle. From z_1, two tangents are drawn to the circle, and the line joining the points of tangency is constructed. This line is the *polar* of z_1, and z_1 is the *pole* of this line. Show that the equation of the polar of z_1 is $\mathbf{R}(\bar{z}_1 z) = R^2$. Locate the inverse to z_1 with respect to the polar of z_1.

4.21   In the terminology of Example 4.20, let z_1 lie inside or on the circle. Then $\mathbf{R}(\bar{z}_1 z) = R^2$ is still called the *polar* of z_1, and z_1 the pole. Locate this line with reference to z_1.

4.22   Given two nonintersecting, nonconcentric circles, locate a pair of points z_1 and z_2 that are inverse points for both circles.

4.23   Let D be the connected domain having the two circles of Example 4.22 as its boundary. Show that D can be mapped onto an annulus $k < |w| < 1$ by a bilinear transformation. Find k, which is uniquely defined.

Solution: Suppose that the mapping is possible. Then, if w_1, w_2 are the images of z_1, z_2, we must have w_1 and w_2 inverse in both circles $|w| = k$ and $|w| = 1$. That is, $w_1\bar{w}_2 = 1$ and $w_1\bar{w}_2 = k^2$ which is meaningless unless $w_1 = 0$ and $w_2 = \infty$. The mapping thus must be of the form

$$w = A\left(\frac{z - z_1}{z - z_2}\right)$$

On the other hand, for any A, this mapping sends the two boundary circles of D onto circles having 0 and ∞ as inverse points (i.e., onto a pair of circles concentric with the origin). In order that the larger of these two circles be $|w| = 1$, $|A|$ is uniquely determined, and hence so is k.

$$k = \frac{R}{r}\left|\frac{a - z_1}{b - z_1}\right|$$

4.24   Map the annular region bounded by $|z| = 1$, $|z - 2/5| = 2/5$ conformally onto the annulus $k < |w| < 1$, and find k.

Solution: See Example 4.23.

4.25   Verify that $w = \sin z$ maps the semi-infinite strip $-\pi/2 < x < \pi/2$, $y > 0$ conformally onto the upper half-plane $v > 0$. Note the nature of the mapping at $z = -\pi/2$, $\pi/2$ where $w'(z) = 0$.

4.26   Show that $w = z/(1 - z)^2$ maps the unit circle $|z| < 1$ conformally onto the whole plane cut from $-\infty$ to $-1/4$. (The mapping is also univalent.)

4.27   The mapping $w = 4/(z + 1)^2$ maps the region $|z| < 1$, $0 < \arg z < \pi/2$ onto a domain D. Find D, the length of its boundary, and show that the mapping is univalent.

4.28 ANALYTICITY AT INFINITY

The nature of an analytic function $f(z)$ near the point at infinity is defined to have the corresponding properties of $f(1/z)$ near $z = 0$. Thus $f(z) = z/(1 + z)$ gives $f(1/t) = 1/(1 + t)$, which is analytic near $t = 0$, so that $f(z) = z/(1 + z)$ is said to be analytic at infinity.

Similarly, $w = f(z)$ is said to map a neighborhood of $z = z_0$ conformally onto a neighborhood of $w = \infty$ if $t = 1/f(z)$ maps a neighborhood of $z = z_0$ onto a neighborhood of $t = 0$.

Which of the following functions are analytic at infinity?

(a) e^z

(b) $\log \left(\dfrac{z + 1}{z - 1} \right)$

(c) $\dfrac{a_0 + a_1 z + \cdots + a_m z^m}{b_0 + b_1 z + \cdots + b_n z^n}$

(d) $\dfrac{\sqrt{z}}{1 + \sqrt{z}}$

4.29   Show that $w = \frac{1}{2}(z + 1/z)$ maps the portion of the upper half-plane exterior to the unit circle conformally onto the upper half-plane. What is the image of the upper half-semicircle $|z| < 1$, $\text{I}z > 0$ under the same mapping?

4.30   Show that $w = z + 1/z$ can be written

$$\frac{w - 2}{w + 2} = \left(\frac{z - 1}{z + 1} \right)^2$$

and so can be considered equivalent to the sequence of mappings

$$w_1 = \frac{z-1}{z+1} \qquad w_2 = w_1{}^2 \qquad w_2 = \frac{w-2}{w+2}$$

Thus show that any circle passing through $z = 1$, -1 is mapped onto a circular arc passing through $w = 2$, -2.

4.31 With the notation of Example 4.30, for the mapping $w = z + 1/z$, discuss the images of circles passing through $z = -1$, and enclosing $z = +1$

(a) By adding the vectors z, $1/z$.
(b) By using the sequence of maps suggested in Example 4.30.

Show that the *exteriors* of such circles are mapped conformally onto the *exteriors* of the cusp-shaped image curves (Figure 4.13).

Figure 4.13 *Joukouski's profile.*

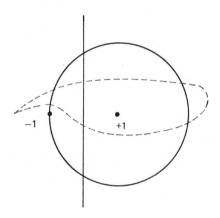

4.32 Generalize the results of Example 4.31 to the mapping

$$\frac{w - kc}{w + kc} = \left(\frac{z - c}{z + c}\right)^k$$

4.33 Show that $w = 2z/(1 - z)^2$ maps one of the four domains into which the circles $|z - 1| = \sqrt{2}$ and $|z + 1| = \sqrt{2}$ divide the plane, conformally onto $|w| < 1$.

4.34 Show that for $0 < c < 1$, $w = z(z - c)/(cz - 1)$ maps $|z| \leq 1$ onto $|w| \leq 1$ (taken twice).

4.35   Find a sequence of conformal maps that transform

(a) The unit upper semicircle to the unit circle.
(b) The region bounded by three circular arcs meeting orthogonally to the upper half-plane.

4.36   Verify the mapping of Example 4.25 by using Euler's formula,

$$w = \sin z = \frac{e^{iz} - e^{-iz}}{2i} \qquad w_1 = e^z$$

and the results of Example 4.29.

4.37   If $u - iv$ is an analytic function of $z = x + iy$, and $F(z) = \phi + i\psi$ satisfies $F'(z) = u - iv$, show that $\psi = $ const gives the general solution of the differential equation (the path curves of the vector-field):

$$\frac{dy}{dx} = \frac{v(x, y)}{u(x, y)}$$

Verify this result for the vector-fields

$$\frac{1}{z}, \qquad 1 - \frac{1}{z^2}$$

4.38   Describe the locus $z^4 + 4z$ as z varies on the unit circle.

Solution: A wheel of unit radius has its center on the rim of a circle of radius 4, and makes four revolutions per unit time turning uniformly while its center moves around the circle in unit time, also uniformly. The locus is the path traced out by a point on the circumference of the smaller circle. The ancient astronomers used such *epicycles* to describe the motion of the planets.

4.39   Show that $f(z) = z + \sum_2^\infty a_n z^n$ is univalent in $|z| < 1$ if $\sum_2^\infty n|a_n| < 1$.

COMMENT 2.   *Conformal mappings by multivalued functions:* $w = \sqrt{z}$.
We recall that for any complex number $z = re^{i\theta}$, there are two square roots: $\sqrt{r}e^{i\theta/2}$ and $-\sqrt{r}e^{i\theta/2}$. Each of these is called a *branch* of the function $\sqrt{z}$. Let us concentrate our attention on one of these, say $\sqrt{z} = re^{i\theta/2}$.

Regarding $w = \sqrt{z}$ as the inverse of the single-valued function $z = w^2$, we see that $w(z)$ is analytic and gives rise to a conformal mapping of each region in which $w'(z)$ exists and is $\neq 0$. We must thus avoid $z = 0$ if our mapping is to be conformal.

$$z = \frac{\xi + i\eta}{1 - \zeta} \qquad |z|^2 = \frac{\xi^2 + \eta^2}{(1 - \zeta)^2} = \frac{1 + \zeta}{1 - \zeta}$$

Hence

$$1 - \zeta = \frac{2}{1 + |z|^2} \qquad \xi + i\eta = \frac{2z}{1 + |z|^2}$$

$$|P_1 P_2|^2 = (\xi_2 - \xi_1)^2 + (\eta_2 - \eta_1)^2 + (\zeta_2 - \zeta_1)^2$$

$$= [\xi_2 + i\eta_2 - (\xi_1 + i\eta_1)] \times [\xi_2 - i\eta_2 - (\xi_1 - i\eta_1)]$$
$$+ [(1 - \zeta_1) - (1 - \zeta_2)]^2$$

$$= \left(\frac{2z_2}{1 + |z_2|^2} - \frac{2z_1}{1 + |z_1|^2} \right) \left(\frac{2\bar{z}_2}{1 + |z_2|^2} - \frac{2\bar{z}_1}{1 + |z_1|^2} \right)$$

$$+ \left(\frac{2}{1 + |z_2|^2} - \frac{2}{1 + |z_1|^2} \right)^2$$

$$= \frac{4 |z_2 - z_1|^2}{(1 + |z_1|^2)(1 + |z_2|^2)}$$

Next, in the triangles $NP_1 P_2'$, $NP_2 P_1$, the side

$$|NP'|^2 = x^2 + y^2 + 1 = 1 + |z|^2$$

$$|NP|^2 = \xi^2 + \eta^2 + (\zeta - 1)^2$$

$$= \frac{4 |z|^2 + 4}{(1 + |z|^2)^2} = \frac{4}{1 + |z|^2}$$

$$|NP| \, |NP'| = 2$$

Finally,

$$|NP_2| = \frac{2 |NP_1'|}{\sqrt{1 + |z_1|^2} \sqrt{1 + |z_2|^2}}$$

$$|NP_1| = \frac{2 |NP_2'|}{\sqrt{1 + |z_1|^2} \sqrt{1 + |z_2|^2}}$$

$$|P_1 P_2| = \frac{2 |P_1' P_2'|}{\sqrt{1 + |z_1|^2} \sqrt{1 + |z_2|^2}}$$

4.1.6 P_1, P_2 ends of a diameter gives $|P_2 - P_1| = 2$

Hence

$$\frac{|z_2 - z_1|^2}{(1 + |z_1|^2)(1 + |z_2|^2)} = 1$$

$$|z_1|^2 + |z_2|^2 + |z_1 z_2|^2 + 1 = |z_1|^2 + |z_2|^2 - z_1 \bar{z}_2 - \bar{z}_1 z_2$$

or

$$(z_1\bar{z}_2 + 1)(\bar{z}_1 z_2 + 1) = 0 = |z_1\bar{z}_2 + 1|^2$$

4.2.1 $$w \cdot z = \frac{\xi^2 + \eta^2}{1 - \zeta^2} = 1$$

4.2.2 $$w = i\frac{z - i}{z + i}$$

4.4.12 $(z_1 - a)\overline{(z_2 - a)} = R^2$ says that $z_1 - a$ and $z_2 - a$ are on the same half-line (or z_1, z_2 are on the same "radius" of the circle), and also that the product of their lengths is R^2.

4.3 $(z - a)(\bar{z} - \bar{a}) = R^2$

$$\frac{\xi^2 + \eta^2}{(1 - \zeta)^2} - a\frac{\xi - i\eta}{1 - \zeta} - \bar{a}\frac{\xi + i\eta}{1 - \zeta} + |a|^2 = R^2$$

$$(a + \bar{a})\xi + (i\bar{a} - ia)\eta + (|a|^2 - R^2 - 1)\zeta = 1 + |a|^2 - R^2$$

Annex to Section 4.1   For an elementary proof showing that "a perfect map is impossible" see G. Polya, Mathematical Discovery, vol. 2, pp. 129–132 (John Wiley & Sons, 1968).

INTEGRATION:
CAUCHY'S THEOREM

The subject of this chapter is the central theorem of the theory of analytic functions, due to Cauchy. The theorem is stated in Section 5.2.

5.1 WORK AND FLUX

We saw in Chapter 3 an interpretation of analytic functions that gives rise to a sourceless and irrotational two-dimensional vector-field. In this section we extend the "point" or local property of such vector-fields, which is expressed by the Cauchy-Riemann equations, to an "overall" or global property.

We imagine a vector $\bar{w} = u + iv$ attached to each point (x, y) of a two-dimensional domain. In this field, we draw a curve C with initial point a and endpoint b.

We may consider the vector $\bar{w}$ either as a force, or as a current density in the two-dimensional flow of a fluid. Each interpretation of the vector-field carries its own interpretation of the curve C. Thus, considering $\bar{w}$ as a force, it is expedient to think of C as a *path*, along which a material particle may move. The direction of such a motion, represented by the *unit tangent vector* to the curve, $e^{i\tau}$, is of importance here (Figure 5.1); τ is the angle between the tangent and the positive x-axis.

We can describe the curve in either of two directions. For definiteness (and to conform to usual practice) we single out the *counterclockwise direction* as the preferred, or positive, direction, when dealing with closed curves. (Here and later, the term "closed curve" is used in the same meaning as "simple closed curve"; see Section 5.9.)

If we interpret $\bar{w}$ as a current density, then it is natural to consider C as a *boundary* across which a material point may move. For such a motion the *unit normal vector* to the curve $e^{i\nu}$, is important (Figure 5.2); ν is the angle between the normal and the positive x-axis. We take as preferred normal direction the *outer normal* for the closed curves.

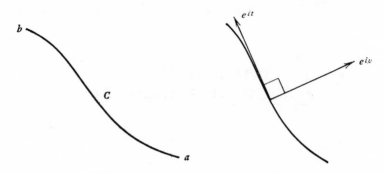

Figure 5.1

Figure 5.2

Accordingly, we have the following relation between our two unit vectors (compare Figures 5.1 and 5.2):

$$(1) \qquad\qquad e^{i\tau} = ie^{iv}$$

and the corresponding real relations

$$(2) \qquad\qquad \cos\tau = -\sin v \qquad \sin\tau = \cos v$$

For all curves, closed or not, we choose the direction along the normal so that the relations (1) and (2) remain valid.

The projections of $\bar{w}$ on the tangent and normal of C that we call w_τ and w_v, respectively, are real numbers, and are useful in calculating various physical quantities.

First, regarding $\bar{w}$ as a force, let us compute the *work* done in transporting a material particle along C from a to b. Let ds be the differential of arc of C at the point z. The contribution of this element of arc to the work done is $w_\tau\, ds$, that is, the projection of force times the distance, so that the total work is given by

$$(3) \qquad\qquad \int_C w_\tau\, ds$$

The symbol $\int_C$ indicates that we have to include into the sum all the elements of the curve C.

Second, let us regard $\bar{w}$ as a current density. The amount of matter crossing the line element ds per unit time in $w_v\, ds$, or the normal component times the length of line crossed. The total amount of matter crossing C per unit time, the *flux* across C, is thus:

$$(4) \qquad\qquad \int_C w_v\, ds$$

Now, see Figure 5.3,

$$w_\tau = \mathbf{R}\,\bar{w}e^{-i\tau} = |w| \cos (\tau - \arg \bar{w})$$

$$= u \cos \tau + v \sin \tau$$

$$w_\nu = \mathbf{R}\,\bar{w}e^{-i\nu} = |\bar{w}| \cos (\nu - \arg \bar{w})$$

$$= u \cos \nu + v \sin \nu$$

(5)

$$\int_C w_\tau \, ds = \int_C (u \cos \tau + v \sin \tau) \, ds \qquad \int_C w_\nu \, ds = \int_C (u \cos \nu + v \sin \nu) \, ds$$

Figure 5.3   **Figure 5.4**

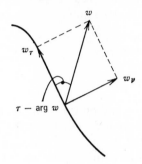

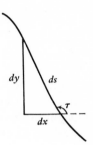

From Figure 5.4, we see that

$$\frac{dx}{ds} = \cos \tau \qquad \frac{dy}{ds} = \sin \tau$$

and therefore, see (2),

$$\cos \tau \, ds = dx \qquad \sin \tau \, ds = dy$$

$$\cos \nu \, ds = dy \qquad \sin \nu \, ds = -dx$$

Incorporating these into (5), we obtain

(6) $$\int_C w_\tau \, ds = \int_C u \, dx + v \, dy \qquad \int_C w_\nu \, ds = \int_C u \, dy - v \, dx$$

If we condense the two real equations (6) into one complex equation, we obtain

(7) $$\int_C w_\tau \, ds + i \int_C w_\nu \, ds = \int_C (u - iv)(dx + i \, dy) = \int_C w \, dz$$

where we have put

(8) $$dx + i \, dy = dz$$

What we have found is important: a physical interpretation of the complex line integral. We can rephrase it in the suggestive, if unorthodox, form

(9)
$$\int_C w\, dz = \text{work} + i\,\text{flux}$$

5.2 THE MAIN THEOREM

We consider the particular case in which a and b, the initial and final points of C, coincide, and C is a "simple" closed curve, that is, a closed curve without multiple points (does not cross itself) surrounding a domain D of the z-plane. If an integral is extended around a closed curve, we emphasize this circumstance by writing $\oint$ instead of $\int$. With this symbol, we have

(1)
$$\oint_C w\, dz = \oint_C w_\tau\, ds + i \oint_C w_\nu\, ds$$

$$= \text{work} + i\,\text{flux}$$

We recall (see Section 5.1) that each real line integral on the right hand side of (1) has to be viewed in a different physical context. The real part is the line integral of the tangential component of the vector $\bar{w}$. In this case it is convenient to interpret $\bar{w}$ as a force, C as a closed path, and the line integral as the work done by the field of force $\bar{w}$ as a particle makes the round trip along the closed path C. In the imaginary part of (1) we interpret C as the boundary of a domain D, and $\bar{w}$ as a current density, so that the line integral represents the total flux per unit time crossing the boundary C.

We now assume that the function w is analytic throughout the domain D, including the boundary C. Then, as we have seen in Section 3.5, the field of the vector $\bar{w}$ is both sourceless and irrotational. Since the field is irrotational, the work done along the closed curve C vanishes. Since the field is sourceless, the total flux across the boundary vanishes. That is, both the real and the imaginary parts of the right side of Section 5.2 (1) vanish, so that the left side is similarly zero. We have thus obtained an intuitive proof of Cauchy's theorem, the central theorem of the theory of analytic functions: If the function $w = f(z)$ is analytic at each point of the domain D bounded by the closed curve C, and on this curve itself, then

$$\oint_C f(z)\, dz = 0$$

This result, so intuitively attained, is of such extreme importance that it is desirable to investigate more closely the underlying notion of a complex line integral.

5.3 COMPLEX LINE INTEGRALS

Complex line integrals are closely analogous to ordinary real integrals. Therefore, let us recall first the concept of integrals in the real domain.

We focus our attention on the particular case in which the real function

(1) $$y = f(x)$$

of the real variable x is continuous and takes positive values between the limits a and b; $a < b$. We may interpret the integral of $f(x)$ from a to b as the area under the curve in Figure 5.5 and conceive of it as the limit of a sum of rectangles

(2) $$\int_a^b f(x)\, dx = \lim_{n \to \infty} \sum_{k=1}^n f(\xi_k)(x_k - x_{k-1})$$

Figure 5.5

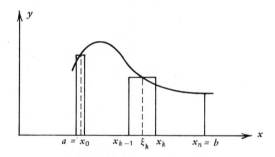

Let us recall the precise meaning of the sum on the right hand side of (2); see Figure 5.5. We subdivide the interval from a to b into n subintervals by inserting the successive points $x_1, x_2, \ldots, x_{n-1}$. For the sake of uniformity, we set $x_0 = a$, $x_n = b$. We choose any point ξ_k in the subinterval (x_{k-1}, x_k). The situation is expressed by the inequalities

$$a = x_0 < \xi_1 < x_1 < \xi_2 < \cdots < x_{k-1} < \xi_k < x_k < \cdots < x_n = b$$

(In fact, we could admit $\leq$ instead of $<$ as long as $x_{k-1} < x_k$.) We approximate the strip under the curve that lies above the subinterval (x_{k-1}, x_k) by a rectangle of base $x_k - x_{k-1}$, the length of the subinterval, and height $f(\xi_k)$, the ordinate corresponding to the intermediate abscissa ξ_k chosen in the subinterval. The area of the rectangle is $f(\xi_k)(x_k - x_{k-1})$. The areas of all the n rectangles of this kind make up the sum on the right of Section 5.3 (2). It is intuitive that such a sum of rectangles becomes arbitrarily close to the true value of the area under the curve if the bases of all the rectangles (the lengths of all the subintervals) become sufficiently small, and this is precisely what Equation 2 states.

We now consider a complex function

(3) $$w = f(z)$$

of the complex variable z. Being given two complex numbers a and b, we regard them as two points, a and b, in the complex plane. Since these points can be joined by infinitely many different curves, we must specify a definite curve C, starting from a and ending at b. We set $a = z_0$, $z_n = b$, and choose the remaining $(n + n - 1 = 2n - 1)$ points in the sequence $a = z_0$, ζ_1, z_1,

Figure 5.6

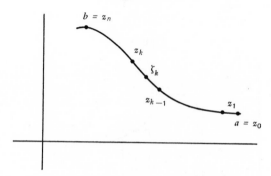

ζ_2, z_2, . . . , z_{k-1}, ζ_k, z_k, . . . ζ_n, $z_n = b$; we encounter these points in just this order when we move steadily from a to b along the curve C (Figure 5.6). We *define* the complex line integral of $w = f(z)$ along the curve C in strict analogy with Section 5.3 (2) as a limit:

(4) $$\int_C w \, dz = \lim_{n \to \infty} \sum_{k=1}^{n} f(\zeta_k)(z_k - z_{k-1})$$

We pass to the limit by letting n tend to infinity, and making all the subarcs of C, such as (z_{k-1}, z_k), tend to zero. The existence of this limit is less intuitive than in (2); we might have expected this, since the geometric meaning of (4) is less obvious than the meaning of the integral in (2) (area). However, since the rules of operating with complex numbers are the same as the rules for real numbers, and the notions of limit and continuity are defined in the same terms in the complex and the real domains, if the existence of one of the limits (2) and (4) can be proved by a formal argument, then that of the other can be proved by the same argument. Moreover, since the line integral itself has an intuitive meaning (see Section 5.1 (9)) we can contrive to give an intuitive meaning to the approximating sum; see Example 5.9.

A few concrete illustrations may help the understanding of these general ideas.

Example 1.   Let c denote a given constant. Find, using its definition as the limit of a sum, the integral

$$\int_C c \, dz$$

Wherever we choose the intermediate points $\zeta_1, \zeta_2, \ldots, \zeta_n$ in their respective intervals, the given integrand assumes the same value c at all these points. Therefore we have to consider the sum

$$\sum_{k=1}^{n} c(z_k - z_{k-1}) = c[(z_1 - z_0) + (z_2 - z_1) + \cdots + (z_n - z_{n-1})]$$

$$= c(z_n - z_0) = c(b - a)$$

Since changing n does not alter this expression, we let n tend to infinity, obtaining

$$\int_C c \, dz = c(b - a)$$

It is worth noticing that the value of the integral depends only on the endpoints a and b, and is independent of the choice of the curve C joining a and b. If the curve C is a closed curve, the points a and b coincide, and the integral vanishes. This, of course, is a very special example of the general theorem of Cauchy, since c, considered as a function of z, is analytic everywhere.

Example 2.   Find, using its definition as the limit of a sum, the integral

$$\int_C z \, dz$$

We choose the two extreme positions for ζ_k, first $\zeta_k = z_k$, and then $\zeta_k = z_{k-1}$. These extreme positions are, of course, admissible, although, for the sake of brevity, we did not mention them explicitly in the foregoing general formulation. In this way we obtain

$$\int_C z \, dz = \lim \sum_{k=1}^{n} z_k(z_k - z_{k-1})$$

$$= \lim \sum_{k=1}^{n} z_{k-1}(z_k - z_{k-1})$$

Complex Line Integrals   **149**

If we do not see how to evaluate these limits, we may try to find their arithmetic mean:

$$\int_C z\,dz = \lim \sum_{k=1}^{n} \frac{z_k + z_{k-1}}{2}(z_k - z_{k-1})$$

$$= \lim \sum_{k=1}^{n} \tfrac{1}{2}(z_k^{\,2} - z_{k-1}^2)$$

$$= \lim \tfrac{1}{2}[(z_1^{\,2} - z_0^{\,2}) + (z_2^{\,2} - z_1^{\,2}) + \cdots + (z_n^{\,2} - z_{n-1}^{2})]$$

$$= \lim \tfrac{1}{2}(z_n^{\,2} - z_0^{\,2})$$

$$= \lim \tfrac{1}{2}(b^2 - a^2)$$

$$= \tfrac{1}{2}(b^2 - a^2)$$

Again, the value of the integral depends only on the endpoints. Letting a and b coincide, we confirm Cauchy's theorem in another particular case.

Example 3.   Let C be the unit circle, described in the positive direction. Find

$$\int_C z^{-1}\,dz$$

as the limit of a sum.

The nth roots of unity divide C, the path of integration, into n equal arcs. Taking z^{-1} at the last point of each arc, we obtain the sum

$$\sum_{k=1}^{n} \exp\left(-\frac{2\pi i k}{n}\right)\left(\exp\frac{2\pi i k}{n} - \exp\frac{2\pi i(k-1)}{n}\right)$$

$$= \sum_{k=1}^{n}\left(1 - \exp\left(-\frac{2\pi i}{n}\right)\right)$$

$$= n\left(1 - \exp\left(-\frac{2\pi i}{n}\right)\right)$$

$$= n\left(1 - \cos\frac{2\pi}{n} + i\sin\frac{2\pi}{n}\right)$$

$$= n\left(2\sin^2\frac{\pi}{n} + i\sin\frac{2\pi}{n}\right)$$

$$\sim n\left(\frac{2\pi^2}{n^2} + i\frac{2\pi}{n}\right)$$

$$\to 2\pi i$$

We have used the fact that for an infinitesimally small angle α, $\sin \alpha \sim \alpha$. This means, more precisely expressed,

$$\frac{\sin \alpha}{\alpha} \to 1 \quad \text{as} \quad \alpha \to 0$$

We obtain

$$\oint_C \frac{dz}{z} = 2\pi i$$

if C is the unit circle described in the counterclockwise direction.

Observe that this result does *not* contradict Cauchy's theorem. In fact, Cauchy's theorem supposes that the integrand is analytic at all points of the domain enclosed by the closed curve, C, of integration, without exception. In the present case, however, there is an exception: the function z^{-1} ceases to be analytic at the origin, which is the center of the circle. In fact, z^{-1} is not even defined at $z = 0$, and becomes infinite as z approaches the exceptional point. We observe that $z = 0$ is the *only* exceptional point; at any other point z^{-1} is analytic. We see that violation of the hypothesis of Cauchy's theorem at even a single point may invalidate the conclusion of the theorem. One fly may spoil the milk.

5.3.1   Evaluate $\int_C (x + iy^2)\,dz$ along two paths joining $(0, 0)$ to $(1, 1)$:

(a) C consists of a segment of the real axis from 0 to 1, and of a segment of the line $x = 1$ from $y = 0$ to $y = 1$.

(b) C is the straight line joining $(0, 0)$ and $(1, 1)$.   (S)

5.3.2   Evaluate $\int_{-1}^{1} z^{1/2}\,dz$ taken around the upper unit semicircle, where $z^{1/2}$ has its principal value.   (S)

5.3.3   What is the value of the integral in Example 5.3.2, if the path is the lower unit semicircle?

5.3.4   Evaluate

$$\int_{-1}^{1} \frac{dz}{z}$$

taken

(a) Along the upper unit semicircle.
(b) Along the lower unit semicircle.

and compare the results with $\oint dz/z$ taken around the unit circle.

5.3.5 Evaluate $\oint z^n\, dz$ for n an arbitrary integer, the path of integration being the unit circle. (S)

5.3.6 Evaluate $\oint (z - z_0)^n\, dz$ for n an arbitrary integer, the path of integration being a circle centered at z_0.

5.3.7 Evaluate $\oint e^z/(z - z_0)\, dz$ taken around a circle centered at z_0.

Solution:

$$\oint \frac{e^z\, dz}{z - z_0} = \oint \frac{e^{z-z_0}}{z - z_0}\, dz\, e^{z_0}$$

$$= e^{z_0}\oint \left[\frac{1}{z - z_0} + 1 + \frac{z - z_0}{2!} + \frac{(z - z_0)^2}{3!} + \cdots \right] dz$$

$$= 2\pi i\, e^{z_0}$$

To justify the interchange of summation and integration we can write the integral as

$$e^{z_0} \oint \frac{e^{z-z_0} - 1}{z - z_0}\, dz + e^{z_0} \oint \frac{dz}{z - z_0}$$

and

$$\frac{e^{z-z_0} - 1}{z - z_0} = 1 + \frac{z - z_0}{2!} + \frac{(z - z_0)^2}{3!} + \cdots$$

is analytic inside and on C.

5.3.8 $f(z) = \sum_0^\infty a_n z^n$ is analytic for $|z| < R$, $R > 1$. Evaluate $\oint f(z)/(z^{k+1})\, dz$, the path of integration being the unit circle.

5.3.9 Evaluate $\oint dz/z$ about the square with vertices at $x = \pm 1$, $y = \pm 1$ and show that the value is $2\pi i$.

5.4 RULES FOR INTEGRATION

Some of the rules for ordinary real integrals follow so immediately from the definition of the integral as the limit of a sum that they can at once be extended to complex line integrals. They are the following:

If p and q are constants

(1)
$$\int_C [pf(z) + qg(z)]\, dz = p\int_C f(z)\, dz + q\int_C g(z)\, dz$$

Let a be the initial point of the curve C, b the final point of C and the initial point of the curve C', and c the final point of C'. Let $C + C'$ stand for the compound curve consisting of C and C' described in succession; $C + C'$

starts at a, passes through b and ends at c. Then

(2)
$$\int_C f(z)\,dz + \int_{C'} f(z)\,dz = \int_{C+C'} f(z)\,dz$$

Let C^{-1} denote the curve consisting of the same points as C, but described in the opposite direction. Then

(3)
$$\int_C f(z)\,dz + \int_{C^{-1}} f(z)\,dz = 0$$

The rule (3) can be regarded as the extreme case of (2) in which $C' = C^{-1}$, $c = a$ and the curve $C + C'$ passes through each of its points twice, first in one direction and then in the opposite direction (so that the two increments dz belonging to the two passages cancel).

Let L denote the length of the curve C, and M denote a (positive) number such that

$$|f(z)| \leqq M$$

along the whole curve C (M is the maximum of $|f(z)|$, or any greater number). Then

(4)
$$\left| \int_C f(z)\,dz \right| \leqq M \cdot L$$

This result is particularly important; it follows from Section 5.3 (4), and the triangle inequality.

Thus
$$\left| \sum_{k=1}^n f(\zeta_k)(z_k - z_{k-1}) \right| \leqq \sum_{k=1}^n |f(\zeta_k)|\,|z_k - z_{k-1}|$$

$$\left| \int_C f(z)\,dz \right| = \lim_{n \to \infty} \left| \sum_{k=1}^n f(\zeta_k)(z_k - z_{k-1}) \right|$$

$$\leqq \lim_{n \to \infty} \sum_{k=1}^n |f(\zeta_k)|\,|z_k - z_{k-1}|$$

$$= \int_C |f(z)|\,|dz|$$

$$\leqq M \int_C |dz|$$

$$= M \int_C \sqrt{dx^2 + dy^2}$$

$$= M \int_C ds$$

$$= M \cdot L$$

Still other familiar rules of the integral calculus remain valid in the complex domain if they are suitably interpreted.

5.4.1 $f(z)$, $g(z)$ are analytic; show that integration by parts is valid for complex line integrals

$$\int_C f(z)g'(z)\,dz = f(z)g(z)\Big|_C - \int_C g(z)f'(z)\,dz$$

where $f(z)g(z)\big|_C$ means $f(z)g(z)$ evaluated at the terminal end of C less its value at the beginning of C.

5.4.2 $z = g(t)$ is analytic. Show that.

$$\int_C f(z)\,dz = \int_{C'} f[g(t)]g'(t)\,dt$$

where C is the image of C' under the mapping $z = g(t)$.

5.4.3   Estimate $I = \oint dz/(1 + z^2)$ about $|z| = R$, and show that

$$|I| \leq \frac{2\pi R}{R^2 - 1} \qquad R > 1$$

Solution: $$\left|\frac{1}{z^2 + 1}\right| = \frac{1}{|R^2 e^{2i\theta} - 1|} = \frac{1}{|R^2 - e^{-2i\theta}|}$$

Geometrically, the points $R^2 - e^{-2i\theta}$ lie on a circle, centered at $z = R^2$ with unit radius. The maximum value of $|1/(1 + z^2)|$ occurs at the point nearest the origin, and is

$$\frac{1}{R^2 - 1}$$

5.4.4   Evaluate $\int_\pi^i e^{iz}\,dz$

   (a) Along the straight line joining the limits.
   (b) Along segments of the coordinate axes joining the limits.
   (c) Estimate the integrals using $|\int| \leq M \cdot L$ in each case, and compare the estimates with the values.

5.4.5   Show that $\oint_{|z|=1} \bar{z}\,dz = 2\pi i$.

5.4.6   Evaluate $\oint \bar{z}\,dz$ taken in the positive sense about the square with vertices at $x = \pm 1$, $y = \pm 1$.

5.5 THE DIVERGENCE THEOREM

The object of this section is to prove two parallel formulas

(1)
$$\oint_C u\,dx + v\,dy = \iint_D \left(\frac{\partial v}{\partial x} - \frac{\partial u}{\partial y}\right) dx\,dy$$

(2)
$$\oint_C u\,dy - v\,dx = \iint_D \left(\frac{\partial u}{\partial x} + \frac{\partial v}{\partial y}\right) dx\,dy$$

C stands for a closed curve without multiple points described in the positive direction and D for the domain surrounded by C; the functions u and v and their partial derivatives involved are supposed to be continuous in D and on C. We may regard u and v as the components of the vector

$$\bar{w} = u + iv$$

Remembering the formulas 5.1 (6) and the definitions given in Sections 3.5 and 3.6, we can transform (1) and (2) into

(3)
$$\oint_C w_\tau\,ds = \iint_D \operatorname{curl} \bar{w} \cdot dx\,dy$$

(4)
$$\oint_C w_v\,ds = \iint_D \operatorname{div} \bar{w} \cdot dx\,dy$$

respectively. In this new form, we can perceive an intuitive meaning. In fact, remember that curl $\bar{w}$ has been defined as the limit of a ratio: the ratio of the work along a small closed curve to the area surrounded by that curve. Multiplying by an infinitesimal area, which we denote by $dx\,dy$, we see that

$$\operatorname{curl} \bar{w}\,dx\,dy = \text{infinitesimal work}$$

Similarly,

$$\operatorname{div} \bar{w}\,dx\,dy = \text{infinitesimal flux}$$

And now we can express (3) and (4) (or, which is the same, (1) and (2)) as follows:

The total work along a closed path C equals the sum of the works around all elements of the area surrounded by C. The outflow across a boundary C equals the total output of all sources within the area surrounded by C.

We shall call both theorems jointly the *divergence theorem*. In the literature, these theorems and their space analogues are variously connected with the names of *Green* and *Gauss*, or *Riemann* and *Ostrogradsky*. The divergence theorem can be proved in various ways; for a usual short proof see Example 5.1.

We sketch here another proof that deepens and explains more fully the intuitive meaning emphasized by the verbal formulation.*

Draw two series of equidistant straight lines, parallel to the x-axis and to the y-axis, respectively. These lines divide the domain D into several portions; some of these portions are squares, others have a mixed boundary, consisting of straight line segments and of certain arcs of C (in the simplest case of just

Figure 5.7

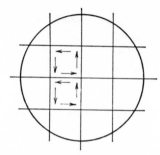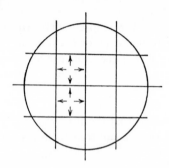

one small arc of C); see Figure 5.7. Let A_k denote the area of the kth portion and P_k its boundary curve described in the positive direction, We say that

$$(5) \qquad \oint_C w_r \, ds = \sum \oint_{P_k} w_r \, ds$$

$$(6) \qquad \oint_C w_v \, ds = \sum \oint_{P_k} w_v \, ds$$

the sums on the right hand side are extended over all the portions into which D has been subdivided.

In fact, consider the various parts of the curve P_k. Such a part may be a straight segment separating the portion considered from one of its neighbors whose boundary curve is P_j: the segment in question belongs both to P_k and to P_j, but described in opposite directions and yields, therefore, *exactly opposite contributions* to the line integrals extended along P_k and P_j in (5) or in (6); see Figure 5.7 and consider both work and flux; *cf.* Section 5.4 (3). Thus, the contribution of any such straight line segment to the sums considered is exactly nil, and nothing remains of any of these sums but the contributions of the several arcs into which the boundary curve C has been cut by the subdivision (any such arc belongs to *just one* portion). Yet these arcs make up the whole curve C and so (5) and (6) become obvious.

---

* The idea of the following proof goes back to Ampère who used it in investigating the magnetic action of electric currents; *cf.* J. C. Maxwell, *A Treatise on Electricity and Magnetism*, Vol. 2, Sect. 483, 484.

Now, by the definition of curl and divergence (see Sections 3.5 and 3.6) we have the approximate equations

$$(7) \qquad \oint_{P_k} w_\tau \, ds \sim A_k \, \text{curl } \bar{w}$$

$$(8) \qquad \oint_{P_k} w_\nu \, ds \sim A_k \, \text{div } \bar{w}$$

which tend to become precise as the dimensions of the portion, and so its area A_k, tend to 0. By virtue of (7) and (8), we can recognize the right-hand sides of (5) and (6) as approximations to the double integrals on the right-hand sides of (3) and (4), respectively, and we obtain these latter formulas in the limit as the distance between the neighboring equidistant parallels of the subdivision tends to 0.

5.6 A MORE FORMAL PROOF OF CAUCHY'S THEOREM

We obtained in Section 3.3 a necessary and sufficient condition that $w(z)$ be analytic; this condition can be written

$$\frac{\partial w}{\partial x} = \frac{\partial w}{i \, \partial y}$$

the real and imaginary parts yielding the Cauchy-Riemann equations. We further assumed that the derivatives of the real functions u, v in this expression were continuous. If we combine this result with 5.5 (1) and (2) we obtain

$$\int_C w \, dz = \int_C (u - iv)(dx + i \, dy)$$

$$= \int_C u \, dx + v \, dy + i \int_C u \, dy - v \, dx$$

$$= \iint_D (v_x - u_y) \, dx \, dy + i \iint_D (u_x + v_y) \, dx \, dy$$

$$= 0$$

since

$$\frac{\partial}{\partial x}(u - iv) = \frac{\partial}{i \, \partial y}(u - iv)$$

so that

$$v_x - u_y = 0 \qquad u_x + v_y = 0$$

at each point of D.

This proof may appear to be more different from our former proof than it actually is. The reader should realize that the foregoing more formal proof

A More Formal Proof of Cauchy's Theorem   **157**

in fact renders the intuitive proof of Section 5.2 just more explicit and explains it more fully.

5.7 OTHER FORMS OF CAUCHY'S THEOREM

Up to now, we have relied on the intuition of the reader in describing domains D, and their boundary curves C. It now becomes necessary to have

Figure 5.8

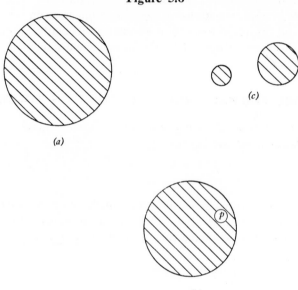

(a)

(c)

(b)

more precise ideas of these concepts, since modifications of the results obtained and their extension to more general cases are often needed in the applications.

The "simplest" domain we can think of is a set of points such as the interior of a circle, see Figure 5.8 (a), or of a square. There are also more complicated domains; in Figure 5.8, (b) is a ring-shaped domain between two concentric circles: it consists of such points as the ones inside the larger (encompassing) circle *and* outside the smaller (encompassed) circle. Figure 5.8 (c) consists of the points belonging to either of two nonoverlapping circles. These three examples illustrate very important differences.

Domains (a) and (b) are *connected*; any two points can be joined by a curve wholly within the domain, that is, passing through interior points of the domain. Clearly this is not so for (c), which is disconnected.

Let us focus our attention on connected domains.

The essential difference between the connected domains (a) and (b) is that while every simple closed curve in (a) encloses only points of the set, there are simple closed curves fully contained in (b) that enclose points not belonging to (b). We say that (a) is *simply* connected, while other connected domains such as (b) are called *multiply* connected.

We are now in a position to give another form to Cauchy's theorem, which will be useful in developing the integral calculus of analytic functions.

Theorem 1. If $w = f(z)$ is analytic in a simply connected domain D, and if a and b are any two points of D, then the integral

$$\int_a^b f(z)\, dz$$

has the same value along every path of integration connecting the endpoints a and b and lying entirely within D.

We can see intuitively that this form of Cauchy's theorem is equivalent to the form stated in Section 5.2; we shall use some simple facts listed in Section 5.4. Thus, let C_1 and C_2 be two paths joining a and b and lying entirely in D. We restrict ourselves to the simplest case in assuming that the path $C_1 + C_2^{-1}$ is a simple closed curve. (The reader should free himself from this restriction.) Since D is simply connected, all points inside $C_1 + C_2^{-1}$ belong to D and so $f(z)$ is analytic at all these points—we shall use this fact.

We now obtain, from Section 5.4 (2) and (3),

$$\left(\int_{a_{C_1}}^b - \int_{a_{C_2}}^b \right) f(z)\, dz = \int_{C_1 + C_2^{-1}} f(z)\, dz$$

$$= \oint f(z)\, dz = 0$$

by virtue of Cauchy's theorem, and hence

$$\int_{a_{C_1}}^b f(z)\, dz = \int_{a_{C_2}}^b f(z)\, dz$$

Conversely, if we assume the proposition just proved we can derive, retracing our steps, the form given in Section 5.2.

Cauchy's theorem, either in its new form or in its original form found in Section 5.2, is the starting point for almost all deeper investigations concerning analytic functions. Several immediate consequences, important in their own right, follow below.

Let C_1 and C_2 be simple closed curves, C_2 lying entirely in the interior of C_1. The points that are both interior to C_1 and exterior of C_2 form a domain that is not simply but *doubly connected*, and is called the *annular* domain determined by C_1 and C_2.

Figure 5.9

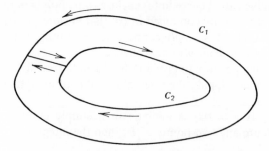

Theorem 2. If $f(z)$ is single valued and analytic at all points of the annular domain determined by C_1 and C_2, boundaries included, and both C_1 and C_2 are described in the positive sense, then

$$\int_{C_1} f(z)\, dz = \int_{C_2} f(z)\, dz$$

To prove this result, we connect the two paths C_1 and C_2 by an auxiliary path R (a "cut") lying entirely in the annular domain (Figure 5.9). Imagine now the right and left "banks" of the path R (the "river") pulled apart somewhat, as in Figure 5.10.

We have "cut" the annular domain along the path R, and by cutting it we have changed it into a simply connected domain: the path *abcdefa* is a non-self-intersecting simple closed curve, enclosing only points for which $f(z)$ is analytic. It follows from Cauchy's theorem that

(1)
$$\oint_{abcdefa} f(z)\, dz = 0$$

the integral being taken in the positive sense. We can decompose (1) into

$$\left[\int_{C_1} + \int_c^d - \int_{C_2} + \int_f^a \right] f(z)\, dz = 0$$

Figure 5.10

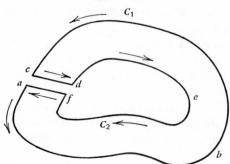

the first and third integrals being taken in the positive (counterclockwise) sense. Since the paths involved actually coincide, but are described in opposite directions, we see that the sum of the two integrals

$$\left[\int_c^d + \int_f^a\right] f(z)\, dz = 0$$

cf. Section 5.4 (3), leaving

(2) $$\int_{C_1} f(z)\, dz - \int_{C_2} f(z)\, dz = 0$$

and this proves Theorem 2.

Imagine a moving rubber band that may change its length, shape, and location. From its initial position where it coincides with the curve C_1, the rubber band moves, in sweeping across the intervening ring-shaped domain, to its final position where it coincides with the curve C_2. This rubber band suggests the concept of "continuous deformation" of a curve. To express Theorem 2 intuitively, we say: the continuous deformation of the line of integration leaves the value of the integral unchanged, provided that the continuously deformed line sweeps across only such points at which the integrand is analytic. Imagine the singular points at which the integrand ceases to be analytic as nails sticking out from the plane in which the rubber band moves and blocking its passage.

Example 4.   From Example 3, Section 5.3,

$$\oint_{C_1} \frac{dz}{z} = 2\pi i$$

where C_1 is the unit circle, $|z| = 1$. From the theorem just proved, since $f(z) = 1/z$ is analytic at each finite $z \neq 0$, we see first that

(3) $$\oint_C \frac{dz}{z} = 2\pi i$$

where C is any curve encircling $|z| = 1$. Especially the result is true for any sufficiently large circle, which will encircle any *simple closed curve C enclosing the point $z = 0$*; and so, by a second application of Theorem 2, the result (3) is true for any such C.

Observe that Theorem 2 supposes that $f(z)$ is analytic in the annulus between C_1 and C_2, but supposes nothing about $f(z)$ inside the interior curve C_2, where $f(z)$ can be analytic or otherwise.

In very much the same way, we establish Theorem 3.

Theorem 3.   Let C_0 be a simple closed curve. Let each of the simple closed curves $C_1, C_2, \ldots, C_n$ lie completely in the interior of C_0, but exterior to each other (especially, they do not intersect). Then

$$\int_{C_0} f(z)\,dz = \int_{C_1} f(z)\,dz + \int_{C_2} f(z)\,dz + \cdots + \int_{C_n} f(z)\,dz$$

provided $f(z)$ is analytic inside and on the boundary of the multiply connected region formed by C_0 as the exterior boundary curve and $C_1, \ldots, C_n$ as interior boundary curves. The integrals are all taken in the positive sense.

The proof is very similar to the proof of Theorem 2, and is clearly suggested by Figure 5.11.

Figure 5.11

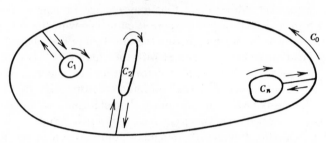

5.8 THE INDEFINITE INTEGRAL IN THE COMPLEX DOMAIN

We are now in a position to investigate the "indefinite integral" of an analytic function $f(z)$.

Let us consider a *simply connected* domain D, and any two points, z_0 and z of D. We may connect z_0 and z by infinitely many curves that lie completely in D. From Theorem 1 of the preceding section,

$$\int_{z_0 \atop C_1}^{z} f(z)\,dz = \int_{z_0 \atop C_2}^{z} f(z)\,dz$$

for *any* two such paths C_1 and C_2, see Figure 5.12. That is, the integral $\int_{z_0}^{z} f(z)\,dz$ has a meaning that does not depend on the path of integration, so that $F(z) = \int_{z_0}^{z} f(z)\,dz$ is a well-determined single-valued function of z (we can evaluate it by choosing any curve between z_0 and z). Let us add the important remark that $F(z)$ is analytic, and $F'(z) = f(z)$. To prove this statement, we proceed from the definition of the derivative, and construct $[F(z + \Delta z) - F(z)]/\Delta z$ (Figure 5.13).

$$F(z + \Delta z) - F(z) = \int_{z_0}^{z+\Delta z} f(t)\,dt - \int_{z_0}^{z} f(t)\,dt = \int_{z}^{z+\Delta z} f(t)\,dt$$

and since we may take any path connecting z and $z + \Delta z$, we choose the simplest path, which is the straight line joining the two points (this choice is

Figure 5.12　　　　　　　　　　　**Figure 5.13**

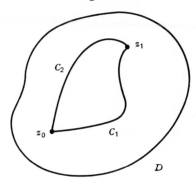

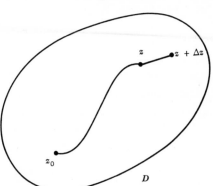

possible if $|\Delta z|$ is sufficiently small). Then

$$
\frac{F(z + \Delta z) - F(z)}{\Delta z} - f(z) = \frac{1}{\Delta z} \int_{z}^{z+\Delta z} f(t)\, dt - f(z)
$$

(1)

$$
= \frac{1}{\Delta z} \int_{z}^{z+\Delta z} [f(t) - f(z)]\, dt
$$

Since $f(z)$ is continuous, $|f(t) - f(z)|$ is small for suitably small $|t - z|$. In more precise form, given any $\varepsilon > 0$, however small, it is possible to find a $\delta > 0$, so that $|f(t) - f(z)| < \varepsilon$ for all $|t - z| < \delta$. Using this fact, we may estimate the right side of (1) as follows:

$$
\left| \frac{1}{\Delta z} \int_{z}^{z+\Delta z} [f(t) - f(z)]\, dt \right| \leqq \frac{1}{|\Delta z|} \int_{z}^{z+\Delta z} |f(t) - f(z)|\, |dt|
$$

(2)

$$
\leqq \frac{1}{|\Delta z|} \varepsilon\, |\Delta z| = \varepsilon
$$

Compare (1) and (2) and observe that ε is arbitrarily small. Thus, the difference $(\Delta F/\Delta z) - f(z)$ becomes arbitrarily small: it tends to zero as $\Delta z \to 0$, and so $F'(z) = f(z)$.

The result we have just obtained is important: *the integral of a given analytic function is itself an analytic function of the upper limit of integration and its derivative is the given function.*

We may use now for the function discussed, $F(z)$, the usual terminology of calculus (of the theory of real variables): the term "indefinite integral" and the notation

$$
F(z) = \int f(z)\, dz
$$

The Indefinite Integral in the Complex Domain　　**163**

The importance of the indefinite integral is twofold. First, we may express the definite integral

$$\int_a^b f(z)\,dz = \int_{z0}^b f(z)\,dz - \int_{z0}^a f(z)\,dz$$
$$= F(b) - F(a)$$

as the difference of two values of the indefinite integral. Second, indefinite integration is the inverse operation to differentiation, the indefinite integral is the "antiderivative." Or, stated still differently, the indefinite integral $F(z)$ of a given function $f(z)$ is the solution of the differential equation

$$(3) \qquad \frac{d\,F(z)}{dz} = f(z)$$

We need only the simplest rules for differentiation to convince us of this fact: if $F(z)$ satisfies (3), so does $F(z) + C$ where C is any constant. Yet is the converse true? Is $F(z) + C$ the most general solution of (3)?

To examine this question let $G(z)$ be an analytic function satisfying

$$(4) \qquad \frac{d\,G(z)}{dz} = f(z)$$

Set $G(z) - F(z) = w$; then, from (3) and (4),

$$w' = 0$$

If $w = u + iv$, where u and v are real as usual, it follows that

$$\frac{\partial w}{\partial x} = \frac{\partial w}{i\,\partial y} = 0$$

$$u_x = v_x = u_y = v_y = 0$$

and hence, by the theory of the functions of two real variables, we see that

$$u = \text{const} \qquad v = \text{const}$$

Therefore, with some (complex) constant C

$$w = C$$
$$G(z) - F(z) = C$$
$$G(z) = F(z) + C$$

and so we have found the most general solution of the differential equation (3). Expressed in other words, the difference between any two indefinite integrals or antiderivatives is necessarily a constant.

We are now in a position to explain Examples 1 and 2 of Section 5.3. We found that $\int_C c\, dz = c(b-a)$. Since $f(z) = c$ is an analytic function for all z, and $F(z) = cz$ is an indefinite integral, it follows that

$$\int_a^b c\, dz = [cz]_a^b = cb - ca = c(b-a)$$

Analogously

$$\int_a^b z\, dz = [\tfrac{1}{2}z^2]_a^b = \tfrac{1}{2}(b^2 - a^2)$$

We consider now Example 3 of Section 5.3, which needs more care since $f(z) = 1/z$ is not analytic inside $|z| \leq 1$: it has a singularity at $z = 0$. We have

Figure 5.14

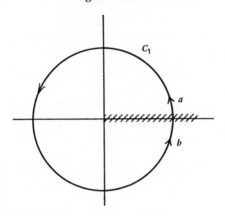

to avoid this singularity, and, moreover, we have to avoid it so that we obtain a *simply connected* domain: our preceding investigation of the indefinite integral was restricted to such a domain.

To accomplish this we "delete" or "cut out" the positive half of the real axis, the origin included; that is, we consider that the subset of the set of all complex numbers that remains when the nonnegative real numbers are excluded. We call this remaining subset the "cut plane"; it is a simply connected domain: a circle, or any other simple closed curve entirely contained in the cut plane, cannot encircle the origin or any other point that does not belong to the cut plane.

As in Example 3 of Section 5.3, we have to integrate $1/z$ over the unit circle $|z| = 1$. We know that any branch of $\log z$ is an antiderivative of $1/z$; yet the line of integration $|z| = 1$ is not entirely contained in the cut plane. Therefore, we consider an arc of this circle, from a to b, which is still inside the cut plane, although a and b are arbitrarily close to the point 1 and the arc is almost a full circle (Figure 5.14).

Now we consider $\int_a^b dz/z = \log b - \log a$; observe that we arrive at the value $\log b$ by starting from the value $\log a$ and following the logarithmic function as it changes *continuously* along the arc from a to b. Finally, we pass to the limit letting a and b tend to 1 and the line of integration to the entire unit circle C:

$$\int_C \frac{dz}{z} = \lim \int_a^b \frac{dz}{z} = \lim \,[\log z]_a^b$$

$$= (\log e^{2\pi i} + 2n_1\pi i) - (\log 1 + 2n_1\pi i)$$

$$= 2\pi i$$

In the first line, $\log z$ denotes any determined branch of the logarithmic function. In the second line, however, log refers to the principal branch as far as a is concerned, and to the limiting value of the principal branch with regard to b so that

$$\log 1 = 0 \qquad \log e^{2\pi i} = 2\pi i$$

The value of the integral turns out (as it should) independent of our arbitrary choice: it does not matter which one of the infinitely many branches we choose for $\log z$.

This example is instructive in several respects; it shows especially that the indefinite integral of $f(z)$ may be multivalued if $f(z)$ ceases to be analytic even at a single point of the region considered.

5.8.1   In the following integrals, C is the unit circle, $|z| = 1$, described in the positive (counterclockwise) sense. Show that the value of each of these integrals is zero.

(a) $\displaystyle\int \frac{e^z \, dz}{z - 2}$

(b) $\displaystyle\int \frac{\sin z}{z} \, dz$

(c) $\displaystyle\int \frac{dz}{\cos z}$

(d) $\displaystyle\int \sqrt{z + 4} \, dz$

(e) $\displaystyle\int \frac{\sinh z}{z^2 + 2z + 2}$

5.8.2   Evaluate each of the following integrals along any curve joining the end points of integration.

(a) $\displaystyle\int_{-1}^{i} \cosh 3z \, dz$

(b) $\displaystyle\int_{-\pi i}^{\pi i} e^{2z} \, dz$

(c) $\displaystyle\int_{-R}^{R} (z^3 - iz) \, dz$

5.8.3   Deforming the path of integration, transform the integral $\oint_{|z|=1} dz/z$ into one taken along the real axis from $z = -R$ to $z = -\delta$, then along the lower semicircle $|z| = \delta$, then along the real axis from δ to R, and finally along the upper semicircle $|z| = R$; suppose $R > \delta$.

5.8.4   Show that $\int_{-1}^{1} dz/z$ has the same value for every path joining $-1, 1$, which remains completely in the upper half-plane.

5.8.5   Evaluate $\int_{1}^{z} \log t \, dt$ in the cut plane, cut from 0 to ∞ on the positive real axis, where $\log t$ has its principal value.

Solution: $z \log z - z + 1$.

5.9   GEOMETRIC LANGUAGE

Until the beginning of this chapter we have used certain general geometric terms, such as "curve," "region," and "domain" somewhat loosely. For the reader who paid sufficient attention to the examples at hand, the danger of misunderstanding was not great. Yet we need a stricter terminology if we wish to deal efficiently with the geometric concepts, the importance of which became manifest in this chapter.

(a) *Curves.*   As in Section 2.10, we consider a complex valued function

$$z = x + iy = f(t)$$

of a real variable t. We assume now that $f(t)$ is nonconstant and defined in an interval including its endpoints, in a *closed* interval. We may even choose, without real loss of generality, a specific interval. We assume that

$$0 \leq t \leq 1$$

is the interval of definition of the function $f(t)$. We assume further that $f(t)$ is continuous. Then the set of values (points) assumed by $f(t)$ forms a *curve*. This curve is described in one *sense* or the opposite accordingly as we consider t as increasing from 0 to 1 or as decreasing from 1 to 0. The curve is *closed* if

(1) $f(0) = f(1)$

A curve that is not closed may be called an *arc*.

Observe that these definitions are general; they say nothing about tangents or continuous change of direction, and so on. Thus, we regard the boundary of a square as a closed curve, and other polygonal lines as curves, closed curves or arcs.

A curve that has no multiple points (which does not intersect itself) is called a *simple* curve. The full definition follows: an arc is called simple if for any two real numbers t_0 and t_1

(2) $\qquad\qquad 0 \leqq t_0 < t_1 \leqq 1 \qquad$ implies $\qquad f(t_0) \neq f(t_1)$

A closed curve is called simple if (2) is valid with the exception of the single pair $t_0 = 0$, $t_1 = 1$ for which we have (1). Thus, there is a one-to-one continuous correspondence between the points of a simple arc and the points of a closed interval, which is the "simplest arc." Similarly, there is such a correspondence between the points of any simple closed curve and the points of the simplest closed curve, the circle. (This follows from the definition that postulates such a correspondence between the simple closed curve and the interval, the two endpoints of which are identified, considered as "coincident.")

A simple closed curve separates the plane into two domains: its interior and its exterior. This is obvious in familiar cases: for the circle, the boundary of a square, the boundary of a convex polygon, and so on. The reader is advised to visualize intuitively the surrounded interior and the unbounded exterior in such accessible cases, and postpone worrying about a general proof of the fact for the most general simple closed curve (which may be horribly complicated.)

A curve is *oriented* when it is described in a definite sense (there are only two different senses to choose from, as we have seen above). A simple closed curve is *positively oriented* if it is described counterclockwise, as by a person walking along it and leaving the interior to his left.

(b) *Point sets, regions, and domains.* The plane (the complex plane, the z-plane) is the set of all complex numbers $z = x + iy$. Any subset of this set is a *point set.* A point set is well defined for us if we know an unambiguous rule by which we can decide of any point of the plane whether it does belong or does not to the point set.

We wish to characterize those kinds of point sets that are most useful in our study.

A point set S is called *open* if each point that belongs to S is the center of a (sufficiently small) circular disc, each point of which also belongs to S. (Each point of an open set is completely surrounded by points of the same kind. No point of an open set can be the limit of a sequence of points not belonging to that open set.) For instance, the interior of a square (to which the points on the boundary of the square do *not* belong) is an open set.

An open set S is *connected* if any two points z_1 and z_2 of S can be joined by an arc in S, that is, are endpoints of an arc each point of which belongs to S. We could say "polygonal arc, consisting of a finite number of straight line-segments" instead of "arc"; (the restriction to the more elementary kind of arc turns out inessential on closer consideration.) For instance, the open set consisting of the interiors of two nonoverlapping circles, see Figure 5.8(c), is not connected.

A connected open set is called a *domain.*

A *region* is a point set consisting of a domain D and of a subset of the boundary points of the domain D. (The term "subset" is used here inclusively, including the two extreme cases, the improper subsets: the empty set and the full set.) A good example is the region of convergence of a power series with a positive finite radius of convergence: it contains the domain interior to the circle of convergence, and all points or no points or some points on the circle of convergence itself.

(c) *Connectivity.* The concept of a simply connected domain was introduced in Section 5.7. (Read again the explanation and pay attention to the terminology.)

The interior of a simple closed curve is a simply connected domain. (Visualize intensively several accessible particular cases; do not worry about a general proof.)

A domain with one hole, or annular domain, is doubly connected; see Figures 5.8(b) and 5.9. A domain with n holes is multiply connected for $n \geqq 1$; for $n = 3$ see Figure 5.11.

(d) The terminology explained in the foregoing is useful. For instance, a line of integration, arc or closed curve, must be oriented; remember the rule 5.4 (3). A simply connected domain must be carefully distinguished from a multiply connected domain as the foregoing sections show, and so on.

We shall take the liberty to use the terminology introduced somewhat "colloquially." We shall not insist on a long complete description when a short incomplete expression can be used with little danger of misunderstanding. For instance, again and again we must consider integrals along "positively oriented simple closed curves." We prefer positively oriented curves to clockwise described curves, simple curves to self-intersecting curves. When there is no particular reason to be especially careful or to suspect a deviation from our general preference we may drop "positively oriented" or even "simple" and say just "closed curve" instead of the longer phrase fully stated before although, of course, only the longer phrase is fully precise.

(e) In reviewing former chapters, the reader may notice that the terminology just explained has been already used, although not with rigid consistency. For instance, in conforming to the prevalent usage, we employed the term "domain of definition" of a function although a function of a complex variable may be defined in any point set and not just in an open connected set. (For analytic functions, however, domains in the technical sense of the term play a special role.)

Additional Examples and Comments on Chapter Five

COMMENT 1. Although we often lean somewhat heavily on geometric intuition in the development of our theorems, all of these results can be given in precise terms. Thus, in Chapter 2 we defined what we meant by a simple closed curve (a Jordan curve), and in this chapter we discussed various types of domains bounded by such curves.

The curves that we admit for the purposes of integration must, of necessity, be reasonably smooth. It is sufficient for our purpose to require only that these curves have continuously turning tangents, or be composed of a finite number of arcs with such tangents. In this case, each component arc may be represented parametrically in the form

$$x = \phi(t)$$
$$y = \psi(t) \qquad 0 \leq t \leq 1$$

where $\phi(t)$ and $\psi(t)$ have a continuous derivative. It follows then that each such arc has a finite length, is rectifiable.

$$L = \int ds = \int_0^1 \sqrt{\left(\frac{dx}{dt}\right)^2 + \left(\frac{dy}{dt}\right)^2}\, dt \text{ exists}$$

In the following, it will be understood that each finite path of integration is a rectifiable arc, or a finite combination of such arcs, that is, is piecewise rectifiable, or piecewise smooth.

5.1   Establish the divergence theorem for a simply connected domain D bounded by a smooth convex curve C; that is

$$\iint_D \left(\frac{\partial P}{\partial x} + \frac{\partial Q}{\partial y}\right) dx\, dy = \int_C P\, dy - Q\, dx$$

5.2   Using $z = x + iy \qquad \bar{z} = x - iy$

$$\frac{\partial}{\partial x} = \frac{\partial}{\partial z} + \frac{\partial}{\partial \bar{z}} \qquad \frac{\partial}{\partial y} = i\left(\frac{\partial}{\partial z} - \frac{\partial}{\partial \bar{z}}\right)$$

show that the divergence theorem can be written in the form

$$\iint_D \frac{\partial f}{\partial \bar{z}}\, dx\, dy = -\frac{i}{2}\int_C f\, dz$$

where f is obtained from $\partial f/\partial \bar{z}$ by "formal integration," that is, keeping z fixed.

5.3   Using the divergence theorem, express the area and polar moment of inertia of the area, bounded by a smooth simple closed curve, as a contour integral.

Solution: Area $= A = \iint_D dx\, dy = \iint_D \frac{\partial f}{\partial \bar{z}}\, dx\, dy$

where $\partial f/\partial \bar{z} = 1$. Then $f = \bar{z} + g(z)$, and

$$A = -\frac{i}{2} \int_C [\bar{z} + g(z)]\, dz = -\frac{i}{2} \int_C \bar{z}\, dz$$

since $g(z)$ is analytic.

$$I_0 = -\frac{i}{4} \int_C z\bar{z}^2\, dz \qquad \left[= \iint_D (x^2 + y^2)\, dx\, dy \right]$$

5.4   Show that for the epicycloids formed by rolling a circle of radius a/n around a fixed circle of radius a,

$$\frac{nz}{a} = (n + 1)t - t^{n+1} \qquad \text{where} \qquad t = e^{i\theta}$$

When n is an integer, these epicycloids are closed curves with n cusps. Show also that

$$\text{area} = A = \frac{(n + 1)(n + 2)}{n^2}\,\pi a^2$$

$$\text{arc length} = L = \frac{8(n + 1)}{n}\,a$$

$$\text{polar moment of inertia} = I_0 = \frac{A^2}{2} + \frac{Aa^2}{2n^2}$$

Solution: From Figure 5.15, the length of the arc RQ is $a\theta = $ length of the arc $QP = (a/n)\psi$. Hence $\psi = n\theta$. Let z be the coordinates of P. Then we

Figure 5.15

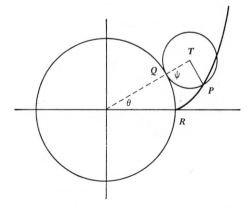

reach P by going from $z = 0$ to the center of the small circle $(ae^{i\theta} + a/n\, e^{i\theta})$ and then adding the vector TP, which is TQ rotated through an angle ψ, that is $(-a/n\, e^{i\theta})\, e^{i\psi}$.

Thus

$$z = a\left(\frac{n+1}{n}\right)e^{i\theta} - \frac{a}{n}\, e^{i(n+1)\theta} = \frac{a}{n}[(n+1)t - t^{n+1}]$$

$$A = \iint_D dx\, dy = -\frac{i}{2}\int \bar{z}\, dz$$

$$= -\frac{i}{2}\int_{|t|=1} \frac{a}{n}\left[\frac{(n+1)}{t} - \frac{1}{t^{n+1}}\right]\frac{a}{n}\,[(n+1)(1-t^n)]\, dt$$

$$= \frac{\pi a^2}{n^2}(n+1)(n+2)$$

See Example 5.3.

5.5 Show, as in Example 5.4, that the hypocycloids, obtained by rolling a circle of radius a/n inside a fixed circle of radius a, satisfy

$$\frac{nz}{a} = (n-1)t + t^{1-n} \qquad t = e^{i\theta}$$

Find L, A, I_0 for these curves for n an integer.

5.6 $f(z) = \sum_0^\infty a_n z^n$ is analytic and univalent for $|z| \le 1$. Consider the curve C, the image of the unit circle under the mapping. Compute the area bounded by C. (See Examples 5.3 and 5.4.)

5.7 Show that

$$\iint_D \overline{f'(z)}\, g(z)\, dx\, dy = -\frac{i}{2}\int_C \overline{f(z)}\, g(z)\, dz$$

where $f(z)$, $g(z)$ are analytic inside and on C. Thus show that if $P(z)$ is a polynomial

$$P(z) = \sum_0^n a_k z^k$$

and D is the circle $|z| < R$,

$$\iint_D |P(z)|^2\, dx\, dy = \pi \sum_0^n \frac{|a_k|^2\, R^{2k+2}}{k+1}$$

5.8   Let $P(z) = \sum_1^n a_k z^k$ be a polynomial with $P(x)$ real. Integrate $\log z$ $[P(z)]^2$ along the "keyhole" contour in the z plane (cut along the positive real axis) consisting of the unit circle, the two edges of the real axis from ε to 1, and a small circle of radius ε about the origin. Show that

$$\int_0^1 [P(x)]^2\, dx = \sum_1^n \sum_1^n \frac{a_\mu a_\nu}{\mu + \nu + 1} \leqq \pi \sum_1^n |a_\nu|^2$$

REMARK.   Since this inequality is valid for every polynomial, it remains valid for all real infinite sequences $[a_\nu]$ for which

$$\sum_1^\infty |a_\nu|^2 < \infty \text{ (Hilbert's inequality)} \qquad [S]$$

5.9   Recalling (Example 1.9.4) that the area enclosed by the polygon with vertices $z_1, z_2, \ldots z_n$ is

$$\tfrac{1}{2}\mathbf{I}(\bar{z}_1 z_2 + \bar{z}_2 z_3 + \cdots + \bar{z}_n z_1)$$

show that the area enclosed $= \tfrac{1}{2}\mathbf{I} \sum_{k=1}^n \bar{z}_k(z_{k+1} - z_k)$. Interpret this sum as part of the approximating sum in the definition of $\oint_C \bar{z}\, dz$.

5.10   If C is a curve joining z_1, z_2, show that

$$\tfrac{1}{2}\mathbf{I}\int_C \bar{z}\, dz = A - \tfrac{1}{2}\mathbf{I}\bar{z}_2 z_1$$

where A is the area enclosed by C and the chord through z_1, z_2.

Solution: See Example 5.9.

5.11   Let $z = z(t)$ be the parametric representation of a curve in the x-y plane (t may conveniently be interpreted as time, and $\cdot$ as d/dt). Show that

$$\dot{z} = \frac{dz}{ds}\frac{ds}{dt} = e^{i\tau}\frac{ds}{dt} \qquad \text{where } e^{i\tau} \text{ is the unit}$$

tangent vector, and that

$$\ddot{z} = e^{i\tau}\frac{d^2 s}{dt^2} + e^{i\nu}\left(\frac{ds}{dt}\right)^2 \frac{d\tau}{ds}$$

Here $e^{i\nu} = ie^{i\tau}$ is the unit inner normal, and $d\tau/ds$ is the curvature of the curve.

5.12 The real functions $x(t)$, $y(t)$ satisfy

$$\dot{x} = (k \sin t \cos t)x + (k \sin^2 t - \tfrac{1}{2})y$$

$$\dot{y} = (\tfrac{1}{2} + k \sin^2 t)x - (k \sin t \cos t)y$$

By examining $z(t) = x + iy = e^{it/2}\,\zeta(t)$, solve for $x(t)$, $y(t)$.

5.13 (z_n) is a sequence of complex numbers. Show that

$$\underline{\lim}\left|\frac{z_{n+1}}{z_n}\right| \leqq \underline{\lim} \sqrt[n]{|z_n|} \leqq \overline{\lim} \sqrt[n]{|z_n|} \leqq \overline{\lim}\left|\frac{z_{n+1}}{z_n}\right|$$

where $\underline{\lim}$, $\overline{\lim}$ stand for the smallest and largest of the set of limit points.

Solution:
$$\left|\frac{z_2}{z_1}\cdot\frac{z_3}{z_2}\cdots\frac{z_{n+1}}{z_n}\right| = \left|\frac{z_{n+1}}{z_1}\right|$$

Thus

$$\frac{1}{n}\sum_1^n \log\left|\frac{z_{k+1}}{z_k}\right| = \frac{n+1}{n}\left(\frac{1}{n+1}\log|z_{n+1}|\right) - \frac{1}{n}\log|z_1|$$

If $m = \underline{\lim}\,|(z_{k+1})/z_k|$, only a finite number of terms of the sequence $[\log |(z_{k+1})/z_k|]$ are less than $\log m - \delta$ for any $\delta > 0$. Similarly, if $M = \overline{\lim}\,\sqrt[n]{|z_n|}$, there will be an infinite number of terms of the sequence $(1/n \log |z_n|)$ not exceeding $\log M + \varepsilon$. Hence, for all n sufficiently large, we have the average of $\left(\log\left|\dfrac{z_{k+1}}{z_k}\right|\right)$

$$\frac{1}{n}\sum_1^n \log\left|\frac{z_{k+1}}{z_k}\right| > \log m - \delta$$

and also for some arbitrarily large n

$$< \frac{n+1}{n}(\log M + \varepsilon) - \frac{1}{n}\log|z_1|$$

and so

$$(\log m) - \delta < \frac{n+1}{n}(\log M + \varepsilon) - \frac{1}{n}\log|z_1|$$

Since δ, ε are arbitrary, provided n is sufficiently large, letting $n \to \infty$ gives $\log m \leqq \log M$, $m \leqq M$.

5.14 From Example 5.13, show that

$$f(z) = \sum_0^\infty a_n z^n$$

converges absolutely if

$$\lim_{n \to \infty} \left| \frac{a_{n+1} z^{n+1}}{a_n z^n} \right| = |z| \lim \left| \frac{a_{n+1}}{a_n} \right| < 1$$

and more generally, if

$$|z| < \frac{1}{\overline{\lim} \left| \dfrac{a_{n+1}}{a_n} \right|}$$

This is the ratio test.

5.15   In the notation of Example 5.2, show that the divergence theorem yields

$$\iint_D \Delta u \, dx \, dy = 4 \iint_D u_{z\bar{z}} \, dx \, dy = -2i \int_C u_z \, dz = \int_C \frac{\partial u}{\partial v} \, ds$$

where $\partial/\partial v$ is the directional derivative in the direction of the outer normal. Generalize this result to obtain *Green's identity*

$$\iint_D (v \, \Delta u - u \, \Delta v) \, dx \, dy = \int_C \left(v \frac{\partial u}{\partial v} - u \frac{\partial v}{\partial v} \right) ds$$

SOLUTIONS FOR CHAPTER FIVE

5.3.1(b)

$$\int_c (x + iy^2) \, dz = \int_0^1 (x + ix^2)(dx + i \, dx) = \frac{1 + 5i}{6}$$

5.3.2

$$\int_{-1}^1 z^{1/2} \, dz = -\int_0^\pi e^{i\theta/2}(ie^{i\theta} \, d\theta) = \tfrac{2}{3}(1 + i)$$

5.8

$$0 = \int_C \log z [P(z)]^2 \, dz$$

$$= \int_\varepsilon^1 \log x [P(x)]^2 \, dx + \int_0^{2\pi} i\theta [P(e^{i\theta})]^2 i e^{i\theta} \, d\theta$$

$$- \int_\varepsilon^1 (\log x + 2\pi i)[P(x)]^2 \, dx - \int_0^{2\pi} (i\theta + \log \varepsilon)[P(\varepsilon e^{i\theta})]^2 i \varepsilon e^{i\theta} \, d\theta$$

Let $\varepsilon \to 0$.

$$2\pi \int_0^1 P^2(x)\, dx = \left| \int_0^{2\pi} \theta P^2(e^{i\theta}) i e^{i\theta}\, d\theta \right|$$

$$\leqq \int_0^{2\pi} \theta P(e^{i\theta}) P(e^{-i\theta})\, d\theta$$

$$= \sum_1^n a_\nu^{\,2} \int_0^{2\pi} \theta\, d\theta + \sum_{\mu \neq \nu}\sum a_\mu a_\nu \int_0^{2\pi} \theta \cos (\mu - \nu)\theta\, d\theta$$

$$= 2\pi^2 \sum_1^n a_\nu^{\,2}$$

Chapter | CAUCHY'S INTEGRAL FORMULA
SIX | AND APPLICATIONS

In this chapter we derive a representation formula for single-valued analytic functions, called the Cauchy integral formula. As a consequence of this integral formula we find one of the most useful properties of analytic functions: every analytic function possesses a convergent power series expansion about each point of analyticity.

6.1 CAUCHY'S INTEGRAL FORMULA

In Chapter 5 we studied various forms of Cauchy's theorem and learned how to deform the contours of integration without changing the value of the integral of an analytic function. We learned especially that we must avoid the singular points of the analytic function in such a deformation.

Our next task is to handle integrals whose integrands contain a finite number of singular points inside the contour, and the principal tool that we can use for this task is the Cauchy integral formula.

Let $f(z)$ be any (single-valued) analytic function, free from singular points in a region D of the z-plane, and let C be any simple (smooth) closed curve in D, enclosing the point z_0. We consider the integral

$$(1) \qquad I = \oint_C \frac{f(z)\,dz}{z - z_0}$$

which certainly exists, since the integrand is continuous on C, but whose value is not necessarily zero, since $f(z)/(z - z_0)$ is not analytic inside C. We have already considered a special case of this integral for which $f(z) = 1$, and found $I = 2\pi i$; see Section 5.3.

Before proceeding with the evaluation of (1) let us try to gain some insight into the problem by deforming the contour, as in Section 5.7, to a small circle C_δ of radius δ, centered at z_0 (Figure 6.1). Then

$$I = \oint_C \frac{f(z)\,dz}{z - z_0} = \oint_{C_\delta} \frac{f(z)}{z - z_0}\,dz$$

Since this relation is valid for all δ sufficiently small, it is clear that the value of the integral can involve the values of $f(z)$ only in the immediate vicinity of $z = z_0$. Now we introduce into the integral over C_δ the angle ϕ as variable of integration, where $z = z_0 + \delta e^{i\phi}$; we obtain

$$dz = i\, \delta e^{i\phi}\, d\phi = i(z - z_0)\, d\phi$$

(2)
$$I = i \int_0^{2\pi} f(z_0 + \delta e^{i\phi})\, d\phi$$

Yet the value of I is independent of δ (provided that δ is small enough so that

Figure 6.1

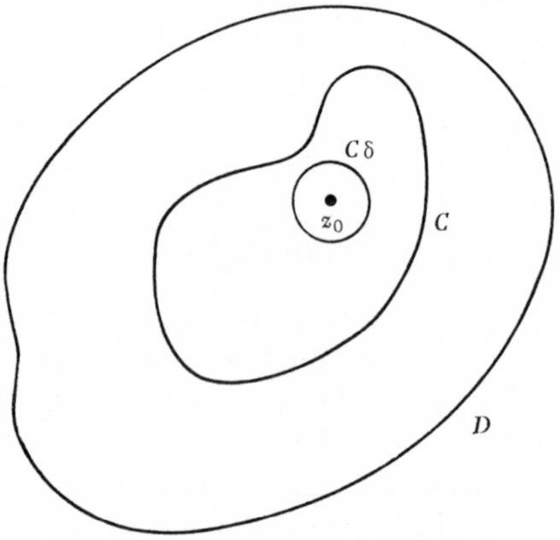

C_δ lies inside C), and continuity suggests that (2) can be evaluated by inspection for $\delta = 0$, giving

$$I = 2\pi i\, f(z_0)$$

Thus we have established *Cauchy's integral formula*: if $f(z)$ is single valued and analytic inside and on the simple closed curve C, and if z_0 is any point inside C,

(3)
$$\frac{1}{2\pi i} \oint_C \frac{f(z)}{z - z_0}\, dz = f(z_0)$$

From this formula we obtain yet another illustration how much the nature of a function $f(z)$ of a complex variable is restricted by the requirement to be analytic: the values of $f(z)$ inside any curve in the region of analyticity are already determined by the values of $f(z)$ on the curve itself.

We may call (3) a "representation formula": it represents $f(z)$ at points inside C in terms of the values of $f(z)$ on C. Let us change our notation slightly: instead of z_0 we write z, and use t as the (complex) variable of integration. Then we can write (3) as

$$(4) \qquad f(z) = \frac{1}{2\pi i} \oint \frac{f(t)}{t - z}\, dt$$

Because of the importance of this formula, and because of the deductions we wish to make from it, we give an alternate, more detailed derivation as follows:

$$I = \oint_C \frac{f(t)\, dt}{t - z}$$

$$= \oint_C \frac{f(t) - f(z)}{t - z}\, dt + f(z) \oint_C \frac{dt}{t - z}$$

$$= 2\pi i f(z) + \int_{C_\delta} \frac{f(t) - f(z)}{t - z}\, dt$$

$$|I - 2\pi i f(z)| \leqq \int_{C_\delta} \left| \frac{f(t) - f(z)}{t - z} \right| |dt|$$

$$= \int_0^{2\pi} |f(t) - f(z)|\, d\phi$$

where $t = z + \delta e^{i\phi}$ on C_δ.

Now $f(t)$ is analytic, and hence continuous, at $t = z$. Thus to each $\varepsilon > 0$ corresponds $\eta > 0$ so that $|f(t) - f(z)| < \varepsilon$ for $|t - z| = \delta < \eta$. In other words, no matter how small $\varepsilon > 0$ is, we have

$$|I - 2\pi i f(z)| < 2\pi\varepsilon$$

and since I does not depend on ε, we must have

$$|I - 2\pi i f(z)| = 0$$

$$I = 2\pi i f(z)$$

6.1.1 From Cauchy's formula applied to a circle of radius r centered at z, show that

$$f(z) = \frac{1}{2\pi} \int_0^{2\pi} f(z + re^{i\theta})\, d\theta$$

and hence, the value of an analytic function at a point is the average of its values on any circle centered at that point (provided, of course, that $f(z)$ is analytic inside and on the circle).

Cauchy's Integral Formula **179**

6.2 A FIRST APPLICATION TO THE EVALUATION OF DEFINITE INTEGRALS

Example 1.
$$I = \oint_C \frac{dz}{z^2 + 1} = \oint_C \frac{dt}{t^2 + 1}$$

$$= \oint_C \frac{dt}{(t + i)(t - i)}$$

where C is any simple closed curve enclosing $z = i$, but not $z = -i$ (Figure 6.2). We can bring this integral in the form 6.1(3) by identifying $f(t)$ with

Figure 6.2

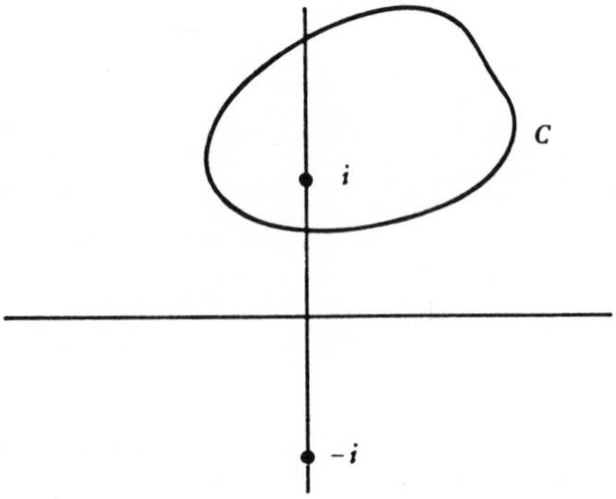

$1/(t + i)$, and z with i. The value is then

$$2\pi i f(z) = 2\pi i \left(\frac{1}{t + i} \right)_{t=i} = \pi$$

Example 2. If C enclosed $z = -i$, but not $z = i$, the value of the integral would be $-\pi$ [identify $f(t)$ and z in this case].

Example 3. Suppose C encloses both $z = i$ and $z = -i$ (Figure 6.3). Then

$$I = \oint_C \frac{dt}{t^2 + 1} = \oint_{C_1} \frac{dt}{t^2 + 1} + \oint_{C_2} \frac{dt}{t^2 + 1}$$

$$= \pi + (-\pi) = 0$$

from examples 1 and 2.

Figure 6.3

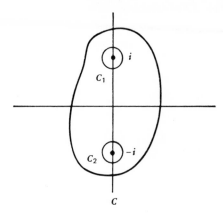

Example 4. Now let us consider the particular curve C consisting of the real axis from $-R$ to R, and the upper semicircle $|z| = R$, $0 \leq \arg z \leq \pi$. This gives

$$\oint_C dt/(1 + t^2) = \pi \text{ from (1) provided } R > 1$$

$$= \int_{-R}^{R} \frac{dx}{1 + x^2} + \int_{0}^{\pi} \frac{iRe^{i\theta}\, d\theta}{1 + R^2 e^{2i\theta}}$$

$$= A + B$$

What about the limiting forms of these integrals as $R \to \infty$?

$$A = \int_{-R}^{R} \frac{dx}{1 + x^2}$$

and

$$\lim_{R \to \infty} \int_{-R}^{R} \frac{dx}{1 + x^2} = \int_{-\infty}^{\infty} \frac{dx}{1 + x^2}$$

while B clearly tends to zero since the integrand tends to zero as $R \to \infty$, and the path of integration is finite. More explicitly, let us estimate B in the usual way.

$$|B| \leq \int_{0}^{\pi} \left| \frac{iRe^{i\theta}}{1 + R^2 e^{2i\theta}} \right| \leq \int_{0}^{\pi} \frac{R\, d\theta}{R^2 - 1} = \frac{\pi R}{R^2 - 1} \to 0$$

as $R \to \infty$

A First Application to the Evaluation of Definite Integrals **181**

COMMENT 1. These inequalities are typical; we shall often meet with similar inequalities, and so an additional comment is in order. We are using

$$\left| \int_0^\pi \frac{iRe^{i\theta}\,d\theta}{1 + R^2e^{2i\theta}} \right| \leqq \text{Max of integrand} \cdot \text{Length of path}$$

$$= M \cdot L$$

See Section 5.4.

In our case, $L = \pi$, clearly, and

$$\left| \frac{Re^{i\theta}}{1 + R^2e^{2i\theta}} \right| = \frac{R}{|1 + R^2e^{2i\theta}|}$$

$$= \frac{R}{|R^2 + e^{-2i\theta}|}$$

Figure 6.4

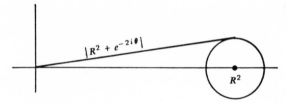

Consider the totality of points

$$R^2 + e^{-2i\theta}$$

These points lie on a circle centered at R^2 with unit radius (Figure 6.4). The largest value of $R/(R^2 + e^{-2i\theta})|$ occurs, for fixed R, when the denominator is smallest, which occurs at the point of the circle nearest the origin; that is,

$$|R^2 + e^{-2i\theta}| \geqq R^2 - 1 \quad \text{and} \quad M = \frac{R}{R^2 - 1}$$

COMMENT 2. The special symmetric limit occurring in A,

$$\lim_{R \to \infty} \int_{-R}^{R} f(x)\,dx$$

is called the *Cauchy principal value* of the integral, and is usually written $P \int_{-\infty}^{\infty} f(x)\,dx$, or $\int_{-\infty}^{\infty} f(x)\,dx$.

In cases where (as here), $\int_{-\infty}^{\infty} f(x)\,dx$ exists, the Cauchy principal value obviously exists, and equals the integral.

$$\int_{-\infty}^{\infty} f(x)\,dx = \lim_{\substack{a\to\infty \\ b\to-\infty}} \int_b^a f(x)\,dx$$

However it often happens that even though $\int_{-\infty}^{\infty} f(x)\,dx$ does not exist in the strict sense, the Cauchy principal-value integral does exist. For instance,

$$\fint_{-\infty}^{\infty} x\,dx = \lim_{R\to\infty} \int_{-R}^{R} x\,dx$$

$$= \lim_{R\to\infty} \left.\left(\frac{x^2}{2}\right)\right|_{-R}^{R}$$

$$= \lim_{R\to\infty} \left(\frac{R^2}{2} - \frac{R^2}{2}\right)$$

$$= \lim_{R\to\infty} 0 = 0$$

COMMENT 3.   The final form our answer has taken is of considerable interest and importance.

$$\oint_C \frac{dt}{1+t^2} = \pi = \int_{-R}^{R} \frac{dx}{1+x^2} + \int_0^{\pi} \frac{iRe^{i\theta}\,d\theta}{R^2 e^{2i\theta}+1} \to \int_{-\infty}^{\infty} \frac{dx}{1+x^2} \qquad \text{as } R\to\infty$$

The answer itself, that is, $\int_{-\infty}^{\infty} dx/(1+x^2) = \pi$ is not at all surprising, but we have been able to find it *without using the indefinite integral*. This suggests that Cauchy's method may enable us to evaluate definite integrals even when we do not know the indefinite integral explicitly. Indeed, such evaluations were the first applications of the theory of analytic functions that Cauchy developed.

6.2.1   In the following integrals, identify the appropriate $f(t)$, and z_0 [of 6.1(3)], and hence evaluate the integrals. The curve C is a large semicircle of radius R in the upper half-plane.

(a) $\displaystyle\int_C \frac{dt}{t^2+t+1}$

(b) $\displaystyle\int_C \frac{e^{it}\,dt}{t^2+a^2} \qquad a>0$

(c) $\displaystyle\int_C \frac{dt}{(t^2+a^2)(t^2+b^2)} \qquad 0<a<b$

6.2.2 Investigate the limiting forms of the integrals in Section 6.2.1 as $R \to \infty$. Show, by estimating the contributions to the integrals around the semicircular arc that they tend to zero as $R \to \infty$.

6.2.3 What is the value of the integral in Cauchy's formula $\oint_C f(t)\, dt/(t - z)$ if z lies outside C? Why?

6.3 SOME CONSEQUENCES OF THE CAUCHY FORMULA: HIGHER DERIVATIVES

There is a connection between the representation of $f(z)$ by means of Cauchy's formula, and the differentiability of $f(z)$. In fact, as

$$f(z) = \frac{1}{2\pi i} \oint_C \frac{f(t)\, dt}{t - z}$$

and

$$\frac{f(z + h) - f(z)}{h} \to f'(z)$$

as $h \to 0$, we are led to

$$\frac{f(z + h) - f(z)}{h} = \frac{1}{2\pi i h} \oint_C \frac{f(t)\, dt}{t - z - h} - \frac{1}{2\pi i h} \oint_C \frac{f(t)\, dt}{t - z}$$

provided both z and $z + h$ are inside C, as in Figure 6.5, and this we can guarantee by taking h sufficiently small. Setting

$$\frac{\Delta f}{\Delta z} = \frac{f(z + h) - f(z)}{h}$$

and noting that

$$\frac{1}{t - z - h} - \frac{1}{t - z} = \frac{h}{(t - z - h)(t - z)}$$

Figure 6.5

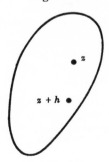

we have

$$\frac{\Delta f}{\Delta z} = \frac{1}{2\pi i} \oint_C \frac{f(t)\,dt}{(t-z)(t-z-h)}$$

The right-hand side clearly tends to a limit, to

$$\frac{1}{2\pi i} \oint_C \frac{f(t)\,dt}{(t-z)^2}$$

as $h \to 0$, and must equal $f'(z)$.

To show this in detail, we form

(1) $\displaystyle \frac{\Delta f}{\Delta z} - \frac{1}{2\pi i} \oint_C \frac{f(t)\,dt}{(t-z)^2} = \frac{1}{2\pi i} \oint_C \frac{f(t)\,dt}{(t-z)(t-z-h)} - \frac{1}{2\pi i} \oint_C \frac{f(t)\,dt}{(t-z)^2}$

$$= \frac{h}{2\pi i} \oint_C \frac{f(t)\,dt}{(t-z)^2(t-z-h)}$$

Hence we find

(2) $\displaystyle \left| \frac{\Delta f}{\Delta z} - \frac{1}{2\pi i} \oint_C \frac{f(t)\,dt}{(t-z)^2} \right| \leqq \frac{|h|}{2\pi} \oint_C \frac{|f(t)|\,|dt|}{|t-z|^2\,|t-z-h|}$

$$\leqq \frac{|h|}{2\pi} \cdot \frac{M \cdot L}{d^3}$$

where

$$M = \max_{t \text{ on } C} |f(t)|$$

$$L = \text{length of } C$$

$$d = \text{minimum distance of } z, z+h \text{ to } C$$

At first glance, it may appear as if d depends on h, but we can easily remove this difficulty. If $d' = $ min distance from z to C, then for $|h| < 1/2(d')$, the right side of (2) becomes

$$\frac{|h|}{\pi} \frac{M \cdot L}{(d')^3}$$

which tends to zero as $h \to 0$.

We have thus established the following extension of Cauchy's formula

(3) $$f'(z) = \frac{1}{2\pi i} \oint_C \frac{f(t)\,dt}{(t-z)^2}$$

valid under the same conditions as the original formula. The result is easy to remember: we can differentiate the integral formula under the integral sign to obtain the derivative of $f(z)$.

And now, we can repeat the whole process. Equation 3 gives us a formula for $f'(z)$, and so leads to a formula for the difference quotient

$$\frac{\Delta f'}{\Delta z} = \frac{f'(z+h) - f'(z)}{h}$$

$$= \frac{1}{2\pi i} \oint_C \frac{f(t)(2t - 2z - h)\,dt}{(t-z)^2(t-z-h)^2}$$

and, precisely as above, the right-hand side tends to a limit as $h \to 0$, giving

$$f''(z) = \frac{2}{2\pi i} \oint_C \frac{f(t)\,dt}{(t-z)^3}$$

We obtain similarly, for each nonnegative integer n, that

(4)
$$f^{(n)}(z) = \frac{n!}{2\pi i} \oint_C \frac{f(t)\,dt}{(t-z)^{n+1}}$$

For a detailed derivation we use mathematical induction.

When we refer to the Cauchy integral formula, we shall include the derived cases (4).

The results we have just obtained have far-reaching consequences. All that we require of an analytic function is that it have one continuous derivative in a region D. Cauchy's formula shows, however, that the existence of a single continuous derivative implies the existence of derivatives of all orders. Thus, if $f(z)$ is analytic, all derivatives $f'(z), f''(z), \ldots f^{(n)}(z), \ldots$ exist, and so they are analytic. It is remarkable that this differentiability property of analytic functions that follows so readily from the integrability properties does not appear obtainable from the differential calculus alone.

We see also a marked difference between real functions and functions of a complex variable; the existence of any finite number of derivatives of a function of a real variable cannot guarantee the existence of higher derivatives.

Let us now estimate the integral (4). We assume that the path of integration is a circle of radius r, and center z (so that $|t - z| = r$), and that

(5)
$$|f(t)| \leq M \quad \text{for} \quad |t - z| = r$$

The perimeter of the circle is $2\pi r$, and so, by the standard inequality (Section 5.4) that we have already used in this section, we immediately obtain from (4) and (5) that

(6)
$$|f^{(n)}(z)| \leq \frac{n!\,M}{r^n}$$

for $n = 0, 1, 2, \ldots$. This is *Cauchy's inequality*. It is valid, in particular, when

(7) $$M = \max_{|t-z|=r} |f(t)|$$

6.3.1 Show that every harmonic function possesses continuous partial derivatives of all orders.

6.4 MORE CONSEQUENCES OF THE CAUCHY FORMULA: THE PRINCIPLE OF MAXIMUM MODULUS

This section is more "theoretical." Although its contents are very important, they will not be needed in the remainder of the present chapter.

(a) The reader should recall the definition of an entire function: a function analytic in the whole plane (for each finite value of z) is called an entire function. We wish to prove *Liouville's theorem: a bounded entire function is necessarily a constant*.

A function $f(z)$ is called *bounded* in a region if there exists a positive number (a bound) M, such that at each point z of the region

(1) $$|f(z)| \leqq M$$

An entire function is called bounded, if it is bounded in its whole domain of existence, in the whole plane; that is, if (1) is valid for all complex numbers z.

The case $n = 1$ of Cauchy's inequality 6.3(6), which we have just derived, states that

(2) $$|f'(z)| \leqq \frac{M}{r}$$

In our case, for a bounded entire function, (2) is valid for any given z with the same value of M, *independent of r*. As r can be made arbitrarily large, we must have $f'(z) = 0$ for any z, and so $f(z)$ is a constant: we have proved Liouville's theorem (see Example 3.3.6).

(b) To arrive at the fact that is the principal concern of this section, we consider a function $f(z)$ that is analytic at each point inside and on the simple closed curve C. Let us observe that $[f(z)]^n$, where n is any positive integer, is analytic to the same extent that $f(z)$ is. Therefore, we can apply Cauchy's formula to $[f(z)]^n$ instead of to $f(z)$; we obtain for any point z interior to C that

(3) $$[f(z)]^n = \frac{1}{2\pi i} \int_C \frac{[f(t)]^n \, dt}{t - z}$$

Let M denote the maximum of $f(t)$ on C, L the length of C, and d the minimum distance from z to C. Then, by our standard estimate, there follows from (3) that

$$|f(z)|^n \leqq \frac{M^n L}{2\pi d}$$

and so

(4)
$$|f(z)| \leqq M\left(\frac{L}{2\pi d}\right)^{1/n}$$

Let n tend to infinity; we obtain in the limit that

(5)
$$|f(z)| \leqq M$$

Remember the definition of M: it is the maximum of $|f(z)|$ on C. Yet (5) tells us that no value of $|f(z)|$ inside C can exceed its maximum value on C: what is maximum on the boundary curve is also maximum in the surrounded closed region. In other words: *the maximum of the absolute value of an analytic function in a closed region is attained on the boundary.*

This is the *principle of the maximum modulus.*

(c) In fact, more is true than we have proved in the foregoing: the region need not be surrounded by a simple curve, it may be multiply connected. Moreover, if the function is not a constant, the maximum of the absolute value is attained *only* on the boundary. (In the trivial case when the function is a constant, the maximum modulus is attained at each point.) For proofs, see Example 6.45.

6.4.1 $f(z)$ is entire, and $M(r) = \max_{|z|=r} |f(z)|$. Show that $M(r)$ is monotone increasing as r increases.

6.4.2 Deduce the maximum principle directly from Example 3.21.

6.4.3 Prove the following generalization of Liouville's theorem: if $f(z)$ is entire, and $|f(z)| \leqq M_1 |z|^k$ for all z (or merely for all z sufficiently large), then $f(z)$ is a polynomial of degree $\leqq k$. [Use the Cauchy inequality for $f^{(n)}(z)$ for any $n > k$.]

6.4.4 Prove the *fundamental theorem of algebra*, which states that every polynomial, $P(z) = a_0 + a_1 z + \cdots + a_n z^n$, with complex coefficients, has a complex zero, that is, there is a number $\zeta = \xi + i\eta$ such that $P(\zeta) = 0$. [If there exists no such ζ, then $1/P(z)$ is analytic everywhere, and so is entire. Show that this contradicts Liouville's theorem.]

6.5 TAYLOR'S THEOREM, MACLAURIN'S THEOREM

We now examine $f(z)$ in the neighborhood of a point of analyticity that, for simplicity, we suppose to be at the origin. Then

(1)
$$f(z) = \frac{1}{2\pi i} \oint_C \frac{f(t)\, dt}{t - z}$$

Figure 6.6

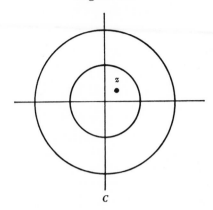

$$C$$

where C is any suitable curve enclosing both points $t = 0$ and $t = z$ and lying in the region in which $f(z)$ is analytic.

In fact, we take C to be a circle centered at the origin (Figure 6.6), restricting z if necessary. Let C be the circle $|t| = R$, and $|z| = r < R$. Then

(2)
$$f(z) = \frac{1}{2\pi i} \oint_C \frac{f(t)}{t} \frac{dt}{1 - \dfrac{z}{t}}$$

Now let

$$\frac{1}{1 - \dfrac{z}{t}} = \frac{1}{1 - \alpha}$$

Then

$$|\alpha| = \left| \frac{z}{t} \right| = \frac{r}{R} < 1$$

for each t on C; it is well known that

$$\frac{1}{1 - \alpha} = 1 + \alpha + \alpha^2 + \cdots + \alpha^k + \cdots$$

and so it is plausible that the following manipulation is valid:

$$f(z) = \frac{1}{2\pi i} \oint_C \frac{f(t)}{t} \frac{dt}{1 - \dfrac{z}{t}}$$

$$= \frac{1}{2\pi i} \oint_C f(t) \sum_0^\infty \frac{z^k}{t^{k+1}} \, dt$$

$$= \sum_0^\infty a_k z^k$$

where

(3)
$$a_k = \frac{1}{2\pi i} \oint_C \frac{f(t)}{t^{k+1}} \, dt$$

In order to justify the term by term integration of an infinite series, we must establish that, along with the remainder after n terms, the integral of the remainder after n terms tends to zero. Thus

(4)
$$\frac{1}{1-\alpha} = 1 + \alpha + \alpha^2 + \cdots + \alpha^{n-1} + \frac{\alpha^n}{1-\alpha} \qquad \alpha \neq 1$$

so that

$$f(z) = \frac{1}{2\pi i} \oint_C \frac{f(t)}{t} \left[1 + \frac{z}{t} + \frac{z^2}{t^2} + \cdots + \left(\frac{z}{t}\right)^{n-1} + \frac{\left(\frac{z}{t}\right)^n}{1 - \frac{z}{t}} \right] dt$$

$$= \sum_0^{n-1} a_k z^k + R_n$$

where

$$R_n = \frac{1}{2\pi i} \oint_C \frac{z^n f(t) \, dt}{t^n (t - z)}$$

If the infinite series is to converge, R_n must tend to zero as n tends to infinity; in fact, we can estimate R_n as follows:

$$|R_n| \leq \frac{1}{2\pi} \oint_C \frac{|f(t)| \, |dt|}{|t| \left(1 - \left|\frac{z}{t}\right|\right)} \left|\frac{z}{t}\right|^n$$

$$\leq \frac{MR}{R\left(1 - \frac{r}{R}\right)} \left(\frac{r}{R}\right)^n = M_1 \left(\frac{r}{R}\right)^n$$

Thus we see that $R_n \to 0$ as $n \to \infty$ for *each* z inside C, so that

$$f(z) = \sum_0^\infty a_k z^k$$

for each z such that $|z| < R$, when $f(z)$ is analytic for $|z| < R$. It is clear from our derivation that the series remains valid (and unchanged) if for C we take a larger and larger circle (restricted only by the requirement that $f(z)$ be analytic inside and on C). Furthermore, from 6.3(4) we can express the coefficients a_k of the series in terms of $f(z)$ itself:

(5)
$$a_k = \frac{1}{2\pi i} \oint_C \frac{f(t)}{t^{k+1}} \, dt = \frac{f^{(k)}(0)}{k!}$$

Thus we have established what we wish to call *MacLaurin's theorem*: if $f(z)$ is analytic in a neighborhood of $z = 0$,

$$f(z) = \sum_0^\infty \frac{f^{(n)}(0)}{n!} z^n$$

the series converging absolutely in $|z| < R$, where R is the radius of the largest circle centered at $z = 0$, inside which $f(z)$ is analytic. (In MacLaurin's time the concept of an analytic function was unknown.)

A slightly different form of this result may be called *Taylor's theorem*: if $f(z)$ is analytic in a neighborhood of a point z_0, then $f(z)$ can be expressed as a series in powers of $z - z_0$; that is, $f(z)$ can be expanded in a power series about $z = z_0$ with the result

$$f(z) = \sum_0^\infty \frac{f^{(n)}(z_0)}{n!} (z - z_0)^n$$

the series converging absolutely in $|z - z_0| < R$, where R is the radius of the largest circle centered at z_0 inside which $f(z)$ is analytic.

Taylor's theorem can be deduced from MacLaurin's theorem by writing $f(z) = f(z_0 + z - z_0)$, and setting $z - z_0 = \zeta$, $f(z_0 + \zeta) = f_1(\zeta)$. Then

$$f_1(\zeta) = \sum_0^\infty \frac{f_1^{(n)}(0)\zeta^n}{n!} = \sum_0^\infty \frac{f^{(n)}(z_0)}{n!} (z - z_0)^n$$

Alternately,

$$f(z) = \frac{1}{2\pi i} \oint_{|t-z_0|=r'} \frac{f(t)\,dt}{t - z} \qquad |z| < r' < R$$

$$= \frac{1}{2\pi i} \oint_{|t-z_0|=r'} \frac{f(t)\,dt}{(t - z_0)\left(1 - \dfrac{z - z_0}{t - z_0}\right)}$$

and from here we may follow the lines of the argument that started from (2).

COMMENT 1. The power series obtained in Taylor's and MacLaurin's theorems are commonly called Taylor series and MacLaurin series, respectively.

We have also *defined* a function

(6) $$f(z) = \sum_0^\infty a_n(z - z_0)^n$$

by giving as its definition the power series in (6), converging in $|z - z_0| < R$. Furthermore, from this definition, we found that $f'(z)$ exists in $|z - z_0| < R$, that $f(z)$ so defined is analytic in $|z - z_0| < R$. Thus we may apply Taylor's theorem to this $f(z)$, and obtain

(7) $$f(z) = \sum_0^\infty b_n(z - z_0)^n \qquad |z - z_0| < R$$

where the numbers $b_0, b_1, \ldots$ are the Taylor series coefficients of $f(z)$. The obvious question now is "are the sequences $a_0, a_1, \ldots$ and $b_0, b_1, \ldots$ identical?" The answer appears to be "obviously yes," and gives us the *uniqueness theorem* for *power series*.

Thus $\sum_0^\infty a_n(z - z_0)^n$ and $\sum_0^\infty b_n(z - z_0)^n$ converge to the same values for each z, $|z - z_0| < R$. If we set $a_n - b_n = c_n$, then $\sum_0^\infty c_n(z - z_0)^n$ converges to 0 for each $|z - z_0| < R$. Let $z \to z_0$. We have $c_0 = 0$. Then $c_1(z - z_0) + c_2(z - z_0)^2 + \cdots = (z - z_0)[c_1 + c_2(z - z_0) + c_3(z - z_0)^2 + \cdots]$ is similarly identically zero for $|z - z_0| < R$, so that $c_1 + c_2(z - z_0) + c_3(z - z_0)^2 + \cdots$ also is identically zero in $|z - z_0| < R$. Again, let $z \to z_0$; then $c_1 = 0$, and we may repeat the process yet again. We have thus $c_0 = a_0 - b_0 = 0$, $c_1 = a_1 - b_1 = 0$, and it is clear that $c_n = a_n - b_n = 0$, for each positive integer n. (A strict proof can clearly be constructed by the use of mathematical induction.)

COMMENT 2.   We have yet another characterization of the collection of analytic functions: it is identical with the collection of convergent power series.

It is thus clear that many of the manipulations involving analytic functions in a given application may well involve performing these manipulations with power series. Indeed, every theorem that is valid for analytic functions implies a corresponding theorem for power series, and conversely. If we keep these thoughts in mind, the power series manipulations become considerably clearer. Consider the following examples.

(a) Suppose $f(z)$, $g(z)$ are both analytic for $|z| < R$,

$$f(z) = \sum_0^\infty a_n z^n \qquad g(z) = \sum_0^\infty b_n z^n$$

then $f(z) \cdot g(z)$ is analytic in $|z| < R$, and

$$f(z) \cdot g(z) = \sum_0^\infty c_n z^n = \left(\sum_0^\infty a_n z^n\right)\left(\sum_0^\infty b_n z^n\right)$$

Of course,

$$c_n = \frac{1}{n!} \frac{d^n}{dz^n} [f(z) \cdot g(z)]\Big|_{z=0}$$

from MacLaurin's theorem, which appears a formidable computation (but it is not, in reality, so difficult if we recall Leibniz' rule for differentiating a product). Alternately, we can, with confidence, consider the problem of multiplying together the two infinite series $\sum_0^\infty a_n z^n$, $\sum_0^\infty b_n z^n$, since we know that the resulting power series will be absolutely convergent for $|z| < R$. We find the coefficient of z^n in $(a_0 + a_1 z + a_2 z^2 + \cdots)(b_0 + b_1 z + b_2 z^2 + \cdots)$ is

(8)
$$c_n = \sum_0^n a_r b_{n-r}$$

(b) If, furthermore, $g(0) \neq 0$, then $g(z) \neq 0$ for $|z| < R_1$, for some $R_1 > 0$, and

$$\frac{f(z)}{g(z)} = \sum_0^\infty d_n z^n \qquad |z| < \min(R_1, R)$$

since the left side is analytic at $z = 0$. Thus, we can divide power series (provided we avoid zeros of the denominator).

One way to obtain the coefficients d_n in this example, or, more realistically, to obtain the *first few* terms of the series, is to rearrange the bookkeeping in the form

$$f(z) = g(z) \sum_0^\infty d_n z^n$$

use (8), and solve for the "undetermined coefficients," $d_0, d_1, d_2, \ldots$. Thus $\sin z / \cos z = \tan z = \sum_0^\infty d_n z^n$, gives $\sin z = \cos z \sum_0^\infty d_n z^n$, gives

$$z - \frac{z^3}{3!} + \cdots = d_0 + d_1 z + \left(d_2 - \frac{d_0}{2}\right)z^2 + \left(d_3 - \frac{d_1}{2}\right)z^3 + \cdots$$

so that

$$d_0 = 0 \qquad d_1 = 1 \qquad d_2 - \frac{d_0}{2} = 0 \qquad d_3 - \frac{d_1}{2} = -\tfrac{1}{6}$$

$$\tan z = z + \frac{z^3}{3} + \cdots$$

6.5.1 Verify MacLaurin's theorem for the particular elementary functions $f(z) = e^z$, $\cos z$, $\sinh z$.

6.5.2 Expand $\sin z$ about $z = \pi/2$.

6.5.3 Find the first three nonvanishing terms of the MacLaurin series for $\sin(\sin z)$.

6.5.4   Since an analytic function of an analytic function is analytic, then $f[g(z)]$ is analytic if f and g are. That is, with

$$g(z) = b_0 + b_1 z + b_2 z^2 + \cdots$$
$$f(z) = a_0 + a_1 z + a_2 z^2 + \cdots$$
$$f[g(z)] = c_0 + c_1 z + c_2 z^2 + \cdots$$
$$= a_0 + a_1[b_0 + b_1 z + b_2 z^2 + \cdots]$$
$$+ a_2[b_0 + b_1 z + b_2 z^2 + \cdots]^2 + \cdots$$

and we are assured of the validity of collecting the coefficients of like powers of z. Use this technique to solve Example 6.5.3.

6.5.5 $f(z)$ is analytic in $|z| < R$ and $f(0) = 0$. Show that a number $R_1 < R$ exists for which $f(z) \neq 0$ in $0 < |z| < R_1$.

6.5.6   If $f(z)$ is analytic in $|z| < R$, and $f(z) = 0$ at $z_1, z_2, \ldots z_n, \ldots$ where $\lim_{n \to \infty} z_n = 0$, show that $f(z) \equiv 0$.

6.5.7   Find the MacLaurin series for the principal value branch of $f(z) = (1 + z)^\alpha$ $[f(0) = 1]$, and compare it with the binomial expansion of $f(z)$. See also Example 3.24.

6.5.8   Find the domain of convergence of

$$f(z) = 1 + \frac{z}{1 + z} + \left(\frac{z}{1 + z}\right)^2 + \left(\frac{z}{1 + z}\right)^3 + \cdots$$

What is the value of $f(z)$ in this domain?

6.5.9   Define the coefficients B_n by

$$f(z) = \frac{z}{e^z - 1} = \sum_0^\infty \frac{(-1)^n B_n}{n!} z^n$$

Show that $B_1 = 1/2$, $B_{2n+1} = 0$, $n > 0$, and find the radius of convergence of the series.

6.5.10   Using the function of Example 6.5.9 show that

$$\coth z = \frac{1}{z} + \sum_2^\infty \frac{(-1)^n B_n 2^n z^{n-1}}{n!}$$

6.5.11   Find the power series expansion of $\tan z$ from the results of Example 6.5.10.

6.5.12 Find the region of convergence of the following series (the ratio test, Example 2.2, is often very useful).

(a) $\displaystyle\sum_{1}^{\infty} \frac{(z-2)^n}{n}$ (b) $\displaystyle\sum_{1}^{\infty} n^{17}(z+i)^n$

(c) $\displaystyle\sum_{1}^{\infty} n^{\log n}\, z^n$ (d) $\displaystyle\sum_{1}^{\infty} \frac{n!}{n^n}\, z^n$

(e) $\displaystyle\sum_{1}^{\infty} \left(1 - \frac{1}{n}\right)^n z^n$ (f) $\displaystyle\sum_{1}^{\infty} n!\, z^n$

(g) $1 + 7z + 5^2 z^2 + 7^3 z^3 + 5^4 z^4 + 7^5 z^5 + 5^6 z^6 + \cdots$

(h) $\displaystyle\sum_{2}^{\infty} \frac{z^n}{(\log n)^n}$ (i) $\tan z = z + \dfrac{z^3}{3} + \cdots$

6.5.13 Establish the validity of l'Hopital's rule for analytic functions, that is, if $f(z_0) = g(z_0) = 0$, then

$$\lim_{z \to z_0} \frac{f(z)}{g(z)} = \frac{f'(z_0)}{g'(z_0)}$$

if this limit exists. If also $f'(z_0) = g'(z_0) = 0$, we may continue:

$$\lim_{z \to z_0} \frac{f(z)}{g(z)} = \lim_{z \to 0} \frac{f''(z)}{g''(z)}$$

6.6 LAURENT'S THEOREM

Suppose now that $f(z)$ is single valued and analytic at all points in a domain D except possibly for the single point $z = z_0$. We make no assumptions here about the behavior of $f(z)$ at z_0. For simplicity we take $z_0 = 0$, and assume $f(z)$ to be analytic for $z \neq 0$, and inside a domain that includes the origin.

Now we construct two circles, $|z| = r_1$ and $|z| = r_2$, $r_2 > r_1$, centered at the origin, and lying inside the domain in which $f(z)$ is analytic. Then $f(z)$ is analytic in the *annulus*, the ring shaped region $r_1 \leqq |z| \leqq r_2$ (Figure 6.7).

We now use a form of Cauchy's formula to obtain a representation for $f(z)$ in this annular region (doubly connected region, see Section 5.7). First we introduce the auxiliary line L connecting the inner and the outer circles, and consider the simply connected domain so formed from C_1, C_2, and L (see Figure 6.8). For this simply connected domain, we may apply Cauchy's integral formula, and obtain

$$f(z) = \frac{1}{2\pi i} \oint_{C_2} \frac{f(t)}{t-z}\, dt + \frac{1}{2\pi i} \int_{L} \frac{f(t)\, dt}{t-z} + \frac{1}{2\pi i} \oint_{C_1} \frac{f(t)}{t-z}\, dt - \frac{1}{2\pi i} \int_{L} \frac{f(t)\, dt}{t-z}$$

Figure 6.7

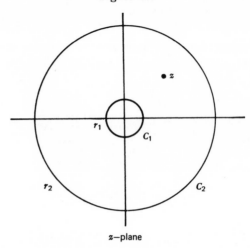

z—plane

The two integrals on L are in opposite directions, and so cancel; the integral

$$\oint_{C_1} \frac{f(t)}{t-z}\, dt = -\oint_{C_1} \frac{f(t)}{t-z}\, dt$$

Thus

(1)
$$f(z) = \frac{1}{2\pi i}\oint_{C_2} \frac{f(t)\, dt}{t-z} - \frac{1}{2\pi i}\oint_{C_1} \frac{f(t)\, dt}{t-z}$$

$$= f_2(z) + f_1(z)$$

where both line integrals are taken in the positive (counterclockwise) sense.

Figure 6.8

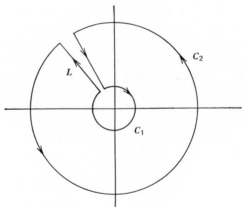

Let us first consider

$$f_2(z) = \frac{1}{2\pi i} \oint_{C_2} \frac{f(t)\, dt}{t - z}$$

On the path of integration, $|t| = r_2$, and z, being inside the circle, satisfies $|z| = r < r_2$. Thus, precisely as in MacLaurin's theorem, we obtain

$$f_2(z) = \frac{1}{2\pi i} \oint_{C_2} \frac{f(t)}{t\left(1 - \dfrac{z}{t}\right)}\, dt$$

$$= \frac{1}{2\pi i} \oint_{C_2} f(t) \sum_{0}^{\infty} \frac{z^n}{t^{n+1}}\, dt$$

$$= \sum_{0}^{\infty} a_n z^n \qquad |z| < r_2$$

with

(2) $$a_k = \frac{1}{2\pi i} \oint_{C_2} \frac{f(t)}{t^{k+1}}\, dt$$

The series converges absolutely for $|z| < r_2$, $f_2(z)$ is analytic in $|z| < r_2$. Now let us look at

$$f_1(z) = -\frac{1}{2\pi i} \oint_{C_1} \frac{f(t)\, dt}{t - z}$$

In this integral, $|t| = r_1$, and $|z| = r > r_1$. Thus

$$f_1(z) = \frac{1}{2\pi i} \oint_{C_1} \frac{f(t)}{z\left(1 - \dfrac{t}{z}\right)}\, dt$$

$$= \frac{1}{2\pi i} \oint_{C_1} f(t)\left\{\frac{1}{z} + \frac{t}{z^2} + \cdots + \frac{t^{n-1}}{z^n} + \frac{t^n}{z^n(z - t)}\right\} dt$$

$$= \sum_{1}^{n} \frac{1}{z^k} \frac{1}{2\pi i} \oint_{C_1} t^{k-1} f(t)\, dt + R_n$$

using Section 6.5 (4). Our usual estimate of the integral for R_n gives

$$|R_n| \le \frac{1}{2\pi} \oint_{C_1} \frac{|f(t)|\,|dt|}{|z|\left(1 - \left|\dfrac{t}{z}\right|\right)} \left|\frac{t}{z}\right|^n$$

$$\le \frac{1}{2\pi} \frac{M}{\left(1 - \dfrac{r_1}{r}\right)} \left(\frac{r_1}{r}\right)^n \cdot 2\pi r_1$$

$$= M_1 \left(\frac{r_1}{r}\right)^n$$

It is clear that $R_n \to 0$ as $n \to \infty$, giving

$$f_1(z) = \sum_{1}^{\infty} \frac{b_k}{z^k} \qquad |z| > r_1$$

(3)
$$b_k = \frac{1}{2\pi i} \oint_{C_1} t^{k-1} f(t)\, dt$$

Thus $f_1(z)$ is analytic for $|z| > r_1$, for all z outside the inner circle. Since $f(z) = f_1(z) + f_2(z)$, and $f_1(z)$ is analytic for $|z| > r_1$, $f_2(z)$ is analytic for $|z| < r_2$, the sum is clearly analytic in the region common to both $f_1(z)$ and $f_2(z)$, or $r_1 < |z| < r_2$, which verifies our assumptions.

It is convenient to express the formulas we have obtained for the coefficients in a slightly more suggestive form. First, we replace k by $-k$ in (3). We obtain

$$b_{-k} = \frac{1}{2\pi i} \oint_{C_1} \frac{f(t)\, dt}{t^{k+1}} \qquad a_k = \frac{1}{2\pi i} \oint_{C_2} \frac{f(t)\, dt}{t^{k+1}}$$

$$f_1(z) = \sum_{-\infty}^{-1} b_{-n} z^n \qquad f_2(z) = \sum_{0}^{\infty} a_n z^n$$

In place of the expression b_{-n}, we use a_n; this is a consistent notation, since a_n for $n = 0, 1, 2, \ldots$ is a coefficient of $f_2(z)$, and now a_n for $n = -1, -2, \ldots$ is a coefficient of $f_1(z)$. In this way, we may collect our results in the form

(4)
$$f(z) = \sum_{-\infty}^{\infty} a_n z^n \qquad r_1 < |z| < r_2$$

$$a_k = \frac{1}{2\pi i} \oint_{C_2} \frac{f(t)\, dt}{t^{k+1}} \qquad k \geqq 0$$

(5)

$$a_k = \frac{1}{2\pi i} \oint_{C_1} \frac{f(t)\, dt}{t^{k+1}} \qquad k < 0$$

Finally, from Cauchy's theorem, we may deform the contours of integration in (5) as long as we stay in the region in which $f(z)$ is analytic. Thus we may deform C_2 into any convenient circle inside $r_1 < |t| < r_2$, and also deform C_1 into *the same* intermediate circle. In this way, we obtain one formula that is valid for all of the coefficients:

(6)
$$a_k = \frac{1}{2\pi i} \oint_C \frac{f(t)\, dt}{t^{k+1}}$$

where C is any simple closed curve (circle) in the annulus $r_1 < |t| < r_2$, which encloses the inner circle.

The series (4) is called the *Laurent expansion* of the function $f(z)$; it is a representation of $f(z)$ as a power series, with both positive and negative powers of z, and converges absolutely in the annulus $r_1 < |z| < r_2$.

COMMENT 1. The series $f_1(z) = \sum_{-\infty}^{-1} a_n z^n$ converges for $|z| > r_1$; in fact, it must converge in a larger region. Our choice of r_1 was one of convenience; we could have chosen any number $r_1 > 0$ (and sufficiently small), since $f(z)$ was assumed to be analytic for all points z inside the domain D except for $z = 0$. By the same reasoning, the series $f_2(z) = \sum_{0}^{\infty} a_n z^n$ will be convergent inside each circle, $|z| = r$, which does not include a singularity of $f(z)$ except for $z = 0$ (why?). Thus $f(z) = \sum_{-\infty}^{\infty} a_n z^n$, $0 < |z| < R$, and this is the "largest" annulus in which the series converges.

The formula for the coefficients a_k can be expressed

$$a_k = \frac{1}{2\pi i} \oint_C \frac{f(t)}{t^{k+1}} \, dt$$

where C is any simple closed curve in the region $0 < |t| < R$ which encloses $t = 0$.

COMMENT 2. The derivation of the Laurent expansion given in this section suggests an even more general (and equally important) result. Suppose that all we know about $f(z)$ is that it is single valued and analytic inside and on the annulus $r_1 \leq |z| \leq r_2$ (we say nothing about $f(z)$ for $|z| < r_1$ or for $|z| > r_2$). Then our derivation, starting from (1), applies equally well in this case, too. We find

(7) $$f(z) = \sum_{-\infty}^{\infty} a_n z^n \qquad r_1 < |z| < r_2$$

and

(8) $$a_k = \frac{1}{2\pi i} \oint_C \frac{f(t)}{t^{k+1}} \, dt$$

where C is any suitable curve in the annulus (and surrounding $t = 0$). This is still called the Laurent expansion for the annulus $r_1 < |z| < r_2$, and may still converge in a larger annulus. However, we expect singular points of $f(z)$ to occur on the boundaries of the largest annulus.

Example. $$f(z) = \frac{1}{z^2(z-1)(z-2)}$$

Figure 6.9

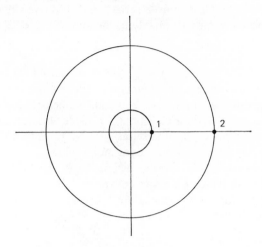

Clearly, $f(z)$ is analytic at all points $z \neq 0, 1, 2$ (see Figure 6.9). Thus, we may consider $f(z)$ in the annulus, $0 < |z| < 1$ with

$$f(z) = \frac{3}{4z} + \frac{1}{2z^2} + \frac{1}{1-z} - \frac{1}{4(2-z)}$$

(9) $\qquad f(z) = \frac{1}{2z^2} + \frac{3}{4z} + \sum_0^\infty z^n - \frac{1}{8}\sum_0^\infty \left(\frac{z}{2}\right)^n \qquad 0 < |z| < 1$

Thus

$$a_n = 0 \qquad n < -2 \qquad a_{-2} = \tfrac{1}{2} \qquad a_{-1} = \tfrac{3}{4}$$

and

$$a_n = 1 - \tfrac{1}{8}(2)^{-n} \qquad n \geqq 0.$$

We may equally well consider $f(z)$ in the annulus $1 < |z| < 2$, where it is also analytic. This gives us

(10) $\qquad f(z) = \frac{1}{2z^2} + \frac{3}{4z} - \sum_0^\infty \left(\frac{1}{z}\right)^{n+1} - \frac{1}{8}\sum_0^\infty \left(\frac{z}{2}\right)^n \qquad 1 < |z| < 2$

Finally, $f(z)$ is likewise analytic in the "annular region" $2 < |z| < \infty$. In this region, we have

(11) $\qquad f(z) = \frac{3}{4z} + \frac{1}{2z^2} - \sum_0^\infty \left(\frac{1}{z}\right)^{n+1} + \frac{1}{4z}\sum_0^\infty \left(\frac{2}{z}\right)^n \qquad 2 < |z| < \infty$

We have here three distinct Laurent series, all in powers of z, but converging in three distinct annuli; all three series converge to $f(z)$. We see that it is always a good idea to keep in mind the specific annulus in which a Laurent series converges.

COMMENT 3. Precisely as in the relation between the Taylor series and the MacLaurin series, we may state the following result: if $f(z)$ is (single valued) and analytic in the annulus $r_1 < |z - z_0| < r_2$, then

$$(12) \qquad f(z) = \sum_{-\infty}^{\infty} a_n(z - z_0)^n \qquad r_1 < |z - z_0| < r_2$$

where

$$(13) \qquad a_k = \frac{1}{2\pi i} \oint_C \frac{f(t)\, dt}{(t - z_0)^{k+1}}$$

with C being any (simple, closed) curve enclosing $t = z_0$ and lying in the annulus. This is Laurent's theorem.

6.6.1 If two Laurent expansions

$$\sum_{-\infty}^{\infty} a_n z^n \qquad \sum_{-\infty}^{\infty} b_n z^n$$

converge in the same annulus, $r_1 < z < r_2$, to the same function $f(z)$, show that $a_n = b_n$ for all n.

6.6.2 Find two Laurent expansions of

$$f(z) = \frac{1}{z^2(1 - z)}$$

in powers of z, and specify the annuli in which they converge.

6.6.3 Find the first two nonvanishing terms of the Laurent series expansions of the following functions, valid for the specified regions

(a) $\tan \pi z \qquad 0 < |z - \tfrac{1}{2}| < 1$

(b) $\dfrac{z}{e^z - 1} \qquad 0 < |z - 2\pi i| < 2\pi$

(c) $\dfrac{\sin z}{(z - \pi)^2} \qquad 0 < |z - \pi|$

6.6.4 The function $f(z) = e^{-(t/2)[z-(1/z)]}$ is analytic in $0 < |z| < \infty$. The Laurent expansion for this range is

$$f(z) = \sum_{-\infty}^{\infty} a_n(t) z^n$$

Show that

$$a_n(t) = \frac{1}{\pi} \int_0^\pi \cos(t \sin \theta + n\theta) \, d\theta$$

and

$$a_{-n}(t) = (-1)^n a_n(t).$$

6.6.5   Consider $f(z)$, periodic with period 2π, and analytic in a domain containing a strip of length 2π in the x-direction, between $y = a$ and $y = b$, $a < b$. Then, from periodicity, $f(z)$ is analytic in the infinite strip $-\infty < x < \infty$, $a < y < b$. Show that the mapping $z = i \log t$ maps this strip onto an annulus in which $f(i \log t) = F(t)$ is single valued. Thus show that the Laurent expansion, expressed in terms of z gives the usual Fourier expansion of a periodic function.

6.6.6   Find the Laurent expansion for

$$f(z) = \frac{1}{(2z + 1)(z - 1)} \quad \text{valid in } \tfrac{1}{2} < |z| < 1$$

and also for the same function in the annulus

$$0 < |z + \tfrac{1}{2}| < \tfrac{3}{2}$$

Where are the two series valid simultaneously?

6.7   SINGULARITIES OF ANALYTIC FUNCTIONS

Laurent's theorem is very valuable to us: it permits us to evaluate many definite integrals of analytic functions in a systematic way. This section, together with the next two sections, will present some of the applications of Laurent's theorem, and suggest many more.

Let us first return to the situation of Section 6.6. Suppose that $f(z)$ is single valued and analytic in a domain D (enclosing the origin) but that $f(z)$ need not be analytic at $z = 0$. Then, from Section 6.6 (4).

$$(1) \qquad f(z) = \sum_{-\infty}^{\infty} a_n z^n \qquad 0 < |z| < r_2$$

$$= f_1(z) + f_2(z)$$

$$f_1(z) = \sum_{-\infty}^{-1} a_n z^n \qquad f_2(z) = \sum_{0}^{\infty} a_n z^n$$

Thus $f(z)$ is decomposed into two parts: $f_2(z)$, analytic inside all of $|z| < r_2$, and $f_1(z)$, analytic for $|z| > 0$.

The function $f_1(z)$, containing the negative powers of z, is called *the principal part* of $f(z)$ *at* $z = 0$, and describes the singularity of $f(z)$ at $z = 0$. We single out three possible cases:

(a) The principal part of $f(z)$ at $z = 0$ is absent; that is, $a_k = 0$, $k = -1$, $-2, -3, \ldots$ clearly $f(z) = f_2(z) = \sum_0^\infty a_n z^n$ for all $|z| < r_2$ except $z = 0$. If we define $f(0) = f_2(0)$, then $f(z)$ is actually analytic inside the full circle $|z| < r_2$.

In this case, we say that $f(z)$ has a *removable singularity* at $z = 0$; the singularity can be removed by correctly defining $f(0)$. This situation arises, for instance, when we consider $f(z) = \sin z/z$, or $f(z) = (e^z - 1)/z$, for which $f(0)$ is not immediately clearly defined. In each case, $f(0) = 1$ makes $f(z)$ analytic at $z = 0$.

(b) There are a finite number of nonvanishing terms in the principal part; thus $a_n = 0$ for $n = -(k + 1), -(k + 2), \ldots$, but $a_{-k} \neq 0$. In this case, we say that $f(z)$ has a *pole of order k at $z = 0$*. Another (and useful) way of viewing this case is that $z^k f(z)$ is analytic at $z = 0$. For example, $f(z) = 1/[z^2(z - 1)(z - 2)]$ has a pole of order two at $z = 0$. (Can you describe the nature of the singularity at $z = 1$? $z = 2$?)

(c) The principal part is a full infinite series, it contains an infinite number of nonvanishing terms. In this case, we say that $z = 0$ is an *essential singularity*. For example

$$f(z) = e^{1/z} = \sum_0^\infty \frac{1}{z^n n!} \qquad |z| > 0$$

has an essential singularity at $z = 0$. Let us now consider

(2)
$$\oint_C f(z)\, dz$$

where C is any contour enclosing $z = 0$. There are several ways of obtaining the value of this integral from the Laurent series. First, we have

$$f(z) = \sum_{-\infty}^\infty a_n z^n$$

and, as we learned from our derivation of this series, Section 6.5 (4), we may evaluate $\oint_C f(z)\, dz$ by integrating $\oint_C \sum_{-\infty}^\infty a_n z^n\, dz$ term by term. This gives

$$\oint_C f(z)\, dz = \sum_{-\infty}^\infty a_n \oint_C z^n\, dz$$

Now, by explicit computation,

$$\oint z^n \, dz = 0 \qquad \text{if } n \geqq 0 \text{ (also, from Cauchy's theorem)}$$

$$= 2\pi i \qquad \text{if } n = -1$$

$$= 0 \qquad \text{if } n < -1$$

Thus

(3) $$\oint_C f(z) \, dz = 2\pi i a_{-1}$$

Alternately, from Section 6.6 (6), the coefficients

$$a_k = \frac{1}{2\pi i} \oint_C \frac{f(t)}{t^{k+1}} \, dt$$

and for $k = -1$, we again obtain (3).

The only term of the Laurent series that gives a nonzero contribution to $\oint_C f(z) \, dz$ is the term a_{-1}/z. *The coefficient a_{-1} is called the residue of $f(z)$ at $z = 0$.* We restate our result,

(4) $$\oint_C f(z) \, dz = 2\pi i \text{ (residue of } f(z) \text{ at } z = 0)$$

6.7.1   Find all of the singular points of $\csc \pi z$, and classify them as to poles and essential singularities, for example, finding the residue at each isolated singularity. (A singular point of an analytic function is called isolated if it is possible to enclose it at the center of a circle so that all other points of the circle are points of analyticity. For example, if $f(z)$ has a finite number of singular points in a bounded region, they are isolated.)

6.7.2   Classify the singular points, and compute the residue at each, for $f(z) =$

(a) $\dfrac{\sin \sqrt{z}}{z \sqrt{z}}$

(b) $\dfrac{1}{\sin \dfrac{\pi}{z}}$

(c) $\dfrac{z}{e^z - 1}$

6.7.3   Prove Riemann's theorem which states that if $f(z)$ is analytic in a domain D with a point z_0 removed, and is bounded in D, then there is a unique extension to $f(z_0)$ that makes $f(z)$ analytic at z_0 as well.

6.7.4   If $f(z)$ is analytic in D, and $f(z_0) = 0$ for some z_0 in D, then

$$f(z) = \sum_{0}^{\infty} \frac{f^{(n)}(z)}{n!} (z - z_0)^n$$

$$= a_p(z - z_0)^p + a_{p+1}(z - z_0)^{p+1} + \cdots$$

where a_p is the first nonvanishing coefficient. We say that $f(z)$ *has a zero of order p* at z_0. Show that if $f(z)$ has a zero of order p at z_0, $1/[f(z)]$ has a pole of order p at z_0, and conversely.

6.7.5 THE NATURE OF $f(z)$ AT $z = \infty$

We define the behavior of $f(z)$ at $z = \infty$ to be precisely the same as the behavior of $f(1/t)$ at $t = 0$. This coincides with our concept of analyticity at infinity. If $f(1/t)$ has a pole or an isolated essential singularity at $t = 0$, then $f(z)$ is said to have the same property at ∞. Show that the sum of the orders of the poles minus the sum of the orders of the zeros of a rational function (including $z = \infty$) is zero. (See Example 6.7.4.)

6.7.6   The integral $-\oint_{|z|=1} f(z)\,dz$ can be viewed as the integral around the unit circle in the negative sense (clockwise), or as the integral about the boundary of the infinite part of the plane (containing $z = \infty$) in the positive sense.

Then, with $z = 1/t$, this becomes

$$\int_{|t|=1} f\left(\frac{1}{t}\right) \frac{dt}{t^2}$$

By analogy with the residue theorem, $-\int_{|z|=1} f(z)\,dz$ should be $2\pi i \cdot$ residue at ∞ (assuming no other singularities). From this point of view, find the residue at $z = \infty$ for $f(z) =$

(a) $\dfrac{z}{1 - z}$ (b) $z + \dfrac{1}{z}$ (c) $\dfrac{1}{1 + z^2}$

6.7.7   Find the residue of $z^n e^{1/z}/(1 + z)$ at each singular point.

6.7.8   Show that every rational function can be decomposed into partial fractions (consider the principal parts of the Laurent series at each pole and show that the remainder is entire and bounded by a power of z; see Example 6.4.3).

6.7.9 $f(z)$ is analytic in $|z| < 1$, continuous in $|z| \leq 1$, with $|f(z)| \leq 1$ there, and $f(0) = 0$. Prove that $|f(z)| \leq |z|$ in $|z| \leq 1$. (Schwarz' lemma).

6.8 THE RESIDUE THEOREM

The results of Section 6.7 are applicable in many cases of interest. We consider the following situation: it is required to evaluate $\oint_C f(z)\,dz$, the definite integral of a single-valued function $f(z)$, analytic inside and on the simple closed curve C, except at a *finite* number of singular points inside C (Figure 6.10). Let us call these points $z_1, \ldots z_p$. Using one of the forms of Cauchy's theorem, we obtain

(1)
$$\oint_C f(z)\,dz = \sum_{k=1}^{p} \oint_{C_k} f(z)\,dz$$

where C_k is a (small) circle, centered at z_k, and lying inside C. Since there are only a finite number of singular points inside C, we may choose the circles C_k

Figure 6.10

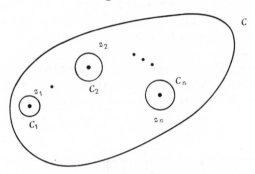

so that they do not intersect any of the other circles; that is, we may *isolate* each singular point inside a circle C_k so that the only singular point of $f(z)$ inside or on each C_k is at the center, z_k. (We can clearly so isolate each point of any finite collection of points but cannot do so for the points $1, 1/2, 1/3, \ldots$ $1/n, \ldots$, for example. In this case, the limit point, 0, is said to be nonisolated.)

Each of the integrals on the right-hand side of (1),

$$\oint_{C_k} f(z)\,dz$$

can be interpreted as a line integral in an annular region, $0 < |z - z_k| < r_k$, where r_k is small enough that $f(z)$ is analytic in the annulus. Applying Laurent's theorem to $f(z)$ in this annulus, we find

(2)
$$f(z) = \sum_{-\infty}^{\infty} a_n^{(k)}(z - z_k)^n \qquad 0 < |z - z_k| < r_k$$

where $[a_n^{(k)}]$ are the coefficients of the Laurent expansion about the point

$z = z_k$. The residue at $z = z_k$ is thus $a_{-1}^{(k)}$. Thus

$$(3) \qquad \oint_{C_k} f(z) \, dz = 2\pi i \text{ (residue of } f(z) \text{ at } z = z_k)$$

We thus obtain the following prescription for the value of the definite integral $\oint_C f(z) \, dz$: *the value of the integral is $2\pi i$ times the sum of the residues at the singular points of $f(z)$ inside the contour C,*

$$\oint_C f(z) \, dz = 2\pi i \sum \text{residues}$$

This result is called the *residue theorem.*

We can readily visualize the value of this theorem for evaluating definite integrals; we do not require the full Laurent series about the singular points; we require only one single coefficient. Frequently we can find these coefficients by inspection. However, let us look for a systematic procedure for obtaining residues.

6.8.1 If C is the unit circle described in the positive sense, evaluate the integrals

(a) $\int_C \dfrac{dz}{\sin z}$ (b) $\int_C e^{1/z} \dfrac{dz}{z}$

(c) $\int_C \dfrac{e^{-z}}{z(z+2)} \, dz$ (d) $\int_C \left(z + \dfrac{1}{z}\right)^{2n} dz$

(e) $\int_C \cot z \, dz$

6.8.2 Evaluate $\displaystyle\int_{|z|=2} \dfrac{dz}{1 + z^2}$

6.8.3 Evaluate $\displaystyle\int_{|z|=e} \pi \cot \pi z \, dz$

6.8.4 Evaluate

$$\int_C \frac{2z^2 + 5}{(z+1)(z^2+1)^2} \, dz$$

where C is the circle $|z + i| = 1$.

6.9   COMPUTATION OF RESIDUES

(a) At a removable singularity, since the principal part is absent altogether, $a_{-1} = 0$, and the residue is zero.

(b) The simplest case occurs when $f(z)$ has a first order pole at $z = z_0$. The Laurent series has the form

$$(1) \qquad f(z) = \frac{a_{-1}}{z - z_0} + a_0 + a_1(z - z_0) + \cdots$$

As we observed earlier [see Section 6.7 (2)],

$$(z - z_0)f(z) = a_{-1} + a_0(z - z_0) + a_1(z - z_0)^2 + \cdots$$

is analytic at $z = z_0$, and now a_{-1} is the leading coefficient of a power-series expansion.

$$(2) \qquad a_{-1} = \lim_{z \to z_0} (z - z_0)f(z)$$

If $f(z)$ has a first order pole at $z = z_0$, the residue of $f(z)$ at $z = z_0$ is $\lim_{z \to z_0} (z - z_0) f(z)$.

For example, in integrating $f(z) = 1/(1 + z^2)$ over a large semicircle in the upper half-plane, we observe that $z = i$ is the only singularity of $f(z)$ inside the contour.

$$f(z) = \frac{1}{(z - i)(z + i)} \qquad (z - i)f(z) = \frac{1}{z + i}$$

is analytic at $z = i$. Hence, without writing down the Laurent series, we find

$$\text{residue of } \frac{1}{z^2 + 1} = \lim_{z \to i} (z - i)f(z) = \frac{1}{2i}$$

of course, the value of the integral, $2\pi i (1/2i) = \pi$, agrees with the result we obtained earlier (using Cauchy's integral formula).

(c) Residues at poles of higher order: suppose $f(z)$ has a pole of order k at $z = z_0$. The Laurent series about $z = z_0$ is

$$(3) \qquad f(z) = \frac{a_{-k}}{(z - z_0)^k} + \cdots + \frac{a_{-1}}{z - z_0} + a_0 + a_1(z - z_0) + \cdots$$

and we wish to find a_{-1}. With the same ideas as in the case of first order poles (but with a little more work), we have

$$(z - z_0)^k f(z)$$

$$= a_{-k} + (z - z_0)a_{-(k-1)} + \cdots + a_{-1}(z - z_0)^{k-1} + a_0(z - z_0)^k + \cdots$$

which is analytic at $z = z_0$, and we wish to find a_{-1}, the coefficient of $(z - z_0)^{k-1}$.

The results of Taylor's theorem suggest the following manipulation:

$$\frac{d^{k-1}}{dz^{k-1}}[(z - z_0)^k f(z)] = (k - 1)!\, a_{-1} + k!\, a_0 (z - z_0) + \cdots$$

If we now let $z = z_0$, the right-hand side reduces to $(k - 1)!\, a_{-1}$. We have

(4)
$$a_{-1} = \frac{1}{(k - 1)!} \lim_{z \to z_0} \frac{d^{k-1}}{dz^{k-1}}[(z - z_0)^k f(z)]$$

Example. $f(z) = 1/[(z^2 + 1)^3]$. Find the residue at $z = i$. We have $f(z) = 1/[(z + i)^3 (z - i)^3]$, and recognize that $z = i$ is a pole of order 3. Thus $(z - i)^3 f(z) = 1/[(z + i)^3]$. Formula 4 gives

$$a_{-1} = \frac{1}{2!} \lim_{z \to i} \frac{d^2}{dz^2} \left[\frac{1}{(z + i)^3} \right]$$

$$= \frac{1}{2!} \lim_{z \to i} \frac{12}{(z + i)^5}$$

$$= \frac{6}{(2i)^5} = \frac{-3i}{16}$$

While (4) is a formula that works in every instance, it may be very cumbersome to apply for other than poles of low order. There are frequently more convenient ways to proceed. Let us illustrate one possibility. Again consider $f(z) = 1/[(z^2 + 1)^3]$, near $z = i$. Since it is easier to manipulate power series in powers of z than those in powers of $z - z_0$, we translate so that $z = i$ corresponds to $t = 0$: let $t = z - i$. Then

$$f(z) = \frac{1}{(z - i)^3 (z + i)^3} = \frac{1}{t^3 (2i + t)^3}$$

We now look for the coefficient of $1/(z - i)$ or, what is the same thing, the coefficient of $1/t$ in $1/[t^3 (2i + t)^3]$. We have

$$\frac{1}{t^3 (2i + t)^3} = \frac{1}{t^3 (2i)^3} \left(1 + \frac{t}{2i} \right)^{-3}$$

$$= \frac{i}{8t^3} \left(1 - \frac{3t}{2i} + \frac{12t^2}{2(2i)^2} + \cdots \right)$$

expanding $(1 + t/2i)^{-3}$ by the binomial expansion and retaining only enough terms as to be able to find the coefficient of $1/t$. We have

$$\frac{i}{8} \frac{12}{2(2i)^2} = -\frac{3i}{16}$$

as previously.

(d) Residues at essential singularities: there is no simple formula, as in the previous cases; the full Laurent series is frequently required. For example,

$$f(z) = e^{z+1/z} = e^z e^{1/z}$$

$$= \sum_0^\infty \frac{z^n}{n!} \sum_0^\infty \frac{1}{z^m m!} \qquad 0 < |z| < \infty$$

The coefficient a_{-1}, of $1/z$, comes from each of the terms $z^n/n!\ 1/z^m m!$ for which $m = n + 1$; $\sum_0^\infty 1/[n!\ (n + 1)!]$. The residue at $z = 0$ is thus $\sum_0^\infty 1/[n!\ (n + 1)!]$.

6.9.1   Since it is generally simpler to expand analytic functions about $z = 0$ than about any other point, we may advantageously use the same ideas in computing residues in general. Find the residue of $f(z) = \log z/(z^2 + 1)^2$ at $z = i$, where $\log z$ is the principal branch.

Solution: Let $z = i + t$. We require the residue of

$$\frac{\log (i + t)}{t^2(t + 2i)^2}$$

at $t = 0$. This is the coefficient of t in

$$\frac{\log (i + t)}{(t + 2i)^2} = \frac{\log i + \dfrac{t}{i} + \cdots}{(2i)^2\left(1 + \dfrac{t}{2i}\right)^2} = \frac{\left(\dfrac{\pi i}{2} - it\right)(1 + it) + \cdots}{(2i)^2}$$

and we expand just far enough to collect the terms in t. The required coefficient is $\frac{1}{4}(\pi/2 + i)$.

6.9.2   Find the residue of $f(z) = 1/(z^2 + 1)^{2n}$ at $z = i$.

6.9.3 $g(z)$ has a double zero at $z = a$, and $f(a) \neq 0$. Show that the residue of $f(z)/g(z)$ at $z = a$ is

$$\frac{6f'(a)g''(a) - 2f(a)g'''(a)}{3[g''(a)]^2}$$

6.10 EVALUATION OF DEFINITE INTEGRALS

In this section we illustrate how the use of Cauchy's theorem, Cauchy's integral formula, and the calculus of residues can be utilized to evaluate certain types of (real) definite integrals. These examples serve only as a guide; they may suggest further applications.

Example 1. If we were to consider

$$I = \int_{-\infty}^{\infty} \frac{dx}{1 + x^2}$$

then guided by the results of the previous sections, we might argue as follows: the integrand, $1/(1 + x^2)$, is the value, on the real axis, of the analytic function $f(z) = 1/(1 + z^2)$. Also, the path of integration is the full real axis. Can we set up a finite integral, so that our required integral is its limiting form? We are led to consider $\oint dz/(1 + z^2)$ around a contour, part of which coincides with the real axis, say from $-R$ to R, and the rest of the contour being an arc either in the upper or the lower half-plane. The simplest such arc is a semicircle. We are thus led to consider $\int_C dz/(1 + z^2)$, where C is the real axis from $-R$ to R, and the upper semicircle centered at $z = 0$. The integral over the semicircle tends to zero as $R \to \infty$. (We can see this result intuitively since the length of the path is πR, and the integrand is "of the order" $1/R^2$ on the path. This argument is not precise, of course, but it frequently serves as a useful guide in choosing contours.)

Let us generalize this result. Consider the definite integral

$$(1) \qquad\qquad I = \int_{-\infty}^{\infty} \frac{p(x)}{q(x)} \, dx$$

where $p(x)$, $q(x)$ are polynomials (the integrand is a rational function) of degrees p and q respectively.

In order that (1) converge, we must require that $q(x) \neq 0$ for all *real* x [we assume there are no common factors in $p(x)$, $q(x)$]. Furthermore, the degree p of $p(x)$ must be at least two less than the degree q of $q(x)$: $p \leq q - 2$. Under these restrictions, (1) is absolutely convergent as an improper integral.

Next, $p(x)/q(x)$ is the specialization to the real axis of the analytic function $f(z) = p(z)/q(z)$; $q(z)$ is a polynomial, and has q complex zeros (counting multiplicity, see Example 6.31). Hence $f(z)$ has a finite number of poles in the full complex plane, $z_1, z_2, \ldots z_q$. We suppose $R > |z_1|, |z_2|, \ldots |z_q|$. By analogy with the preceding example, we now consider the integral

$$(2) \qquad\qquad \oint_C \frac{p(z)}{q(z)} \, dz$$

where C is the upper semicircle of radius R, centered at $z = 0$. From the residue theorem,

$$\oint_C \frac{p(z)}{q(z)} \, dz$$

$= 2\pi i$ (sum of residues of $p(z)/q(z)$ at the poles in the upper half plane)

Specializing the integrand to the contour, we have

$$\int_{-R}^{R} \frac{p(x)}{q(x)}\, dx + \int_{0}^{\pi} \frac{p(Re^{i\theta})}{q(Re^{i\theta})} iRe^{i\theta}\, d\theta = A + B$$

Clearly

$$\lim_{R\to\infty} A = \int_{-\infty}^{\infty} \frac{p(x)}{q(x)}\, dx$$

and we expect $\lim_{R\to\infty} B = 0$. We have

$$|B| \leqq R \int_{0}^{\pi} \frac{|p(Re^{i\theta})|}{|q(Re^{i\theta})|}\, d\theta$$

$$= R \int_{0}^{\pi} \frac{|a_0 + a_1 Re^{i\theta} + a_2 R^2 e^{2i\theta} + \cdots + a_p R^p e^{ip\theta}|}{|b_0 + b_1 Re^{i\theta} + \cdots + b_q R^q e^{iq\theta}|}\, d\theta$$

$$\leqq R \int_{0}^{\pi} \frac{|a_p| R^p + |a_{p-1}| R^{p-1} + \cdots + |a_0|}{|b_q| R^q - |b_{q-1}| R^{q-1} - \cdots - |b_0|}\, d\theta$$

$$= R \int_{0}^{\pi} \frac{|a_p| R^p \left(1 + \dfrac{|a_{p-1}|}{|a_p| R} + \cdots + \dfrac{|a_0|}{|a_p| R^p}\right)}{|b_q| R^q \left(1 - \dfrac{|b_{q-1}|}{|b_q| R} - \cdots - \dfrac{|b_0|}{|b_q| R^q}\right)}\, d\theta$$

We now choose R so large that

$$1 - \frac{|b_{q-1}|}{|b_q| R} - \cdots - \frac{|b_0|}{|b_q| R^q} \geqq \tfrac{1}{2}$$

$$1 + \frac{|a_{p-1}|}{|a_p| R} + \cdots + \frac{|a_0|}{|a_p| R^p} \leqq 2$$

for example. We thus have

$$|B| \leqq R \int_{0}^{\pi} \frac{|a_p|}{|b_q|} \frac{4\, d\theta}{R^{q-p}} = \frac{4\pi\, |a_p|}{|b_q|\, R^{q-p-1}}$$

and since

$$q - p - 1 > 0 \qquad |B| \to 0 \text{ as } R \to \infty$$

We have proved the following result:

$$\int_{-\infty}^{\infty} \frac{p(x)}{q(x)}\, dx$$

(under the convergence hypotheses that $q(x) \neq 0$, degree $p \leqq$ degree $q - 2$) has the value $2\pi i$ (sum of the residues of the integrand in the upper half-plane).

To illustrate the use of this result, consider

$$I = \int_{-\infty}^{\infty} \frac{dx}{x^4 + 1}$$

The zeros of $z^4 + 1$ are $z = e^{i\pi/4}, e^{3\pi i/4}, e^{5\pi i/4}, e^{7\pi i/4}$ of which $e^{i\pi/4}, e^{3\pi i/4}$ are in the upper half-plane

$$\text{Residue of } \frac{1}{z^4 + 1} \text{ at } z = e^{i\pi/4} \text{ is } \lim_{z \to e^{i\pi/4}} \frac{z - e^{i\pi/4}}{z^4 + 1}$$

$$= \lim_{z \to e^{i\pi/4}} \frac{1}{4z^3}$$

$$= \frac{e^{-3\pi i/4}}{4}$$

Similarly, the residue at $e^{3\pi i/4}$ is $e^{-\pi i/4}/4$. Thus

$$I = \frac{2\pi i}{4} \left(e^{3\pi i/4} + e^{-\pi i/4} \right)$$

$$= \frac{\pi i}{2} \left(e^{-\pi i/4} - e^{\pi i/4} \right)$$

$$= \frac{\pi i}{2} \left(-2i \sin \frac{\pi}{4} \right)$$

$$= \pi \sin \frac{\pi}{4} = \frac{\pi}{\sqrt{2}}$$

$$\int_{-\infty}^{\infty} \frac{dx}{1 + x^4} = \frac{\pi}{\sqrt{2}}$$

Example 2.   Consider the integral

$$I = \int_{-\infty}^{\infty} \frac{\cos x \, dx}{x^2 + a^2} \qquad a > 0$$

This integral is sufficiently similar to that of the preceding example that we might try the same finite contour, the upper half semicircle of radius R, with the analytic integrand $f(z) = \cos z/(z^2 + a^2)$. All goes well up to the estimate of the integral

$$B = \int_0^{\pi} \frac{\cos (Re^{i\theta}) i Re^{i\theta}}{R^2 e^{2i\theta} + a^2} \, d\theta$$

Evaluation of Definite Integrals   **213**

While $\cos x$ is bounded, for real x,

$$\cos z = \frac{e^{iz} + e^{-iz}}{2} = \frac{e^{ix-y} + e^{-ix+y}}{2}$$

Thus, as y becomes large and positive, e^{ix-y} is small, but e^{-ix+y} has an exponentially large magnitude. The integral B above does not tend to zero as $R \to \infty$.

We resort to a trick. Instead of writing $I = \int_{-\infty}^{\infty} \cos x \, dx/(x^2 + a^2)$, we write the equivalent statement

$$I = \mathbf{R} \int_{-\infty}^{\infty} e^{ix} \frac{dx}{x^2 + a^2}$$

Now taking the upper half semicircle as our contour, and $e^{iz}/(z^2 + a^2)$ as integrand, we see that $|e^{iz}| = |e^{ix-y}| = e^{-y} \leqq 1$ on the full contour. Accordingly

$$\left| \int_0^\pi \frac{e^{i\,Re^{i\theta}} iR \, d\theta}{R^2 e^{2i\theta} + a^2} \right| \leqq R \int_0^\pi \frac{d\theta}{R^2 - a^2} = \frac{\pi R}{R^2 - a^2} \to 0$$

Hence

$$\int_{-\infty}^{\infty} e^{ix} \frac{dx}{x^2 + a^2} = 2\pi i \left(\text{residue of } \frac{e^{iz}}{z^2 + a^2} \text{ at } z = ai \right)$$

$$= 2\pi i \frac{e^{-a}}{2ia}$$

$$= \pi \frac{e^{-a}}{a}$$

The prescription is precisely the same as in Example 1. Hence

$$\int_{-\infty}^{\infty} \cos x \frac{dx}{x^2 + a^2} = \mathbf{R}\left(\frac{\pi e^{-a}}{a} \right) = \frac{\pi e^{-a}}{a}$$

$[\int_{-\infty}^{\infty} \sin x/(x^2 + a^2) \, dx = 0$, Why?]

Let us consider another illustration.

(3)
$$J = \int_0^\infty \frac{\sin x}{x} \, dx$$

In order to use the ideas introduced above, we would prefer an integral on $(-\infty, \infty)$. This we can do; from symmetry,

$$J = \frac{1}{2} \int_{-\infty}^{\infty} \frac{\sin x}{x} \, dx$$

We are also prepared to replace $\sin x$ by e^{ix} and take the imaginary part of the resulting integral. However e^{ix}/x is singular at $x = 0$. Still, e^{iz}/z

Figure 6.11

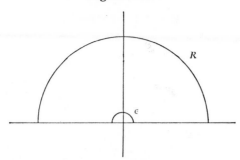

is analytic for $z \neq 0$, and we can utilize it as follows: we take a contour consisting of the upper half semicircle of radius R, and remove a small semicircle of radius ε near the origin; then $\oint e^{iz}\,dz/z = 0$, since the integrand is analytic inside and on the contour (Figure 6.11). Adapting the integrand to the contour we have

$$\int_{-R}^{-\varepsilon} \frac{e^{ix}}{x}\,dx + \int_{\pi}^{0} e^{i\varepsilon e^{i\theta}}i\,d\theta + \int_{\varepsilon}^{R} \frac{e^{ix}}{x}\,dx + \int_{0}^{\pi} e^{i\,Re^{i\theta}}i\,d\theta = 0$$

The first and third integrals combine to give

$$\int_{\varepsilon}^{R} \frac{e^{ix} - e^{-ix}}{x}\,dx = 2i\int_{\varepsilon}^{R} \frac{\sin x}{x}\,dx$$

We can see our desired result by letting $\varepsilon \to 0$, $R \to \infty$. Then

$$\lim_{\varepsilon \to 0} \int_{\pi}^{0} e^{i\varepsilon e^{i\theta}}i\,d\theta = \int_{\pi}^{0} i\,d\theta = -i\pi$$

and we are left with

$$|B| = \left| \int_{0}^{\pi} e^{i\,Re^{i\theta}}i\,d\theta \right| \leqq \int_{0}^{\pi} e^{-R\sin\theta}\,d\theta$$

$$\int_{0}^{\pi} e^{-R\sin\theta}\,d\theta = 2\int_{0}^{\pi/2} e^{-R\sin\theta}\,d\theta$$

and on $(0, \pi/2)$, $\sin\theta/\theta$ decreases steadily, so that

$$\frac{\sin\theta}{\theta} \geqq \frac{\sin\dfrac{\pi}{2}}{\dfrac{\pi}{2}} = \frac{2}{\pi}$$

$$\sin\theta > \frac{2\theta}{\pi} \qquad \text{on } \left(0, \frac{\pi}{2}\right)$$

$$e^{-R\sin\theta} \leqq e^{-2R\theta/\pi} \qquad 0 \leqq \theta \leqq \frac{\pi}{2}$$

Evaluation of Definite Integrals   **215**

Hence

$$|B| \leq 2 \int_0^{\pi/2} e^{-R \sin \theta} \, d\theta$$

$$\leq 2 \int_0^{\pi/2} e^{-2\theta R/\pi} \, d\theta$$

$$= \frac{2\pi}{2R} [1 - e^{-R}] \to 0$$

Finally

$$2i \int_0^\infty \frac{\sin x}{x} \, dx = i\pi$$

$$\int_0^\infty \frac{\sin x}{x} \, dx = \frac{\pi}{2}$$

Without attempting to write a formal statement describing these integrals in the most general terms, we can use the residue calculus to evaluate

$$\int_{-\infty}^\infty e^{ix} Ra(x) \, dx$$

where $Ra(x)$ stands for a rational function, the quotient of two polynomials. Now we require only that the degree of the denominator be at least 1 greater than the degree of the numerator.

Example 3.   Consider the integral

$$I = \int_0^\pi \frac{d\theta}{a + b \cos \theta} = \frac{1}{2} \int_{-\pi}^\pi \frac{d\theta}{a + b \cos \theta} \qquad 0 < b < a$$

We have encountered such integrals when the path of integration has been a circle, or an arc of a circle. Let $e^{i\theta} = z$, $2 \cos \theta = z + 1/z$, $d\theta = dz/iz$. Then

$$I = \frac{1}{2} \oint_{|z|=1} \frac{dz}{iz \left[a + \frac{b}{2} \left(z + \frac{1}{z} \right) \right]}$$

$$= \frac{1}{i} \oint_{|z|=1} \frac{dz}{bz^2 + 2az + b}$$

The denominator vanishes at

$$z = \frac{-a \pm \sqrt{a^2 - b^2}}{b}$$

with

$$z_1 = \frac{-a + \sqrt{a^2 - b^2}}{b}$$

lying inside the circle (the other lies outside).
Thus

$$I = \frac{2\pi i}{i} \left. \text{res} \right|_{z_1} = \frac{2\pi}{2(bz_1 + a)}$$

$$= \frac{\pi}{\sqrt{a^2 - b^2}}$$

Evaluate the following integrals by the methods of Section 6.10.

6.10.1 $\displaystyle\int_{-\infty}^{\infty} \frac{dx}{x^2 + x + 2}$

6.10.2 $\displaystyle\int_{0}^{\infty} \frac{x^2 \, dx}{x^4 + a^4}$ $\quad a > 0$

6.10.3 $\displaystyle\int_{0}^{\infty} \frac{x^{2m}}{1 + x^{2n}} \, dx$ $\quad 0 \le m < n$, integral

6.10.4 $\displaystyle\int_{0}^{\infty} \frac{dx}{(a + bx^2)^n}$

6.10.5 $\displaystyle\int_{0}^{\infty} \frac{x^2 \, dx}{x^6 + 1}$

6.10.6 $\displaystyle\int_{-\infty}^{\infty} \frac{\cos x \, dx}{(x^2 + a^2)(x^2 + b^2)}$

6.10.7 $\displaystyle\int_{0}^{\infty} \frac{\cos x \, dx}{(x^2 + a^2)^2}$

6.10.8 Deduce the integral of Example 6.10.7 from the result of Example 6.10.6.

6.10.9 $\displaystyle\int_{0}^{\infty} \frac{x \sin x \, dx}{x^2 + 1}$

6.10.10 $\displaystyle\int_{-\infty}^{\infty} \frac{\sin x \, dx}{(x + a)^2 + b^2}$ $\quad a, b > 0$

6.10.11 $\displaystyle\int_0^\infty \frac{\sin x \, dx}{x(x^2 - \pi^2)}$

6.10.12   Show that in indenting the contour around a first order pole, the contribution to the integral by the indented semicircle tends to $\frac{1}{2}$ the residue at the pole in question.

6.10.13 $\displaystyle\int_0^{2\pi} \frac{d\theta}{a + i(b \cos \theta + c \sin \theta)}$

6.10.14 $\displaystyle\int_0^{2\pi} \cot \left(\frac{\theta - a - ib}{2} \right) d\theta$

6.10.15 $\displaystyle\int_0^\pi \frac{\cos 2\theta}{1 - 2a \cos \theta + a^2} \qquad a^2 < 1$

6.10.16

$$\int_0^{2\pi} \frac{\cos n\theta \, d\theta}{5 - 3 \cos \theta}$$

Evaluate in two ways. First with $e^{i\theta} = z$, $2 \cos n\theta = z^n + 1/z^n$; second, consider

$$\mathbf{R}\int_0^{2\pi} \frac{e^{in\theta} \, d\theta}{5 - 3 \cos \theta}$$

6.10.17   Show that $\int_{-\pi}^{\pi} Ra \, (\sin \theta, \cos \theta) \, d\theta$ can be evaluated by the methods of this section, after the elementary change of variable $\tan \theta/2 = x$. (This change of variable produces an integral of the form $\int_{-\infty}^{\infty} Ra(x) \, dx$.)

6.10.18   Show that

$$\int_0^\pi \frac{\cos n\phi \, d\phi}{a - ib \cos \phi} = \frac{\pi i^n}{\sqrt{a^2 + b^2}} \left(\frac{\sqrt{a^2 + b^2} - a}{b} \right)^n \qquad a, b > 0$$

(See Example 6.10.16.)

6.10.19   Integrate $z/(a - e^{-iz})$ around the rectangle with vertices at $\pm \pi$, $\pm \pi + in$, and hence show that, for $a > 1$,

$$\int_0^\pi \frac{x \sin x \, dx}{1 - 2a \cos x + a^2} = \frac{\pi}{a} \log \frac{1 + a}{a}$$

6.1   Let $f(z)$ be analytic inside and on the circle $|z| = R$, and $z = a = re^{i\theta}$ be inside the circle. Cauchy's formula yields

$$f(a) = \frac{1}{2\pi i} \int_{|z|=R} \frac{f(z)\,dz}{z - a}$$

$$0 = \frac{1}{2\pi i} \int_{|z|=R} \frac{f(z)\,dz}{z - \dfrac{R^2}{\bar{a}}}$$

Hence

$$f(a) = \frac{1}{2\pi i} \int_{|z|=R} \frac{R^2 - a\bar{a}}{(z - a)(R^2 - \bar{a}z)} f(z)\,dz$$

This result is called *Poisson's formula*. Show that it yields

$$f(a) = f(re^{i\theta}) = \frac{1}{2\pi} \int_0^{2\pi} \frac{(R^2 - r^2)f(Re^{i\phi})\,d\phi}{R^2 - 2Rr\cos(\theta - \phi) + r^2}$$

and if $u(r, \theta)$ is harmonic in $r \leq R$,

$$u(r, \theta) = \frac{1}{2\pi} \int_0^{2\pi} \frac{(R^2 - r^2)u(R, \phi)\,d\phi}{R^2 - 2Rr\cos(\theta - \phi) + r^2}$$

6.2   Imitate the steps of Example 6.1 for the case in which $f(z)$ is analytic in the upper half-plane, and tends to zero as $z \to \infty$ there, so that Cauchy's integral yields

$$f(a) = \frac{1}{2\pi i} \int_{-\infty}^{\infty} \frac{f(x)\,dx}{x - a} \qquad \text{I}a > 0$$

and

$$0 = \frac{1}{2\pi i} \int_{-\infty}^{\infty} \frac{f(x)\,dx}{x - \bar{a}}$$

Thus deduce Poisson's formula for the half plane

$$u(x, y) = \frac{y}{\pi} \int_{-\infty}^{\infty} \frac{u(t, 0)\,dt}{(x - t)^2 + y^2}$$

for $u(x, y)$ harmonic in $y > 0$, and vanishing at ∞.

6.3   Deduce the results of Example 6.2 by mapping the circle $|z| < R$ conformally onto the upper half-plane $\text{I}w > 0$, and use the mapping function to change variables in the integral.

6.4   If $f(z)$ is analytic in $|z| \leq 1$, $|f(z)| \leq M$ and $f(a) = 0$ where $|a| < 1$, show that

$$|f(z)| \leq M \left| \frac{z - a}{1 - \bar{a}z} \right| \quad \text{in} \quad |z| \leq 1$$

See Example 6.6.9.

6.5   If $f(z)$ is analytic in $|z| < 1$, and $|f(z)| \leq 1/(1 - |z|)$ show that

$$|a_n| = \left| \frac{f^{(n)}(0)}{n!} \right| \leq (n + 1)\left(1 + \frac{1}{n}\right)^n < e(n + 1)$$

6.6   Deduce the maximum modulus theorem from Parseval's equation: if $f(z) = \sum_0^\infty a_n(z - z_0)^n$, $|z - z_0| = r$, then

$$\frac{1}{2\pi} \int_0^{2\pi} |f(z_0 + re^{i\theta})|^2 \, d\theta = \sum_0^\infty |a_n|^2 \, r^{2n}$$

6.7   If $f(z)$ has a pole at $z = a$, show that $\lim_{z \to a} |f(z)| = \infty$.

If $f(z)$ has an isolated essential singularity at $z = a$, show the following:

(a) That there is a sequence $(z_n) \to a$ such that $f(z_n) \to \infty$.
(b) That $1/(f(z) - b)$ has an essential singularity at $z = a$.
(c) That for any arbitrary complex b, there exists a sequence $(z_\nu) \to a$ such that $f(z_\nu) \to b$ (Weierstrass).

Exercises 6.8 to 6.26 illustrate a few additional applications of the residue theorem to the evaluation of definite integrals, summation of certain series, and to certain new representations of analytic functions.

6.8   Evaluate $\displaystyle\int_0^\infty \frac{x^{a-1} \, dx}{1 + x} \quad 0 < a < 1$

Solution: The integrand is the specialization to the real axis of $z^{a-1}/(1+z)$; in general, this is infinitely multivalued. Let us choose a specific branch, for example, the principal branch, which is single valued in the plane cut from $z = 0$ to $z = \infty$ on the positive real axis.

In order to use the method of residues, we require a closed curve, and a curve that coincides with the real axis for $0 < x < R$. We can achieve this in several ways. (Figure 6-12). Consider the curve illustrated: from δ to R along the real axis, around the circle $|z| = R$ to the bottom edge of the cut, back along the real axis from R to δ, and then on the circle $|z| = \delta$ to the starting point. Because of its suggestive shape, this is often called a *keyhole* contour.

Figure 6.12

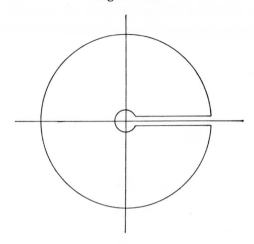

Specializing the integral to the contour, we have

$$\int_{\delta}^{R} \frac{x^{a-1}\, dx}{1 + x} + \int_{0}^{2\pi} \frac{(Re^{i\theta})^{a-1} iRe^{i\theta}\, d\theta}{1 + Re^{i\theta}} + \int_{R}^{\delta} \frac{(e^{2\pi i}x)^{a-1}\, dx}{1 + x} + \int_{2\pi}^{0} \frac{(\delta e^{i\theta})^{a-1} i\, \delta e^{i\theta}\, d\theta}{1 + \delta e^{i\theta}}$$

$$= 2\pi i \text{ (residue at } z = -1\text{)}$$
$$= 2\pi i (e^{\pi i})^{a-1}$$

The first and third of these integrals combine to give

$$(1 - e^{2\pi i a}) \int_{\delta}^{R} \frac{x^{a-1}\, dx}{1 + x}$$

while the second and fourth tend to zero as $R \to \infty$, $\delta \to 0$.

$$\left[\left| \int_{0}^{2\pi} \frac{(Re^{i\theta})^{a-1} iRe^{i\theta}\, d\theta}{1 + Re^{i\theta}} \right| \leq \int_{0}^{2\pi} \frac{R^{a}\, d\theta}{R - 1} = 2\pi \frac{R^{a}}{R - 1} \to 0 \right]$$

Hence

$$\int_{0}^{\infty} \frac{x^{a-1}\, dx}{1 + x} = \frac{2\pi i e^{\pi i a}}{e^{2\pi i a} - 1} = \frac{\pi}{\sin \pi a}$$

6.9 Evaluate the integral of Example 6.8 by integrating over a large semicircle in the upper half-plane, indented at the origin. Thus, also evaluate

$$\fint_{0}^{\infty} \frac{x^{a-1}\, dx}{1 - x} = -\pi \cot \pi a$$

(The integral is a Cauchy principal value integral.)

6.10 Evaluate the integral of Example 6.8 by first making the change of variable $x = e^t$, and then integrating around the rectangle with vertices at $t = \pm R, \pm R + 2\pi i$.

6.11 Show that $\int_0^\infty (\sinh ax)/(\sinh \pi x)\, dx = \frac{1}{2}\tan a/2$, $a^2 < \pi^2$, by integrating around the rectangle with vertices at $\pm R, \pm R + i\pi$, indented at $z = 0, i$.

6.12 By integrating $1/(z \log z)$ around the keyhole contour of Example 6.8, suitably indented, show that

$$\int_0^\infty \frac{dx}{x[(\log x)^2 + \pi^2]} = 1$$

6.13 By integrating around the unit circle, indented at $z = e^{i\phi}, e^{-i\phi}$, show that for positive integral n,

$$\int_0^{2\pi} \frac{\cos n\theta\, d\theta}{\cos \theta - \cos \phi} = \pi \frac{\sin n\phi}{\sin \phi}$$

6.14 Integrate $\log z\, Ra(z)$, where $Ra(z)$ is any suitable rational function, around the keyhole contour of Example 6.8, and thus show how to evaluate $\int_0^\infty Ra(x)\, dx$.

6.15 $f(z)$ is analytic inside and on an ellipse C with foci at ± 1. Show that

$$\int_C f(z) \log \left(\frac{z+1}{z-1}\right) dz = 2\pi i \int_{-1}^{1} f(x)\, dx$$

6.16 Integrate $e^{i\pi z^2} \tan \pi z$ around a large parallelogram, two sides passing through $z = 0, z = 1$ at an angle of $45°$ to the positive real axis, the other two sides being parallel to the real axis. Hence show that

$$\int_{-\infty}^{\infty} e^{-x^2}\, dx = \sqrt{\pi}$$

Solution: $\qquad \oint e^{i\pi z^2} \tan \pi z\, dz = 2\pi i \frac{e^{\pi i/4}}{-\pi}$

$$\int_{/} + \int_{/} \to 2e^{-\pi i/4} \int_{-\infty}^{\infty} e^{-\pi t^2}\, dt$$

and the other two tend to zero. Thus $\int_{-\infty}^{\infty} e^{-\pi t^2}\, dt = 1$. (Note also that $\int_{-\infty}^{\infty} e^{-t^2}\, dt = \int_{-\infty}^{\infty} e^{-(t+a)^2}$ for any complex constant a.)

6.17   Integrate e^{-z^2} around a large circular arc $|z| = R$, $0 \le \theta \le \pi/4$, closed off by radial lines, and show that

$$\int_0^\infty \cos x^2 \, dx = \int_0^\infty \sin x^2 \, dx = \frac{1}{2}\sqrt{\frac{\pi}{2}}$$

6.18   Show that $\pi \cot \pi z$ has residue $+1$ at each of the poles $z = 0, \pm 1, \pm 2, \ldots$, and that it is bounded on the sequence of squares C_N with vertices at $\pm(N + \frac{1}{2}) \pm i(N + \frac{1}{2})$, $N =$ integer, the bound being independent of N.

Consider $\oint_{C_N} [(\pi \cot \pi z)/(z^2)] \, dz$, show that the integral tends to zero as $N \to \infty$, and hence deduce that

$$\sum_1^\infty \frac{1}{n^2} = \frac{\pi^2}{6}$$

6.19   By analogy with Example 6.18, integrate $(\pi \csc \pi z)/z^2$ around C_N, and show that

$$\sum_1^\infty \frac{(-1)^{n+1}}{n^2} = \frac{\pi^2}{8}$$

6.20   If the poles of $f(z)$ are at $b_1, b_2, \ldots b_k$, with residues $r_1, r_2, \ldots r_k$, and if $|zf(z)| \to 0$ as $z \to \infty$, show that

$$\sum_{-\infty}^\infty f(n) = -\sum_1^k (\pi \cot \pi b_\nu) r_\nu$$

(We assume that no b_ν is an integer.) (Integrate $\pi \cot \pi z f(z)$ around C_N and use the residue theorem.)

6.21   Under the hypotheses of Example 6.20, show that

$$\sum_{-\infty}^\infty (-1)^n f(n) = -\sum_1^k (\pi \csc \pi b_\nu) r_\nu$$

6.22   Find the explicit values of the sums

(a) $\sum_0^\infty \frac{1}{n^2 + a^2}$ (c) $\sum_1^\infty \frac{n^2}{n^4 + a^4}$

(b) $\sum_0^\infty \frac{1}{n^4 + a^4}$ (d) $\sum_{-\infty}^\infty \frac{(-1)^n}{n^2 + a^2}$

(See Examples 6.20 and 6.21.)

6.23 The ideas contained in Examples 6.18 through 6.22 suggest the following result, a special case of a more general property of analytic functions that we mentioned in Section 3.9. Consider

$$\frac{1}{2\pi i} \int_{C_N} \frac{\pi \cot \pi z}{z(z-w)} \, dz$$

where $w \neq n$ is considered as a parameter. The value of the integral is

$$\frac{\pi \cot \pi w}{w} + \sum_{-N}^{N}{}' \frac{1}{n(n-w)} - \frac{1}{w^2}$$

and tends to zero as $N \to \infty$ (see Example 6.28). Thus

$$\pi \cot \pi w = \frac{1}{w} + \sum_{-\infty}^{\infty}{}' \left(\frac{1}{n} + \frac{1}{w-n} \right)$$

$$= \frac{1}{w} + \sum_{1}^{\infty} \frac{2w}{w^2 - n^2}$$

The resemblance to the partial fraction expansion of a rational function is clear: the function $\cot \pi w$ is a (uniform) limit of rational functions (the limit of the partial sums).

Let us now integrate

$$\pi \cot \pi w - \frac{1}{w} = \sum_{1}^{\infty} \frac{2w}{w^2 - n^2}$$

from 0 to z, obtaining

$$\log \frac{\sin \pi z}{z} = \sum_{1}^{\infty} \log \frac{(z^2 - n^2)}{(-n^2)} + \log \pi$$

or

$$\frac{\sin \pi z}{\pi z} = \prod_{1}^{\infty} \left(1 - \frac{z^2}{n^2} \right)$$

We observe the entire function $\sin \pi z / \pi z$ in a new light also: the function $\sin \pi z / \pi z$ is a (uniform) limit of polynomials. All of its zeros are displayed in "factored" form. In fact, Euler first guessed this formula by analogy with polynomials.

Show, by similar methods, that

$$\pi^2 \csc^2 \pi z = \sum_{-\infty}^{\infty} \frac{1}{(z+n)^2}$$

6.24 A more general form of the result of Example 6.23 is the following. If $f(z)$ is an analytic function whose only finite singularities are poles at $z = a_1, a_2, \ldots, 0 < |a_1| < |a_2| < \ldots$, with residues $b_1, b_2, \ldots$, and $|f(z)|$

is bounded on a sequence of contours C_N, where C_N contains $a_1, a_2, \ldots a_N$ and no other poles, the length of C_N being of the order R_N, the minimum distance from 0 to C_N ($R_N \to \infty$ as $N \to \infty$), then

$$\frac{1}{2\pi i} \int_{C_N} \frac{f(t)\, dt}{t(t-z)} = \sum_1^N \frac{b_k}{a_k(a_k - z)} + \frac{f(z)}{z} - \frac{f(0)}{z}$$

and passing to the limit,

$$f(z) = f(0) + \sum_{k=1}^{\infty} b_k\left(\frac{1}{z - a_k} + \frac{1}{a_k}\right)$$

This gives us a "partial fractions" expansion of suitably bounded meromorphic (only poles in the finite plane) functions.

Prove the above theorem, and illustrate with $f(z) = \pi \csc \pi z$.

6.25   Find the principal part of the Laurent expansion of $f'(z)/f(z)$ about

(a) A zero of order k of $f(z)$.
(b) A pole of order k of $f(z)$.

6.26   If $f(z)$ is entire, with simple zeros at $a_1, a_2, a_3, \ldots$, and $f'(z)/f(z)$ satisfies the conditions of Example 6.24, show that

$$f(z) = f(0) e^{z[f'(0)/f(0)]} \prod_1^{\infty} \left(1 - \frac{z}{a_k}\right) e^{z/a_k}$$

In particular, show that

$$\frac{e^z - 1}{z} = e^{z/2} \prod_1^{\infty} \left(1 + \frac{z^2}{4n^2 \pi^2}\right)$$

6.27   If $f(z)$ is analytic inside and on a simple closed curve C, except for a finite number of poles inside C, and $f(z) \neq 0$ on C, show that

$$\frac{1}{2\pi i} \int_C \frac{f'(z)}{f(z)} dz = N - P$$

where N is the number of zeros of $f(z)$ (counting multiplicities) inside C, and P is the corresponding number of poles inside C; see Example 6.25.

6.28   For the function $f(z)$ of Example 6.27, let $a_1, a_2, \ldots$ be the zeros, with multiplicities $n_1, n_2, \ldots$ and $b_1, b_2, \ldots$ the poles of orders $m_1, m_2, \ldots$.

Let $g(z)$ be analytic inside and on C. Show that

$$\frac{1}{2\pi i}\int_C g(z)\frac{f'(z)}{f(z)}\,dz = \sum n_k g(a_k) - \sum m_k g(b_k)$$

See Example 6.27.

6.29   In the notations of Example 6.27, show that

$$\frac{1}{2\pi i}\int_C \frac{f'(z)}{f(z)}\,dz = \frac{1}{2\pi}\Delta_C \arg f(z)$$

where $\Delta_C \arg f(z)$ is the change in $\arg f(z)$ when C is traversed in the positive sense.

Solution: $\dfrac{1}{2\pi i}\displaystyle\int_C \frac{f'(z)}{f(z)}\,dz = \frac{1}{2\pi i}\log f(z)\Big|_C$

$$= \frac{1}{2\pi i}\left[\log |f(z)| + i \arg f(z)\right]\Big|_C$$

$$= \frac{1}{2\pi}\Delta_C \arg f(z)$$

This result is known as the *argument principle.*

6.30   If $f(z)$, $g(z)$ are analytic inside and on a simple closed curve C, and $|g(z)| < |f(z)|$ on C, with $f(z) \neq 0$ on C, then prove that $f(z)$ and $f(z) + g(z)$ have the same number of zeros inside C (Rouché's theorem). (Use the argument principle and Example 6.27.)

6.31   Use Rouché's theorem (Example 6.30) to prove the fundamental theorem of algebra. (Let C be a large circle $|z| = R$, $f(z) = a_n z^n$, $f + g = a_0 + a_1 z + \cdots a_n z^n$.)

6.32   If $a > e$, show that $e^z = az^n$ has n zeros in $|z| < 1$. See Example 6.30.

6.33   If $f(z)$ is analytic inside and on a simple closed curve C, and $f(z)$ takes on no value more than once on C, show that $f(z)$ takes on no value more than once inside C, that is, $f(z)$ is univalent in C. (Use the argument principle Example 6.29.)

6.34 C is a simple closed curve given by $|f(z)| = M$, and $f(z)$ is analytic inside and on C. Show that the number of zeros of $f(z)$ inside C exceeds the number of zeros of $f'(z)$ by 1 (Macdonald). In particular, then, $f(z)$ has at least one zero inside C.

6.35 Consider the polynomial $z^3 + z^2 + 6i - 8$. By considering a large quarter-circle in each quadrant, and using the argument principle, show that there is a zero in each of the second, third, and fourth quadrants.

6.36 The coefficients of $Az^4 + Bz^3 + Cz^2 + Dz + E$ are all real. Show that a necessary and sufficient condition that all the roots (zeros) have negative real parts is

1. The coefficients are of one sign (for example, $+$).
2. $BCD - AD^2 - EB^2 > 0$.

These are Routh's conditions.

[(1) expresses the fact that a real factor is $z + a$, $a > 0$ and a complex factor is $(z + b)^2 + c^2$, $b > 0$. In (2), consider $f(z) = z^4 + 4bz^3 + 6cz^2 + 4dz + e$, with $b, c, d, e > 0$, and apply the argument principle to a large semicircle in the right half-plane. To have $\Delta_C \arg f(z) = 0$, it is necessary that $\arg f(iy)$ must increase by 4π as y increases from $-\infty$ to ∞. This gives $6bcd - b^2e - d^2 > 0$.]

6.37 INVERSION OF SERIES

The next three exercises lead up to Lagrange's formula for inverting power series, that is, in essence solving an equation $w = f(z)$ explicitly for $z = g(w)$.

If $f(z)$ is analytic in $|z| < R$, $f(0) \neq 0$, show that $\psi(z) = z - wf(z)$ has precisely one zero inside $|z| = r < R$, provided $|w| \leq \rho$. [Use Rouché's theorem, Example 6.30, with $f(z)$ replaced by z, $g(z)$ by $-wf(z)$; $|g(z)| = |w| |f(z)| < |z|$ if $|w| < \rho$, where $\rho = r/\max_{|z|=r}|f(z)|$.]

6.38 Show that

$$\frac{1}{2\pi i} \int_{|z|=r} z \frac{\dfrac{d\psi}{dz}}{\psi} \, dz = \frac{1}{2\pi i} \int_{|z|=r} z \frac{1 - wf'(z)}{z - wf(z)} \, dz = a$$

where $z = a$ is the zero of $z - wf(z)$ in $|z| < r$. Hence, if $F(z)$ is analytic in $|z| \leq r$,

$$F(a) = \frac{1}{2\pi i} \int_{|z|=r} F(z) \frac{1 - wf'(z)}{z - wf(z)} \, dz$$

and $F(a)$ is analytic in w for $|w| \leq \rho$.

Solution: Examples 6.37 and 6.28.

6.39   From Examples 6.37 and 6.38, we see that $a = wf(a)$ defines $a(w)$ as an analytic function of w, with the restriction that $a(0) = 0$. Let us now replace a by z; then we have

$$F(z) = \frac{1}{2\pi i}\int_{|t|=r} F(t)\frac{1 - wf'(t)}{t - wf(t)}\,dt$$

$$= \frac{1}{2\pi i}\int_{|t|=r} F(t)[1 - wf'(t)]$$

$$\times\left\{\frac{1}{t} + \frac{wf(t)}{t^2} + \frac{[wf(t)]^2}{t^3} + \cdots\right\}dt, \quad\text{and}\quad \left|\frac{wf(t)}{t}\right| < 1$$

$$= \frac{1}{2\pi i}\int_{|t|=r} \frac{F(t)}{t}\,dt + \sum_1^\infty w^n\frac{1}{2\pi i}\int_{|t|=r} F(t)\left[\frac{f^n(t)}{t^{n+1}} - f'(t)\frac{f^{n-1}(z)}{z^n}\right]dz$$

$$= F(0) - \sum_1^\infty w^n\frac{1}{2\pi i}\int_{|t|=r}\frac{d}{dt}\left\{F(t)\left[\frac{f(t)}{t}\right]^n\right\}dt$$

$$F(z) = F(0) + \sum_1^\infty \frac{w^n}{n!}\left\{\frac{d^{n-1}}{dt^{n-1}}F'(t)[f(t)]^n\right\}_{t=0}$$

This is *Lagrange's formula.*
In particular, if $F(z) = z$, we have the *inversion formula*: $w = z/f(z)$,

$$z = \sum_1^\infty \frac{w^n}{n!}\left\{\frac{d^{n-1}}{dt^{n-1}}[f(t)]^n\right\}_{t=0}$$

Show that if

$$w = z - \frac{z^2}{1!} + \frac{2^3}{2!} - \frac{2^4}{3!} + \cdots = ze^{-z}$$

then

$$z = w + \frac{2w^2}{2!} + \frac{3^2w^3}{3!} + \frac{4^3w^4}{4!} + \cdots$$

$$e^z = 1 + w + \frac{3w^2}{2!} + \frac{4^2w^3}{3!} + \cdots$$

6.40   Let z be the root of $z^3 - z^2 - w = 0$ which $\to 1$ as $w \to 0$. Show that

$$z = 1 + \sum_1^\infty (-1)^{n-1}\frac{(3n - 2)!}{n!\,(2n - 1)!}\,w^n$$

for $|w| < \dfrac{4}{27}$ (see Example 6.39)

6.41   Let $w = P(z)$ be a polynomial. Show that the zeros of $P'(z)$ lie in the smallest convex polygon containing the zeros of $P(z)$ (Gauss). (See Example 2.14.18.)

6.42   (a) $f(z)$ is analytic, and with no zeros, in $|z| < R$. Show that

$$\log f(0) = \frac{1}{2\pi i} \int_{|z|=r} \frac{\log f(z)}{z} \, dz \qquad |z| < R$$

(b) Suppose now, that $f(z)$ has a simple zero at $z = a = r_1 e^{i\alpha}$. Apply the result of (a) to $f(z)/(1 - z/a)$, and show that

$$\log |f(0)| = \frac{1}{2\pi} \int_0^{2\pi} \log |f(re^{i\theta})| \, d\theta - \frac{1}{2\pi} \int_0^{2\pi} \log \left| 1 - \frac{re^{i\theta}}{r_1} \right| d\theta$$

and transform the latter integral into

$$\frac{1}{2\pi} \int_0^{2\pi} \log \left| \frac{r}{r_1} - e^{i\theta} \right| d\theta = \log \frac{r}{r_1} - \frac{1}{2\pi i} \mathbf{R} \int_{|z|=1} \log \left(1 - \frac{r_1}{r} z \right) \frac{dz}{z}$$

$$= \log \frac{r}{r_1}$$

(c) Hence deduce Jensen's theorem: $f(z)$ is analytic for $|z| < R$, $f(0) \neq 0$, and $r_1, r_2, \ldots r_n, \ldots$ are the absolute values of the zeros of $f(z)$ in the circle $|z| < R$, arranged in nondecreasing order. Then, for $r_n \leq r \leq r_{n+1}$,

$$\log \frac{r^n |f(0)|}{r_1 r_2 \cdots r_n} = \frac{1}{2\pi} \int_0^{2\pi} \log |f(re^{i\theta})| \, d\theta$$

6.43   Extend the results of Example 6.42 to the case for which $f(z)$ has zeros at $a_1, \ldots a_m$, and poles at $b_1, \ldots b_n$, not exceeding r in modulus, so that

$$\log \left[\left| \frac{b_1 b_2 \cdots b_n}{a_1 a_2 \cdots a_m} f(0) \right| r^{m-n} \right] = \frac{1}{2\pi} \int_0^{2\pi} \log |f(re^{i\theta})| \, d\theta$$

6.44   Combine the results of Examples 6.1, 6.42, and 6.43 and prove the Poisson-Jensen formula: let $f(z)$ have zeros at $a_1, a_2, \ldots a_m$, and poles at $b_1, b_2, \ldots b_n$ inside the circle $|z| \leq R$, and be analytic at all other points inside and on the circle. Then

$$\log |f(re^{i\theta})| = \frac{1}{2\pi} \int_0^{2\pi} \frac{R^2 - r^2}{R^2 - 2Rr \cos (\theta - \phi) + r^2} \log |f(Re^{i\phi})| \, d\phi$$

$$- \sum_{\mu=1}^m \log \left| \frac{R^2 - \bar{a}_\mu re^{i\theta}}{R(re^{i\theta} - a_\mu)} \right| + \sum_{v=1}^n \log \left| \frac{R^2 - \bar{b}_v re^{i\theta}}{R(re^{i\theta} - b_v)} \right|$$

Additional Examples and Comments on Chapter Six   **229**

Note that this formula contains Poisson's formula and Jensen's theorem as special cases.

6.45   If $|f(z)| \leq M$ on a simple closed curve $C, f(z)$ is analytic inside and on C, and if $|f(z_0)| = M$ for some (any) z_0 inside C, show that $f(z) \equiv$ constant. That is $|f(z)| \equiv M$.

SOLUTIONS FOR CHAPTER SIX

6.2.1(b) $\dfrac{\pi e^{-a}}{a}$ (c) $\dfrac{\pi}{ab(a+b)}$

6.5.3 $z - \dfrac{z^3}{3} + \dfrac{z^5}{10} + \cdots$

6.7.9 $f(z)/z$ is analytic in $|z| < 1$, continuous in $|z| \leq 1$, and $|f(z)/z| \leq 1$ on $|z| = 1$. Hence $|f(z)/z| \leq 1$ for $|z| < 1$ as well (maximum principle).

| CONFORMAL MAPPING AND
ANALYTIC CONTINUATION

The main object of this chapter is to derive the Schwarz–Christoffel formula for the conformal mapping of polygonal regions onto the upper half-plane.

7.1 ANALYTIC CONTINUATION

As we observed in Section 6.5, any function of the complex variable z, analytic at $z = a$ (and so, in some region containing $z = a$), can be expanded in a convergent power series about $z = a$,

$$(1) \qquad f(z) = \sum_{n=0}^{\infty} \frac{f^{(n)}(a)}{n!} (z - a)^n \qquad |z - a| < R$$

and every such convergent power series represents an analytic function near $z = a$. This fact is very important: analytic functions and (convergent) power series are synonymous. If we wish to study any arbitrary analytic function near $z = a$, we can do so by studying its power series (1).

For example,

$$(2) \qquad f(z) = \frac{1}{1 - z} = \sum_{0}^{\infty} z^n \qquad |z| < 1$$

furnishes the complete description of $f(z) = 1/(1 - z)$ at the points $|z| < 1$. It is not so evident from this power series, however, that $f(z)$ has a pole of order 1 at $z = 1$, nor that this $f(z)$ is indeed analytic at points outside the unit circle. Of course, there is no need to use the power series in this example, since $f(z) = 1/(1 - z)$ tells the full story for all z, while the power series will only converge up to the singular point nearest to $z = 0$.

On the other hand, suppose that the only information we have about an analytic function is its representation as a power series

$$(3) \qquad f(z) = \sum_{n=0}^{\infty} c_n (z - a)^n \qquad |z - a| < R$$

How can we compute the values of $f(z)$ outside this circle when $f(z)$ is actually analytic in a larger region?

Let us return to the example

$$f(z) = \sum_{n=0}^{\infty} z^n \qquad |z| < 1$$

From this definition, we can compute the value of $f(z)$ at any point z_0, $|z_0| < 1$. In fact, we can further compute the value of $f'(z)$ at z_0, by means of the series

$$f'(z) = \sum_{0}^{\infty} (n + 1)z^n \qquad |z| < 1$$

and so, for $f^{(n)}(z)$ at z_0, for each n. That is, we can compute the Taylor's series expansion for $f(z)$ about a new center, $z = z_0$, $|z_0| < 1$. For example, let us choose $z_0 = -1/2$.

$$f(-\tfrac{1}{2}) = \sum_{0}^{\infty} (-\tfrac{1}{2})^n = \frac{1}{1 + \tfrac{1}{2}} = \frac{2}{3}$$

$$f'(-\tfrac{1}{2}) = \sum_{0}^{\infty} (n + 1)(-\tfrac{1}{2})^n = \frac{1}{(1 + \tfrac{1}{2})^2} = \frac{4}{9}$$

(4) $$f^{(n)}(-\tfrac{1}{2}) = n!\,(\tfrac{2}{3})^{n+1}$$

(The fact that we used $f(z) = 1/(1 - z)$ to evaluate these sums does not actually mean that we must know the answer in advance in order to sum the series: the series in question converge, and we must regard their sums as being computable, even though a limiting operation is involved.)

We now obtain

(5) $$f(z) = \sum_{n=0}^{\infty} (z + \tfrac{1}{2})^n (\tfrac{2}{3})^{n+1} \qquad |z + \tfrac{1}{2}| < \tfrac{3}{2}$$

It is clear that both the series (2) and (5) represent the same function, $f(z) = 1/(1 - z)$. In fact, by the sum of a geometric series,

$$\sum_{0}^{\infty} (z + \tfrac{1}{2})^n (\tfrac{2}{3})^{n+1} = \frac{1}{1 - \tfrac{2}{3}(z + \tfrac{1}{2})} \cdot \frac{2}{3} = \frac{2}{3 - 2z - 1} = \frac{1}{1 - z}$$

However, the series (5) converges at points at which (2) does not converge. We say that the function

$$f(z) = \sum_{0}^{\infty} z^n \qquad |z| < 1,$$

has been *continued analytically* to points outside the original circle of convergence (Figure 7.1).

We may now repeat the whole process, and select a new point z_1, computing

Figure 7.1

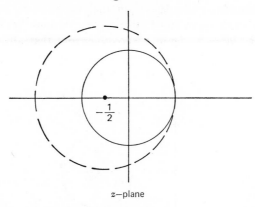

z-plane

$f^{(n)}(z_1)$, and obtain yet another power series. Let us visualize this process being carried on in all possible ways (Figure 7.2).

Clearly we can, by this process, reach every point z contained in any given circle passing through $z = 1$, and since there must be a singularity on each circle of convergence (why?), we find that $f(z)$, *defined only* by the series $\sum_0^\infty z^n$, for $|z| < 1$, can be *continued analytically* to the full complex plane, except for $z = 1$.

It is not clear at this moment that $f(z)$, so obtained, is single valued near $z = 1$. However, the procedure of continuation by power series, illustrated here, provides us with the answer to this problem as well. Consider a path that encircles $z = 1$, and avoids $z = 1$, as illustrated in Figure 7.3.

We compute the power series about z_1 on this path, continue to z_2 on the path, and inside the circle at z_1, and so on, until we return to a neighborhood of z_1 by a circle centered at z_n on the path. That is, we *continue $f(z)$ around the path*, and return to the starting point. Our function $f(z)$ is single valued if, and

Figure 7.2

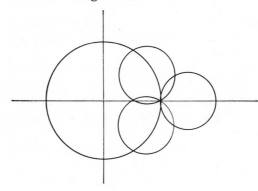

only if, the continued values of $f(z)$ are identical with the given values of $f(z)$.

It is convenient to consider that each analytic function, however defined (by formula, by series, or by some other representation), has been continued analytically to as large a region as possible, and that all of the possible

Figure 7.3

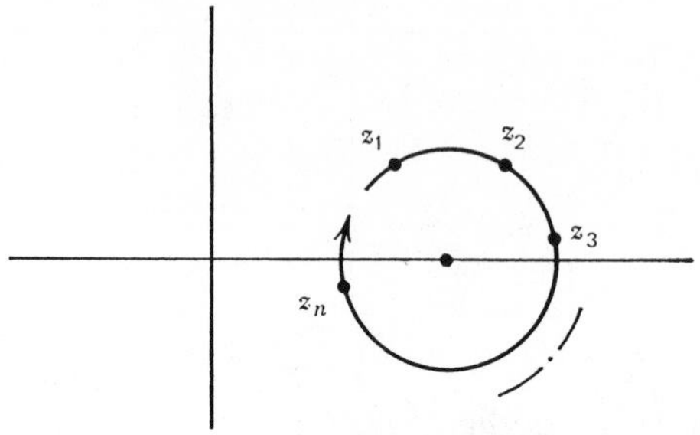

continuations are attached to the definition of $f(z)$. Thus, for example, the function

$$f(z) = \sum_{n=0}^{\infty} (-1)^n \frac{z^{n+1}}{n+1} \qquad |z| < 1$$

represents, for $|z| < 1$, the single-valued analytic function $\log(1+z)$ having the value 0 at $z = 0$, that is, the principal value branch of the logarithm. However, there are infinitely many distinct continuations (around $z = -1$), each differing from the given one by an integral multiple of $2\pi i$. We consider this totality of functional values as representing the actual analytic function $f(z)$.

With these ideas in mind, we have the following useful theorem.

Theorem 7.1   If $f(z)$, $g(z)$ are two analytic functions, and if $f(z) = g(z)$ in a region D where both are analytic, then $f(z) \equiv g(z)$ wherever they are defined. (This theorem includes the uniqueness of power series as a special case.)

COROLLARY.   If $z_1, z_2, \ldots, z_n, \ldots$ is an infinite sequence of points, $f(z_n) = g(z_n)$ for each n, and $f(z)$, $g(z)$ are analytic at $z = \lim_{n\to\infty} z_n$, then $f(z) \equiv g(z)$ everywhere.

COMMENT.   It may well happen that a function defined by a power series cannot be continued across the circle of convergence of the series. For example, consider

$$(6) \qquad f(z) = z + z^2 + z^4 + \cdots = \sum_{0}^{\infty} z^{2^n} \qquad |z| < 1$$

Since $f(z) \to \infty$ as $z \to 1$, the point $z = 1$ is a singularity of $f(z)$. Next, $f(z) = z + f(z^2)$, so that $f(z)$ is singular at $z^2 = 1$, or at $z = -1$ as well as at $z = 1$. In the same manner, $f(z) = z + z^2 + f(z^4)$, and the points $z^4 = 1$ are likewise singular points of $f(z)$. Thus, by similar reasoning, for each n, the points for which $z^{2^n} = 1$ are singular points of $f(z)$; that is, the 2^nth roots of unity. Since each point of the unit circle is either one of these singular points, or is a limit point of such points, it is clear that no circle centered inside $|z| = 1$ can be drawn to include points outside $|z| = 1$ without enclosing singular points of $f(z)$. Thus, this $f(z)$ cannot be continued beyond $|z| = 1$. In this case, the circle of convergence constitutes a barrier beyond which $f(z)$ cannot be defined analytically, and is called a *natural boundary*.

7.1.1   The real exponential function $e^x = \sum_0^\infty x^n/n!$ satisfies the identity $e^{x+a} = e^x \cdot e^a$. Using analytic continuation, show that for the complex exponential $e^z = \sum_0^\infty z^n/n!$

(1) e^z is *the* analytic continuation of e^x to the full z-plane,
(2) $e^{z+a} = e^z \cdot e^a$ for real, fixed a.

7.1.2
$$\arctan x = \int_0^x \frac{dt}{1 + t^2}$$

for real x. Show that *the* analytic continuation of $\arctan x$ to complex values of z is

$$\arctan z = \int_0^z \frac{dt}{1 + t^2} = \frac{1}{2i} \int_0^z \left(\frac{1}{t - i} - \frac{1}{t + i} \right) dt$$

$$= \frac{1}{2i} \log \frac{i - z}{i + z}$$

7.1.3   Show that the two Laurent expansions

$$-\left(\frac{1}{z} + 1 + z + z^2 + \cdots \right) \qquad \frac{1}{z^2} + \frac{1}{z^3} + \frac{1}{z^4} + \cdots$$

$$0 < |z| < 1 \qquad\qquad 1 < |z|$$

are analytic continuations of each other.

7.2   THE GAMMA FUNCTION

In this section we introduce one of the most important of the special functions of mathematical physics. It was introduced by L. Euler, who wanted a continuous function that assumed the value $n!$ at the integer points. Euler's gamma function is

(1)
$$\Gamma(x) = \int_0^\infty t^{x-1} e^{-t} \, dt$$

that exists, as an improper integral, for $x > 0$. Euler also gave the result, obtained by integrating (1) by parts,

$$\Gamma(x+1) = \int_0^\infty t^x e^{-t} \, dt$$

$$= -t^x e^{-t} \Big|_0^\infty + \int_0^\infty x t^{x-1} e^{-t} \, dt$$

(2)
$$\Gamma(x+1) = x\Gamma(x)$$

Since $\Gamma(1) = \int_0^\infty e^{-t} \, dt = 1$, we have immediately,

$$\Gamma(2) = 1\Gamma(1) = 1$$

$$\Gamma(3) = 2\Gamma(2) = 2!$$

.

.

.

(3)
$$\Gamma(n+1) = n\Gamma(n) = n!$$

Thus it is actually $\Gamma(x+1)$ which takes on the value $n!$ for $x = n$.

For our purposes, we consider the complex integral

(4)
$$\Gamma(z) = \int_0^\infty t^{z-1} e^{-t} \, dt$$

Then $|\Gamma(z)| \leq \int_0^\infty |t^{z-1}| \, e^{-t} \, dt = \int_0^\infty t^{x-1} e^{-t} \, dt$ exists for $x > 0$, or $\mathbf{R}z > 0$. Integration by parts is still valid, so that (2) becomes

(5)
$$\Gamma(z+1) = z\Gamma(z) \qquad \mathbf{R}z > 0$$

The question now is whether or not $\Gamma(z)$ is analytic. We note that t^{z-1} is analytic for each t, $0 < t < \infty$, $d/dz(t^{z-1}) = t^{z-1} \log t$. If $\Gamma(z)$ were analytic, and *if* we could compute $\Gamma'(z)$ by differentiating under the integral sign, we would have

$$\Gamma'(z) = \int_0^\infty \log t \, t^{z-1} e^{-t} \, dt$$

These are large "ifs." However, they do suggest to us precisely what we must establish. Since

$$\left| \int_0^\infty \log t \; t^{z-1} e^{-t} \, dt \right| \leq \int_0^\infty |\log t| \; t^{x-1} e^{-t} \, dt$$

and since the integral on the right converges for $0 < x$, we find that we can reverse the above steps by integrating $\int_0^\infty t^{z-1} \log t \, e^{-t} \, dt$ to get $\Gamma'(z)$; $\Gamma(z)$ has a continuous derivative in $\mathbf{R}z > 0$. Hence $\Gamma(z)$ is a single-valued analytic function in $\mathbf{R}z > 0$.

Now we return to (5), $\Gamma(z + 1) = z\Gamma(z)$, the "functional equation" satisfied by the gamma function. Rewriting this equation, we have

$$(6) \qquad \qquad \Gamma(z) = \frac{\Gamma(z + 1)}{z} \qquad \mathbf{R}z > 0$$

The integral definition of $\Gamma(z)$, (4), converges only for $\mathbf{R}z > 0$ (in much the same manner that power series converge only inside circles); we see however, from (6), that $\Gamma(z)$ can be expressed as $\Gamma(z + 1)/z$, and this latter quantity exists and is analytic for $\mathbf{R}z > -1$, $z \neq 0$.

Thus, by analytic continuation, since $\Gamma(z)$ and $\Gamma(z + 1)/z$ are both analytic for $\mathbf{R}z > 0$, and equal to each other there, then they are equal wherever both are defined, which is certainly the case for $\mathbf{R}z > -1$, $z \neq 0$.

We have thus continued $\Gamma(z)$ analytically to the larger region $\mathbf{R}z > -1$, avoiding $z = 0$. Since $\Gamma(1) = 1$, we see further, that $\Gamma(z)$ has a simple pole at $z = 0$, with residue 1.

We may now repeat the whole procedure, writing (5) in the form

$$(7) \qquad \qquad \Gamma(z) = \frac{\Gamma(z + 1)}{z} = \frac{\Gamma(z + 2)}{z(z + 1)}$$

from which we can continue $\Gamma(z)$ to $\mathbf{R}z > -2$, $z \neq 0$, -1, while $z = -1$ is a pole of order 1, residue -1. Repeated application of these methods yields

$$(8) \qquad \qquad \Gamma(z) = \frac{\Gamma(z + n + 1)}{z(z + 1)(z + 2) \cdots (z + n)}$$

for each integer n. Thus $\Gamma(z)$ is analytic in each half plane $\mathbf{R}z > -(n + 1)$, with simple poles at $z = 0, -1, -2, \ldots, -n$. The residue at $z = -n$ is clearly

$$\lim_{z \to -n} (z + n)\Gamma(z) = \frac{1}{-n(1 - n)(2 - n) \cdots (-1)}$$

$$= \frac{(-1)^n}{n!}$$

In other words, $\Gamma(z)$ is meromorphic, with first-order poles at $z = 0, -1,$ $-2, \ldots$ and residue $(-1)^n/n!$ at $z = -n$.

In Example 6.16, we evaluated the particular integral

$$\int_{-\infty}^{\infty} e^{-u^2} \, du = \sqrt{\pi}$$

This result is related to the gamma function in the following way. We write $\int_{-\infty}^{\infty} e^{-u^2} \, du = 2 \int_0^{\infty} e^{-u^2} \, du$ and take $u^2 = t$, a new variable of integration, obtaining

$$\sqrt{\pi} = \int_0^{\infty} t^{-1/2} e^{-t} \, dt = \Gamma(\tfrac{1}{2})$$

(9) $$\Gamma(\tfrac{1}{2}) = \sqrt{\pi}$$

7.2.1   Evaluate $\Gamma(\tfrac{3}{2})$, $\Gamma(-\tfrac{1}{2})$.

7.2.2   Show that the binomial coefficient $\binom{\alpha}{n}$, [the coefficient of z^n in the expansion $(1 + z)^{\alpha}$], can be written

$$\binom{\alpha}{n} = \frac{\Gamma(\alpha + 1)}{\Gamma(n + 1)\Gamma(\alpha - n + 1)}, \qquad \alpha \neq -1, -2, \ldots.$$

7.2.3   Show that

$$\binom{-\tfrac{1}{2}}{n} = \frac{\sqrt{\pi}}{n! \, \Gamma(\tfrac{1}{2} - n)} = \frac{(-1)^n (2n)!}{2^{2n}(n!)^2}$$

(see Example 7.2.2).

7.2.4   Evaluate $\Gamma(\tfrac{1}{2}) = \int_0^{\infty} t^{-1/2} e^{-t} \, dt$ by the following device.

$$[\Gamma(\tfrac{1}{2})]^2 = \left(\int_0^{\infty} t^{-1/2} e^{-t} \, dt \right) \left(\int_0^{\infty} u^{-1/2} e^{-u} \, du \right)$$

Set $t = x^2$, $u = y^2$ obtaining

$$[\Gamma(\tfrac{1}{2})]^2 = 4 \int_0^{\infty} e^{-x^2} \, dx \int_0^{\infty} e^{-y^2} \, dy$$

$$= 4 \iint e^{-(x^2 + y^2)} \, dx \, dy$$

$$= 4 \iint e^{-r^2} r \, dr \, d\theta$$

7.2.5   Following the lines of Example 7.2.4, obtain

$$\Gamma(\alpha)\Gamma(\beta) = \left(\int_0^\infty u^{\alpha-1}e^{-u}\,du \right)\left(\int_0^\infty v^{\beta-1}e^{-v}\,dv \right)$$

$$= 4 \int_0^\infty x^{2\alpha-1}e^{-x^2}\,dx \int_0^\infty y^{2\beta-1}e^{-y^2}\,dy$$

$$= 4 \iint r^{2\alpha+2\beta-1}e^{-r^2}\cos^{2\alpha-1}\theta\,\sin^{2\beta-1}\theta\,dr\,d\theta$$

$$= 2\Gamma(\alpha+\beta) \int_0^{\pi/2} \cos^{2\alpha-1}\theta\,\sin^{2\beta-1}\theta\,d\theta$$

$$= \Gamma(\alpha+\beta) \int_0^1 t^{\alpha-1}(1-t)^{\beta-1}\,dt$$

COMMENT.   These last integrals are of frequent occurrence in applications, and are well worth noting:

(10)*
$$\int_0^1 t^{\alpha-1}(1-t)^{\beta-1}\,dt = \frac{\Gamma(\alpha)\Gamma(\beta)}{\Gamma(\alpha+\beta)}$$

(11)
$$\int_0^{\pi/2} \cos^{2\alpha-1}\theta\,\sin^{2\beta-1}\theta\,d\theta = \frac{\Gamma(\alpha)\Gamma(\beta)}{2\Gamma(\alpha+\beta)}$$

In particular, with $\beta = 1 - \alpha$ in (11) we find

$$\Gamma(\alpha)\Gamma(1-\alpha) = 2 \int_0^{\pi/2} (\cot\theta)^{2\alpha-1}\,d\theta$$

$$= \int_0^\infty \frac{y^{\alpha-1}}{1+y}\,dy$$

where $y = \cot^2\theta$

$$= \frac{\pi}{\sin\pi\alpha}$$

from Example 6.8.

(12)
$$\Gamma(\alpha)\Gamma(1-\alpha) = \frac{\pi}{\sin\pi\alpha}$$

7.2.6   From $\Gamma(z)\,\Gamma(1-z) = \pi/\sin\pi z$, established above for restricted z, $\mathbf{R}z > 0$, $\mathbf{R}(1-z) > 0$, show by analytic continuation that this identity is valid for all complex $z \neq 0, \pm 1, \pm 2, \ldots$. Let $z \to -n$ and equate residues on each side, and verify that the residue of $\Gamma(z)$ at $z = -n$ is $(-1)^n/n!$.

---

* This integral is called the beta function.

7.2.7 Use (12) to transform

$$\binom{-\alpha}{n} = \frac{\Gamma(1-\alpha)}{n!\,\Gamma(1-\alpha-n)} = \frac{(-1)^n\Gamma(\alpha+n)}{\Gamma(\alpha)n!}$$

(see Example 7.2.2).

7.3 SCHWARZ' REFLECTION PRINCIPLE

In our earlier treatment of the elementary analytic functions such as e^x, $\cos x$, ..., we started from a formula, valid for real x, and for which the extension to complex z was immediate. Thus the extension of $e^x = \sum_0^\infty x^n/n!$ to complex arguments, $e^z = \sum_0^\infty z^n/n!$ was natural, even if somewhat arbitrary. We see now, in the light of analytic continuation, that there can be only one way to extend e^x to an analytic function whose real values are e^x.

The real functions that we extended in this fashion possess certain simple properties due to the fact that they are real for real values of z. Their power series expansions, about a point on the real axis, have real coefficients. (All the derivatives of the function, real on the real axis, are real.) Now if

$$f(z) = \sum_{n=0}^\infty a_n(z-x_0)^n \qquad x_0,\, a_n \text{ real}$$

then

$$\overline{f(z)} = \sum_0^\infty a_n\overline{(z-x_0)^n} = \sum_0^\infty a_n(\bar z - x_0)^n$$
$$= f(\bar z)$$

That is, the value of $f(z)$ at the reflected point $\bar z$ is the reflected value of $f(z)$, the reflection being in the real axis. The converse of this property is also true, and we shall make use of this result in Section 7.5. In order to investigate the converse statement, suppose that $f(z)$ is known to be analytic in D, a domain in the upper half-plane, bounded below by a portion L of the real axis, and that $f(z)$ is real and continuous on the open interval consisting of L, endpoints excluded (that is, we say nothing about the behavior of $f(z)$ at the endpoints of L). See Figure 7.4. Now, if $f(z)$ were further known to be analytic in the larger region, consisting of D plus its reflection D', then it would follow that $\overline{f(z)} = f(\bar z)$ in this larger region. We consider the possibility of analytically continuing $f(z)$ into D'; if such continuation is possible, then, clearly, the property $\overline{f(z)} = f(\bar z)$ must hold for the continued function. In other words, we define the proposed continuation by the property

(1) $$f_1(z) = \overline{f(\bar z)}$$

for z in D'. (If z is in D', $\bar z$ is in D, where f is given, so that (1) defines a

Figure 7.4

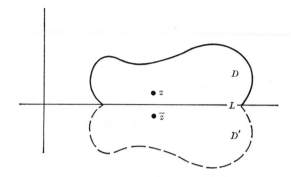

function in D'.) We wish to show that the extended function, $f(z)$, as given in D, and $f_1(z)$, defined in D' is indeed analytic in the combined regions $D + D'$ (Figure 7.5).

Now $(f(z + \Delta z) - f(z))/\Delta z$ tends to a limit as $\Delta z \to 0$ provided z, $z + \Delta z$ are points of D [$f(z)$ is analytic there]. From (1), we now have

$$(2) \qquad \frac{f_1(z + \Delta z) - f_1(z)}{\Delta z} = \frac{\overline{f(\overline{z + \Delta z}) - f(\overline{z})}}{\overline{\Delta z}}$$

where z, $z + \Delta z$ are in D', so that $\overline{z}$, $\overline{z + \Delta z}$ are in D; the right side (2) thus tends to a limit, and so $f_1(z)$ is indeed analytic in D'. We also have from (1)

$$(3) \qquad\qquad\qquad f_1(z) \doteq f(z)$$

for z real, for z on L.

We now have the following situation: $f(z)$ is analytic in D, $f_1(z)$ is analytic in D', and $f_1(z) = f(z)$ on L. Thus, *if* $f(z)$ can be continued, we will have our result; $f_1(z)$ will be *the* continuation to D'.

Figure 7.5

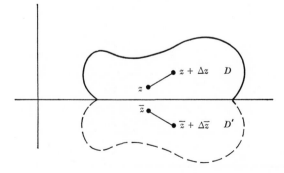

In order to demonstrate that $f(z)$ can, indeed, be continued, we construct an auxiliary function, analytic in $D + D'$, and equal to $f(z)$ for z in D, and to $f_1(z)$ for z in D'.

Define

$$\phi(z) = \begin{cases} f(z) & z \text{ in } D \\ f_1(z) & z \text{ in } D' \end{cases}$$

and consider

(4)
$$F(z) = \frac{1}{2\pi i} \int_C \frac{\phi(t)\, dt}{t - z}$$

where C is a simple closed curve in $D + D'$ (Figure 7.6). Since $\phi(z)$ is continuous inside and on C, then $F(z)$ is analytic for z inside C (see Section 6.1). For z in D, decomposing C into C_1 and C_2, we find

$$F(z) = \frac{1}{2\pi i} \int_{C_1} \frac{\phi(t)\, dt}{t - z} + \frac{1}{2\pi i} \int_{C_2} \frac{\phi(t)\, dt}{t - z}$$

$$= \frac{1}{2\pi i} \int_{C_1} \frac{f(t)\, dt}{t - z} + 0$$

$$= f(z)$$

while for z in D', $F(z) = f_1(z)$, by similar reasoning (Figure 7.7). That is, $F(z)$ is analytic inside C, $F(z) = \phi(z)$ there, and so $\phi(z)$ is analytic inside C, and $f(z)$, $f_1(z)$ are indeed analytic continuations of each other. We have verified *Schwarz' reflection principle: if $f(z)$ is analytic in a domain D, part of whose boundary consists of a portion L of the real axis, and if $f(z)$ is real and continuous on the open segment L, then $f(z)$ can be continued analytically into the reflected domain D', of D in L, where it takes on reflected (conjugate) values at reflected (conjugate) points.*

Figure 7.6

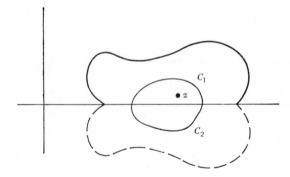

Figure 7.7

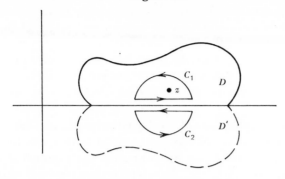

COMMENT. No assumptions are made as to the nature of $f(z)$ at the endpoints of L, where it can well have singularities.

7.3.1 Show that the Schwarz reflection principle is equally valid if $f(z)$ maps a straight line segment L' onto a straight line segment L, then $f(z)$ can be continued from a domain D bounded by L' to the domain D', the reflection of D in L', and $f(z)$ takes on the values which are the reflections of the values of $f(z)$ in L.

7.3.2 If $f(z)$ is analytic in the unit circle, and real and continuous for z on an arc of the unit circle, show that $\overline{f(z)} = f(1/\bar{z})$ holds for the continuation of $f(z)$ outside the unit circle. Observe that Schwarz' principle holds, in this case, if we replace the "reflected" point by the "inverse" point.

7.3.3 If $f(z)$ is analytic in $|z| < 1$, and $|f(z)| = 1$ on an arc of the unit circle, show that $f(z)$ can be continued beyond the unit circle by the rule $f(z)f(1/\bar{z}) = 1$ (consider $i \log f(z)$ and Example 7.3.2).

7.4 THE GENERAL MAPPING PROBLEM: RIEMANN'S MAPPING THEOREM

We have investigated a number of conformal mappings by analytic functions and have learned how to map certain geometrically simple domains onto a half plane, or the unit circle as special cases. The method we adopted was an example of an *inverse* method; that is, we started with a specific, elementary function, and we investigated the images of level lines and other simple lines. Thus we found that the bilinear mapping always maps circles and straight lines onto image circles and straight lines; the bilinear mapping preserves the family of circles with straight lines as limiting cases. These particular mappings have another useful and important property: the inverse of a bilinear mapping is also bilinear, so that if $w = (az + b)/(cz + d)$, there is a one-to-one correspondence between points of the z-plane, and points of the w-plane. Such is not the case for the other elementary mapping functions we have examined. Thus $w = z^2$ maps the unit circle $|z| < 1$ onto the unit circle $|w| < 1$, but each point of the w-plane is the image of two points of the z-plane: the mapping is not one to one: the inverse function is not single valued, $z = \pm\sqrt{w}$.

In most of the applications of conformal mapping we wish to make to problems of mathematical physics, we wish to have a one-to-one correspondence of points, and hence wish to restrict our attention to conformal maps that possess this property: if $w = f(z)$ maps a region D of the z-plane to a region D' of the w-plane, we require the inverse function $z = f^{-1}(w)$ to be single valued for w in D'. (For example $w = z^2$ will have this required property if we suitably restrict the domains D so that $0 \le \arg z < \pi$ holds for the points of D.) Conformal maps that are one to one (both directions) are called *univalent*, or *simple* maps, and our attention will be directed towards the construction of such univalent maps in the following sections.

As previously indicated, one of the applications of complex variable methods in engineering problems is the construction of univalent conformal mappings of a given domain D onto a particularly simple domain, such as the unit circle, or the upper half-plane. That is, we must now find an analytic function $w = f(z)$, which maps a given domain D onto the interior, for example, of the unit circle in the w-plane, so that each point of the w-plane has a unique image point in the z-plane. This direct problem is, of course, much more difficult than the indirect or inverse methods we have used up to now. Thus, it is of considerable importance to us to know in advance that a mapping function actually exists that will perform the desired result, and that indeed we can single out a unique such analytic mapping function. These statements are precisely stated in the *Riemann mapping theorem*: given an arbitrary simply connected domain D in the z-plane, whose boundary ∂D consists of at least two points (and so a whole arc), there exists a unique univalent mapping function $w = f(z)$ which maps D onto the interior of the unit circle, $|w| < 1$, so that (any) prescribed point of D maps onto the origin, $w = 0$, and (any) prescribed direction at this point maps into the direction of the positive real axis.

COMMENT 1. We can conveniently regard this theorem as stating that there is a three-parameter family of mapping functions that will map D onto the unit circle: two degrees of freedom coming from the coordinates of the point of D that we select to map onto $w = 0$, and the third degree of freedom coming from the direction we wish to correspond to the real positive w-axis.

Alternately, we might expect some other set of three restrictions, equivalent to the three described above, that will provide us with a unique mapping (see Section 7.6).

COMMENT 2. The proof of this important theorem involves ideas that would take us too far afield, and so we state the result and will use it as necessary in the following sections, without proof. The interested reader may find a proof of this result in *Complex Analysis* by L. Ahlfors.

7.4.1   Recall Poisson's formula

$$u(r, \theta) = \frac{1}{2\pi} \int_0^{2\pi} \frac{(1 - r^2)u(1, \sigma)\, d\sigma}{1 - 2r \cos(\theta - \sigma) + r^2}$$

for the solution of Laplace's equation in the unit circle, with prescribed values $u(1, \theta)$ on the boundary of this circle; map the unit circle onto the upper half-plane by a bilinear mapping, and use this mapping function to transform Poisson's formula to a form valid for the upper half-plane:

$$u(x, y) = \frac{y}{\pi} \int_{-\infty}^{\infty} \frac{u(\xi, 0)\, d\xi}{(x - \xi)^2 + y^2}$$

for the solution of Laplace's equation $\Delta u = 0$ in terms of the boundary values $u(x, 0)$.

7.4.2   Establish the uniqueness part of Riemann's mapping theorem. If $w = f(z)$ maps the unit circle univalently onto itself, with $f(0) = 0$, then $f(z) = cz$, $|c| = 1$. (See Schwarz' lemma.)

7.5 THE SCHWARZ-CHRISTOFFEL MAPPING

The Riemann mapping theorem guarantees the existence of a conformal mapping function that will map any required domain onto the unit circle (the unit circle, here, is a useful, simple reference domain). Since we know how to map the circle to a half plane and conversely, we may, if convenient, use the upper half-plane as a reference domain. In this section we investigate the univalent conformal mapping function that maps a domain bounded by straight lines, a polygon, onto the upper half-plane. The half plane is more immediately useful here because of the direct applicability of the Schwarz reflection principle.

We consider a simple polygon of n sides (non-self-intersecting), with vertices at $w_1, w_2, \ldots w_n$, in the w-plane, and its image region, the upper half z-plane, with (unknown) mapping function $w = f(z)$, or $z = F(w)$, in its inverse form (Figure 7.8).

Let $x_1, x_2, \ldots x_n$ be the image points corresponding to the vertices $w_1, \ldots$ w_n. (We consider $x_1, \ldots x_n$ to be finite points on the real axis.)

Now, since each side of the polygon, for example the side w_1, w_2, maps onto a segment of the real axis, in this case to x_1, x_2, we have $w = f(z)$ mapping onto a straight line segment for z real. If arg $(w_2 - w_1) = \beta$, then $(w - w_1)e^{-i\beta}$ is real and continuous for z real and on the segment x_1, x_2.

From Schwarz' reflection principle, we observe that $w = f(z)$ can be continued analytically to the full lower half-plane (the reflection of the upper

Figure 7.8

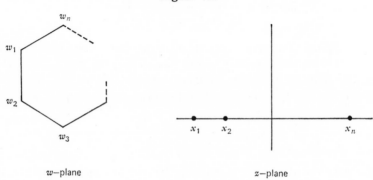

w–plane z–plane

half-plane in the segment x_1, x_2 of the real axis) the values of this continuation are also known in terms of the values of $f(z)$ in the upper half-plane: $(w - w_1)e^{-i\beta}$ takes on conjugate values at conjugate points.

If we repeat this construction for the side w_2, w_3, and the segment of the real axis x_2, x_3, we have $(w - w_2)e^{-i \, \text{arg} \, (w_3 - w_2)}$ takes on conjugate values at conjugate points (Figure 7.9).

Starting from a given point z_1 of the upper half-plane, and reflecting across x_1, x_2, and also across x_2, x_3, we find that the continued value of $w = f(z)$ will be (in general) different in each case, since it is not w that takes on conjugate values, but

$$(w - w_k)e^{-i \, \text{arg} \, (w_{k+1} - w_k)}$$

for $k = 1$ or $k = 2$ in this instance. We must thus anticipate that the points $x_1, \ldots x_n$ will in general be *branch points* of the mapping function.

In order to see more clearly how these branch points affect our mapping function, let us rearrange our point of view concerning the two different reflections across the real axis that we have just discussed. Let us now consider the continuation of $w = f(z)$ to the lower half-plane, along the path C_1 of

Figure 7.9, and return to the starting point z_1 along the path C_2. That is, we first reflect across x_1, x_2, and then reflect back along x_2, x_3. In the w-plane, these reflections correspond to reflecting the given polygon across side w_1, w_2, and then reflecting the new polygon across its side w_2, w_3 (or, more properly, across the side that corresponds to the original side w_2, w_3). The first reflection, across w_1, w_2, moves the whole polygon as a rigid body, but turns the plane of the figure upside down (Figure 7.10). The second reflection returns the plane of the polygon to its original position, although the polygon will now be at a new location; however, the new position of the polygon is very simply related

Figure 7.9   **Figure 7.10**

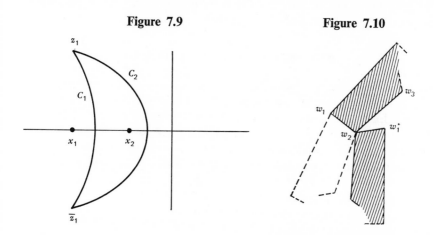

to the old position: it is merely a translation and a rotation of the figure we started from. Thus, if w is any point in the interior of the original polygon, and w^* its location after the two reflections, then

(1)
$$w^* = aw + b$$

where a, b are complex constants, $|a| = 1$. It is also clear that the constants a, b will depend on which of the two sides across which we perform the reflections. These constants (as yet unknown) again describe the multivalued nature of $w = f(z)$ that arises by continuing the function around one of the branch points.

Next, the particularly simple form of (1) provides us with a method of obtaining a *single-valued* function, related to $w = f(z)$.

(2)
$$\frac{dw^*}{dz} = a\frac{dw}{dz}$$

and

$$\frac{d^2w^*}{dz^2} \bigg/ \frac{dw^*}{dz} = \frac{d^2w}{dz^2} \bigg/ \frac{dw}{dz}$$

The Schwarz-Christoffel Mapping   **247**

Thus, even though $w = f(z)$ possesses branch points, the expression w''/w' is single valued in the complete z-plane.

Let us set $w''/w' = [f''(z)]/[f'(z)] = g(z)$. In the next paragraphs we locate the singular points of $g(z)$, which turn out to be simple poles, and then we are able to write down $g(z)$ explicitly.

(a) $g(z)$ is single-valued for all z.

(b) $g(z) = (f''(z))/(f'(z))$ is analytic for all points z for which $f''(z)$, and $f'(z)$ are analytic, except for such points that $f'(z) = 0$. Since $f(z)$ is analytic in the upper half-plane, so are $f'(z)$, $f''(z)$; furthermore, $f(z)$ can be continued analytically to the lower half-plane, so that we have

$$
\text{(3)} \qquad g(z) \text{ is analytic in } \text{I}z > 0, \qquad \text{I}z < 0
$$

except for the points at which $f'(z) = 0$.

(c) Since we require of our mapping that it be conformal in $\text{I}z > 0$, then $f'(z) \neq 0$ for $\text{I}z > 0$, and hence in $\text{I}z < 0$ as well, from the reflection principle. Thus the only possible location of the singular points of $g(z)$ is the real axis.

(d) $g(z)$ is analytic at all points of the real axis with the possible exception of the points $x_1, \ldots x_n$. In order to see this result clearly, let x_0 be an interior point of one of the finite segments (x_k, x_{k+1}), and w_0 its image point on the side of the polygon (w_k, w_{k+1}). Let $\beta_k = \arg(w_{k+1} - w_k)$, so that $(w - w_0)e^{-i\beta_k}$ is real for z real and in the interval (x_k, x_{k+1}).

$$
\text{(4)} \qquad (w - w_0)e^{-i\beta_k} = \sum_{n=p}^{\infty} a_n(z - x_0)^n
$$

for $|z - x_0|$ sufficiently small, and the coefficients a_p, $a_{p+1}, \ldots$ are real. Since the mapping function near x_0 maps a straight line (angle π) onto a straight line (angle π), then $p = 1$, and $a_1 \neq 0$. That is to say, $f'(z) \neq 0$ at $z = x_0$. (For a discussion of the special case $x_0 = \infty$, see Example 7.5.2.)

(e) Collecting our results so far: $g(z)$ is single valued and analytic at all points $z \neq x_1, x_2, \ldots x_n$.

Let α_k be the interior angle of the polygon at the vertex w_k, and $\arg(w_{k+1} - w_k) = \beta_k$ (Figure 7.11).

Then $(w_{k+1} - w)e^{-i\beta_k}$ has argument zero for z on the segment (x_k, x_{k+1}), and argument-α_{k+1} for z on the segment (x_{k+1}, x_{k+2}). Hence

$$
\text{(5)} \qquad [(w_{k+1} - w)e^{-i\beta_k}]^{\pi/\alpha_{k+1}}
$$

is real and continuous for z on the line segment (x_k, x_{k+2}), and analytic for

Figure 7.11

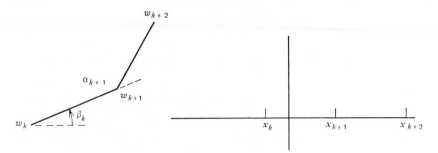

$\mathbf{I}z > 0$. An appeal to the reflection principle yields

(6)
$$[(w_{k+1} - w)e^{-i\beta_k}]^{\pi/\alpha_{k+1}} = \sum_{n=1}^{\infty} c_n(z - x_{k+1})^n$$

with (c_n) real, and $c_1 \neq 0$.
Solving for $w(z)$,

(7)
$$w(z) = w_{k+1} + e^{i\beta_k}(z - x_{k+1})^{\alpha_{k+1}/\pi}\left(\sum_{1}^{\infty} c_n(z - x_{k+1})^{n-1}\right)^{\alpha_{k+1}/\pi}$$

which is of the form

$$w(z) = w_{k+1} + e^{i\beta_k}(z - x_{k+1})^{\alpha_{k+1}/\pi}(\text{analytic and} \neq 0 \text{ at } z = x_{k+1})$$

Thus

$$w'(z) = (z - x_{k+1})^{\alpha_{k+1}/\pi - 1}(\text{analytic and} \neq 0 \text{ at } z = x_{k+1})$$

$$\frac{w''(z)}{w'(z)} = \frac{\dfrac{\alpha_{k+1}}{\pi} - 1}{z - x_{k+1}} + (\text{analytic at } x_{k+1})$$

The function $g(z)$ has a first-order pole at each vertex image point $x_1, \ldots x_n$, the residue at $z = x_k$ being $\alpha_k/\pi - 1$.
We may now write

(8)
$$g(z) = \sum_{k=1}^{n} \frac{\dfrac{\alpha_k}{\pi} - 1}{z - x_k} + E(z)$$

and $E(z)$ is single valued and analytic everywhere (including $z = \infty$, see Example 7.5.1); that is, $E(z)$ is entire.

It certainly seems plausible that $E(z) \equiv 0$. Indeed, we can prove that this is the case (see Example 7.5.3). Let us, however, pursue this simplest possibility,

The Schwarz-Christoffel Mapping   **249**

and consider the resulting formula for $w = f(z)$, assuming $E(z) \equiv 0$.

$$g(z) = \frac{w''(z)}{w'(z)} = \sum_{k=1}^{n} \left(\frac{\alpha_k}{\pi} - 1 \right) \frac{1}{z - x_k}$$

$$\log w'(z) = \sum_{k=1}^{n} \log (z - x_k)^{(\alpha_k/\pi)-1} + \log C$$

$$w'(z) = C \prod_{k=1}^{n} (z - x_k)^{(\alpha_k/\pi)-1}$$

$$(9) \qquad w(z) = C \int_1^z \prod^n (t - x_k)^{(\alpha_k/\pi)-1} \, dt + D$$

This formula is the Schwarz-Christoffel formula for the conformal map of a polygon onto the upper half-plane.

7.5.1　　Take $x_0 = \infty$, on the "segment" of the real axis (x_n, x_1), with w_0 the corresponding point on the side of the polygon (w_n, w_1). Let $\beta = \arg(w_1 - w_n)$ and show that $(w - w_0)e^{-i\beta} = \sum_1^\infty b_n/z^n$, (b_n) real and $b_1 \neq 0$, and hence that

$$\frac{w''(z)}{w'(z)} = -\frac{2}{z} + \cdots$$

so that $g(z)$ is analytic at *the* point at infinity.

7.5.2　　Let the vertex points $w_1, \ldots w_{n-1}$ map onto $x_1, \ldots x_{n-1}$, finite points of the real axis, and w_n map to $z = \infty$. By using the reflection principle, show that

$$\frac{w''(z)}{w'(z)} = g(z) \text{ is analytic at } \infty$$

so that

$$g(z) = \sum_{k=1}^{n-1} \left(\frac{\alpha_k}{\pi} - 1 \right) \frac{1}{z - x_k} + E_1(z)$$

with the same assumption as previously, take $E_1(z) \equiv 0$.

7.5.3　　Deduce that in Section 7.5 (8), $E(z) \equiv 0$;

(a) Show that $g(z)$, analytic at ∞, has a power series expansion $g(z) = -2/z + \cdots$ in powers of $1/z$,

(b) $\sum_1^n (\alpha_k/\pi - 1)/(z - x_k)$ is also analytic at ∞, and has the power series expansion $-2/z + \cdots$. Thus $E(z)$ is analytic at ∞, its power series starting with the $1/z^2$ term.

(c) $E(z)$ is thus bounded, tends to zero at ∞; now use Liouville's theorem.

7.5.4　　Show that $\sum_1^n \alpha_k = (n - 2)\pi$.

7.6 A DISCUSSION OF THE SCHWARZ-CHRISTOFFEL FORMULA

$$w(z) = C \int^z (t - x_1)^{(\alpha_1/\pi)-1}(t - x_2)^{(\alpha_2/\pi)-1} \cdots (t - x_n)^{(\alpha_n/\pi)-1} dt + D$$

(a) The constants $x_1, \ldots x_n$, C, and D must satisfy certain consistency relations for the mapping to send the given polygon onto the upper half-plane. We must also decide on appropriate branches of the multivalued functions $(t - x_k)^{(\alpha_k/\pi)-1}$. The simplest branch, of course, is the principal-value

Figure 7.12

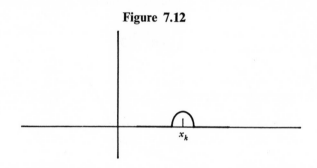

branch, obtained by taking $(t - x_k)^{(\alpha_k/\pi)-1}$ real and positive for t real and $t > x_k$. Then, with

$$t = x_k + re^{i\theta} \qquad (t - x_k)^{(\alpha_k/\pi)-1} = r^{(\alpha_k/\pi)-1}e^{i[(\alpha_k/\pi)-1]\theta},$$

we find the appropriate values for t real and $t < x_k$ for example, by setting $\theta = \pi$:

$$(t - x_k)^{(\alpha_k/\pi)-1} = r^{(\alpha_k/\pi)-1}e^{i(\alpha_k-\pi)}$$

$$= -r^{(\alpha_k/\pi)-1}e^{i\alpha_k}$$

$$= (x_k - t)^{(\alpha_k/\pi)-1}(-e^{i\alpha_k})$$

with the principal value branch again for $(x_k - t)^{(\alpha_k/\pi)-1}$ (Figure 7.12).

A convenient way of proceeding is thus as follows: first choose a particular vertex point, w_1, for example, and evaluate D by integrating from x_1.

$$w(z) = w_1 + C \int_{x_1}^z \prod_1^n (t - x_k)^{(\alpha_k/\pi)-1} dt$$

Next, choose principal value branches for each factor, $(t - x_k)^{(\alpha_k/\pi)-1}$ or $(x_k - t)^{(\alpha_k/\pi)-1}$ accordingly as $t > x_k$ or $t < x_k$ as we integrate along the real axis. Starting, as here, with the left most point x_1, we take $(t - x_1)^{(\alpha_1/\pi)-1}$ near x_1, $t > x_1$, and $(x_k - t)^{(\alpha_k/\pi)-1}$, $k = 2, \dots n$.

$$(1) \quad w(z) = w_1 + C \int_{x_1}^{z} (t - x_1)^{(\alpha_1/\pi)-1}(x_2 - t)^{(\alpha_2/\pi)-1} \cdots (x_n - t)^{(\alpha_n/\pi)-1} \, dt$$

All branches are principal-value branches; any branch differs from the principal value branch by a multiplicative constant; our choice normalizes C.

Now we follow the real axis as closely as possible (not integrating through a singular point, but indenting appropriately in the upper half-plane). As z runs from x_1 toward x_2, w runs along the line (w_1, w_2), with argument

Figure 7.13

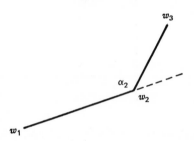

arg $(w_2 - w_1)$. Hence arg $C = $ arg $(w_2 - w_1)$. To reach the segment (x_2, x_3), we indent our path of integration at x_2, and decrease arg $(t - x_2)$ by π; we obtain

$$w(z) = w_1 + C \int_{x_1}^{x_2} (t - x_1)^{(\alpha_1/\pi)-1}(x_2 - t)^{(\alpha_2/\pi)-1} \cdots (x_n - t)^{(\alpha_n/\pi)-1} \, dt$$

$$+ \, e^{-i(\alpha_2 - \pi)} C \int_{x_2}^{z} (t - x_1)^{(\alpha_1/\pi)-1}(t - x_2)^{(\alpha_2/\pi)-1}$$

$$\times (x_3 - t)^{(\alpha_3/\pi)-1} \cdots (x_n - t)^{(\alpha_n/\pi)-1} \, dt$$

$$= w_2 + e^{-i(\alpha_2 - \pi)} C \int_{x_2}^{z} (t - x_1)^{(\alpha_1/\pi)-1}(t - x_2)^{(\alpha_2/\pi)-1}$$

$$\times (x_3 - t)^{(\alpha_3/\pi)-1} \cdots (x_n - t)^{(\alpha_n/\pi)-1} \, dt$$

Hence, as z passes from the segment (x_1, x_2) to the segment (x_2, x_3), arg $(w - w_1) = $ const on (x_1, x_2), arg $(w - w_2) = $ const on (x_2, x_3), and arg $(w - w_2)$ increases by $\pi - \alpha_2$; that is, the polygon angle at w_2 is α_2 as required (Figure 7.13). Thus, regardless of the (finite) values of $x_1, \dots x_n$ and C, the mapping function maps a polygon of interior angles α_k onto the upper half-plane.

Figure 7.14

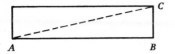

Suppose now that $n = 3$ and our polygon is a triangle. For arbitrary choice of x_1, x_2, x_3, Equation 1 maps the upper half z-plane onto a triangle with the prescribed angles, and such that x_1 maps onto w_1. Since the triangle is similar to our required triangle, we can choose C, involving a rotation (arg C) and a uniform stretching ($|C|$), to bring the triangle to the correct location and size. Thus, in the case of a triangle, we may choose x_1, x_2, x_3 arbitrarily, and perform our required mapping uniquely once we have chosen these points.

The situation is somewhat different for polygons with four, or more sides: the interior angles do not determine the shape.

Yet, if we consider the quadrilaterals in Figure 7.14, and draw in the corresponding diagonals, we see that the single extra requirement that the triangle ABC be similar to triangle $A'B'C'$ (as well as the equality of the interior angles of the quadrilaterals) guarantees similarity of the polygons.

Thus, considering a polygon with interior angles $\alpha_1, \ldots \alpha_n$, we consider all diagonals drawn from any fixed vertex P, obtaining $n - 2$ triangles (Figure 7.15). This polygon will be similar to our given polygon if, and only if, each of the triangles so formed is similar to the corresponding triangles formed from our given polygon. That is, the $n - 2$ diagonals at P, P' must form corresponding equal angles. Since the full angles at P, P' are already assumed equal, we see that $n - 3$ restrictions are necessary to guarantee similarity of the two polygons.

It now appears that we may arbitrarily (conveniently) choose any three of the points $x_1, \ldots x_n$. After this is done, the remainder of $x_1, \ldots x_n$ are determined by the requirement of similarity of the polygons. Finally, a rotation and a scale change determine C, and our mapping problem is solved.

Figure 7.15

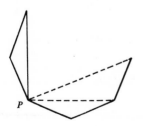

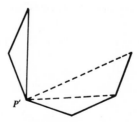

COROLLARY. From Example 7.5.2, we may decide to choose one of the points $x_1, \ldots x_n$ at infinity, for example, $x_n = \infty$. Then we have

(2)
$$w = w_1 + C \int_{x_1}^{z} (t - x_1)^{(\alpha_1/\pi)-1} \cdots (x_{n-1} - t)^{(\alpha_{n-1}/\pi)-1} \, dt$$

having one less factor in the integrand (so it is somewhat simpler). However, in this form, we have used one of our three degrees of freedom, and in using (2) we may only choose two of the remaining points $x_1, \ldots x_{n-1}$ arbitrarily.

7.6.1 Derive Section 7.6(2) from Section 7.6(1) by mapping the upper half z-plane onto an upper half-plane, so that $z = x_n$ maps to infinity $[\tau = 1/(z - x_n)]$.

Figure 7.16

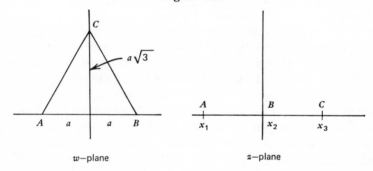

w—plane z—plane

Example 1.   Let us map an equilateral triangle onto the upper half-plane. Suppose the triangle has side $2a$, and is situated as indicated in Figure 7.16. We may choose x_1, x_2, x_3 arbitrarily, provided that we are careful to preserve the correct order, so that interior maps to interior. Let us take $x_1 = -1$, $x_2 = 0$, $x_3 = 1$, as a first case. We may write our mapping function

(3)
$$w(z) = -a + C \int_{-1}^{z} (t + 1)^{-2/3} (-t)^{-2/3} (1 - t)^{-2/3} \, dt$$

with principal value branches for each of the radicals and where we visualize z on the segment $(-1, 0)$. Then, with $z = 0$, $w = a$, and

$$a = -a + C \int_{-1}^{0} (t + 1)^{-2/3} (-t)^{-2/3} (1 - t)^{-2/3} \, dt \qquad t = -s$$

(4)
$$2a = C \int_{0}^{1} s^{-2/3} (1 - s^2)^{-2/3} \, ds$$

evaluates C. The integral (4) is not particularly simple, but we may express its value in terms of the gamma function [see Section 7.2(10)], by taking $s^2 = u$ as a new variable of integration in (4).

$$2a = \frac{C}{2} \int_0^1 u^{-5/6}(1 - u)^{-2/3} \, du$$

$$= \frac{C}{2} \frac{\Gamma(\frac{1}{6})\Gamma(\frac{1}{3})}{\Gamma(\frac{1}{2})}$$

(5)
$$C = \frac{4a\sqrt{\pi}}{\Gamma(\frac{1}{6})\Gamma(\frac{1}{3})}$$

Somewhat surprisingly, the mapping function for so simple a polygon as a triangle is remarkably messy and cannot be expressed in terms of elementary functions.

7.7 DEGENERATE POLYGONS

Our construction of the Schwarz-Christoffel formula shows us that the real z-axis between consecutive points (x_k, x_{k+1}) maps onto a straight line (w_k, w_{k+1}), and that across a point $z = x_k$, the line in the w-plane forms an interior angle α_k. It appears, thus, that our formula, [7.6(1)], is equally applicable to polygons that are limiting forms of finite polygons, such as the limiting "triangle" that we may visualize as being formed from the triangle ABC on the right in Figure 7.17, keeping B and C fixed, and "pulling" the vertex A upwards to infinity. Such limiting forms of polygons are called *degenerate* polygons, and the semi-infinite strip in Figure 7.17 is a degenerate triangle.

Let us examine the Schwarz-Christoffel formula in this case. Let us map A to ∞, B to -1 and C to $+1$ on the real z-axis. We obtain

(1)
$$w(z) = -a + C \int_{-1}^{z} \frac{dt}{\sqrt{1 + t}\sqrt{1 - t}}$$

$$= -a + C \int_{-1}^{z} \frac{dt}{\sqrt{1 - t^2}}$$

Figure 7.17

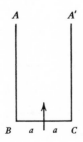

where, as previously, the square root is to be computed as a principal-value branch. Let $z \to 1$,

$$w = a = -a + C \int_{-1}^{1} \frac{dt}{\sqrt{1 - t^2}} = -a + C\pi$$

Thus

$$C = \frac{2a}{\pi},$$

and

$$w = -a + \frac{2a}{\pi} \int_{-1}^{z} \frac{dt}{\sqrt{1 - t^2}}$$

$$= -a + \frac{2a}{\pi} \left(\text{arc sin } z + \frac{\pi}{2} \right)$$

$$= \frac{2a}{\pi} \text{ arc sin } z$$

and

$$z = \sin \frac{\pi w}{2a} \qquad \text{(cf. Section 4.25)}$$

7.7.1   Regarding the infinite strip of width h as a degenerate triangle (Figure 7.18), obtain the explicit mapping function that leaves A at ∞, and maps B onto $z = 0$.

Figure 7.18

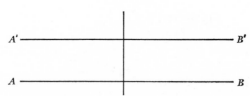

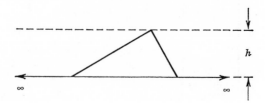

7.7.2   Regarding the upper half w-plane, with the segment $w = iv$, $0 \leq v \leq b$ removed (part of the boundary of the region), find the mapping function that maps this region onto the upper half z-plane (Figure 7.19).

Figure 7.19

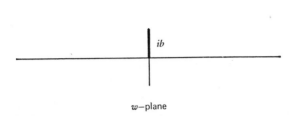

w–plane

7.7.3   Interpret the portion of the upper half w-plane above the line $ABCA'$ as a degenerate polygon (Figure 7.20), and map the region to the upper half z-plane (on CA', $w = ib + \text{real}$).

Figure 7.20

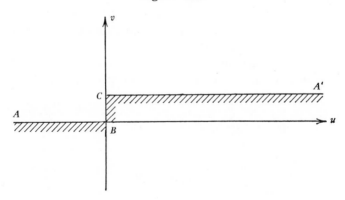

Another example.   As a final example, and to illustrate some of the complications that frequently arise when the Schwarz-Christoffel formula is applied, we consider the problem of mapping the channel of Figure 7.21 onto the upper half-plane. We may interpret this channel as being the degenerate case of a quadrilateral keeping B, D fixed, and letting $K \rightarrow \infty$ along the real axis, $A \rightarrow \infty$ along the imaginary axis (Figure 7.23). There are four vertex image points in this example, so that if we choose three of them, the fourth

Figure 7.21

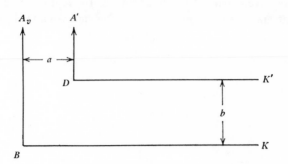

will be determined by the rest of the problem. Let us map A to ∞, B to -1 and K to 0, the origin in the z-plane. Call the image of D the point $z = \alpha$ (Figure 7.22).

Then we have

$$w(z) = C \int_{-1}^{z} \frac{(1 + t)^{-1/2}}{t} (\alpha - t)^{1/2} \, dt$$

with principal-value square roots.

The evaluation of C, α must come from matching the vertex points, so that we choose z on the real axis (when possible) and take the real axis as our path of integration (Figure 7.23).

We have started our integration from $z = -1$, corresponding to B: $w = 0$. Proceeding along the segment $(-1, 0)$, corresponding to BC, $\arg w = 0$, or $\arg C = 0$: C is real.

To reach the segment $(0, \alpha)$, we must avoid $z = 0$, and we do so by indenting the contour of integration at $z = 0$:

Figure 7.22

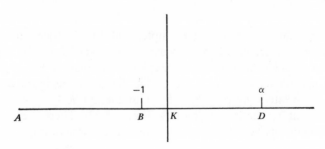

Figure 7.23

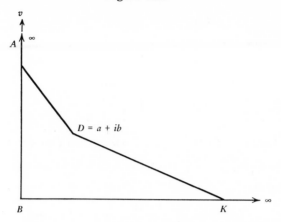

Thus,

$$w = C\int_{-1}^{-\delta}\sqrt{\frac{\alpha - t}{1 + t}}\frac{dt}{t} + C\int_{\pi}^{0}\sqrt{\frac{\alpha - \delta e^{i\theta}}{1 + \delta e^{i\theta}}}\, i\, d\theta + C\int_{\delta}^{z}\sqrt{\frac{\alpha - t}{1 + t}}\frac{dt}{t}$$

Now we have z on the segment $(0, \alpha)$, and may let $z \to \alpha$. This gives $w = a + ib$. If, at the same time, we let $\delta \to 0$, we have

$$(2) \qquad\qquad a + ib = C\fint_{-1}^{\alpha}\sqrt{\frac{\alpha - t}{1 + t}}\frac{dt}{t} + C(-i\pi\sqrt{\alpha})$$

where the integral is a Cauchy principal-value integral. Since C is real, we have $b = -\pi C\sqrt{\alpha}$

$$(3) \qquad\qquad a = C\fint_{-1}^{\alpha}\sqrt{\frac{\alpha - t}{1 + t}}\frac{dt}{t}$$

We now perceive one of the complications that restricts the usefulness of the Schwarz-Christoffel mapping formula. If there are more than three vertices, the equations (determining C and the rest of the points $x_1, \ldots x_n$ that we cannot choose in advance) are in general not solvable by formula, and numerical methods must be used.

The equation (3) does, however, turn out to be solvable. The integral can be transformed into a special case of a more general integral. We set $t = -1 + (1 + \alpha)\cos^2\theta/2$, obtaining

$$\frac{a}{C} = 2\int_{0}^{\pi}\frac{1 - \cos\theta}{\cos\theta - \left(\dfrac{1}{1 + \alpha} - \dfrac{1}{2}\right)} = -\pi$$

(see Example 6.13).

Thus,

$$a = -\pi C, \qquad b = -\pi C \sqrt{\alpha}$$

so that

$$\alpha = \frac{b^2}{a^2} \qquad C = -\frac{a}{\pi} \qquad \text{and}$$

(4)
$$w(z) = -\frac{a}{\pi} \int_{-1}^{z} \sqrt{\frac{b^2/a^2 - t}{1 + t}} \frac{dt}{t}$$

Additional Examples and Comments on Chapter Seven

7.1 Show that $\sum_0^\infty z^{n!}$ has $|z| = 1$ as a natural boundary.

7.2 Show that

$$\Gamma(z) = \sum_0^\infty \frac{(-1)^n}{n! \, (z + n)} + \int_1^\infty t^{z-1} e^{-t} \, dt$$

and use this result to establish the properties of $\Gamma(z)$ obtained in Section 7.2(8).

7.3 Show that for $r > 0$, $s > 0$,

$$0 < a < 1 \qquad 0 < b < 1 \qquad a + b > 1$$

$$\int_{-\infty}^{\infty} \frac{dx}{(r + ix)^a (s - ix)^b} = 2\pi(r + s)^{1-a-b} \frac{\Gamma(a + b - 1)}{\Gamma(a)\Gamma(b)}$$

7.4 $f(z) = \sum_0^\infty a_n z^n$, $g(z) = \sum_0^\infty b_n z^n$ are analytic in a neighborhood of $z = 0$.

Show that

$$\sum_0^\infty a_n b_n z^n = \frac{1}{2\pi i} \int_C f(t) g\left(\frac{z}{t}\right) \frac{dt}{t}$$

for a suitable contour C.

7.5 (a) Take $b_n \equiv 1$ in 7.4 and comment.

(b) Take $f(z) = g(z) = (1 + z)^n$ in 7.4 and so evaluate $\sum_{r=0}^{n} \binom{n}{r}^2$.

(c) If $f(z)$ is analytic for $|z| < R_1$, $g(z)$ for $|z| < R_2$, show that $h(z) = \sum_0^\infty a_n b_n z^n$ converges in $|z| < R_1 R_2$.

7.6 In 7.4, if $f(z)$ is singular only at $\alpha_1, \alpha_2, \ldots$, $g(z)$ singular only at $\beta_1, \beta_2, \ldots$ then $h(z) = \sum_0^\infty a_n b_n z^n$ can be singular only at the points $(\alpha_n \beta_m)$. (Assume that $f(z)$ and $g(z)$ are meromorphic functions.)

7.7   Let $f(x)$ be any continuous function of the real variable x, a any fixed constant, and

$$F(\alpha) = \frac{1}{\Gamma(\alpha)} \int_a^x (x - t)^{\alpha-1} f(t)\, dt$$

(a) Show that $F(\alpha)$ is analytic in α for $\mathbf{R}\alpha > 0$ (for fixed x).

(b) Show that $F(\alpha)$ can be continued to $\mathbf{R}\alpha > -1$ if $f'(x)$ is continuous, and to $\mathbf{R}\alpha > -n$ if $f^{(n)}(x)$ is continuous. In this case $F(-k) = f^{(k)}(x)$ for $k < n$.

(c) $F(k) = \displaystyle\int_a^x \cdots \int_a^x f(t_1)\, dt_1\, dt_2 \cdots dt_k$

is the k-fold iterated integral of $f(x)$.

7.8   If z is a point on the ellipse $x^2/a^2 + y^2/b^2 = 1$, show that

$$(a^2 - b^2)z^2 - 2(a^2 + b^2)z\bar{z} + (a^2 - b^2)\bar{z}^2 + 4a^2b^2 = 0$$

(see Example 4.4.13). Furthermore, define z_1, z_2 as reflected points in the ellipse if

$$(a^2 - b^2)z_1^2 - 2(a^2 + b^2)z_1\bar{z}_2 + (a^2 - b^2)\bar{z}_2^2 + 4a^2b^2 = 0$$

Which points outside the ellipse are reflected points of the interior of the ellipse?

7.9   Referring to 7.8, the ellipse is parametrically represented by

$$x = a \cos t \qquad y = b \sin t \qquad 0 \leq t < 2\pi$$

$$z = a \cos t + ib \sin t$$

In the $s = t + i\tau$ plane, $\cos t$, $\sin t$ are real on the real axis and clearly extend by reflection for complex values of s. Show that for two reflected values of s, s_1 and $\bar{s}_1$, the corresponding values of z are reflected points in the ellipse.

7.10   If $f(z)$ is analytic in a region part of whose boundary consists of an arc C of the ellipse of 7.8, and $f(z)$ is real and continuous on this (open) arc, show that $f(z)$ can be continued analytically to the reflected region (across C) and takes on conjugate (reflected) values at the corresponding reflected points. (From 7.8, the domain may have to be restricted close to C.)

7.11   Generalize the results of 7.10 in the following manner. Let $x = \phi(t)$, $y = \psi(t)$ describe a curve C for $t_0 \leq t \leq t_1$, $\phi(t + i\tau)$, $\psi(t + i\tau)$ be analytic for $|\tau|$ sufficiently small, real for $\tau = 0$. Then $s_1 = t_1 + i\tau_1$ and $\bar{s}_1$ yield, from $z = x(s) + iy(s)$, a pair of "reflected" points in C. Hence show that an analytic function, real on C, can be continued across C by reflection. Show further that the reflected points are independent of the parametrization of C.

7.12   Consider the differential equation for $w(z)$,

$$w'' + p(z)w' + q(z)w = 0$$

for

$$p(z) = \sum_0^\infty p_n z^n \qquad \text{analytic in } |z| < R$$

$$q(z) = \sum_0^\infty q_n z^n \qquad |z| < R$$

In $|z| < R_1 < R$, $p_n R_1{}^n \to 0$ as $n \to \infty$, so that $|p_n| \leqq M/R_1{}^n$ for all n. Thus, $p(z)$ is "majorized" by $\sum_0^\infty M z^n/R_1{}^n = M/(1 - z/R_1)$. (The coefficients for $p(z)$ are, in magnitude, not greater than those of the majorant.) Show that $q(z)$ is majorized by $M_1/(1 - z/R_1)^2$ and hence, that $w(z)$ is majorized by $W(z)$, where $W''(z) - MW'(z)/(1 - z/R_1) - M_1 W(z)/(1 - z/R_1)^2 = 0$.
    $W(z) = M_2(1 - z/R_1)^{-\lambda}$ for suitable $-\lambda$, positive M_2, deduce the result that the original differential equation possesses a solution (existence), having $w(0)$, $w'(0)$ arbitrary, and is uniquely determined if $w(0)$, $w'(0)$ are prescribed. Furthermore, $w(z)$ is analytic in the same circle that $p(z)$, $q(z)$ are.

7.13   From 7.12, show that each solution of

$$w'' + p(z)w' + q(z)w = 0$$

with $p(z)$, $q(z)$ analytic in $|z - z_0| < R$, can be expressed uniquely as

$w = a_1 w_1(z) + a_2 w_2(z)$, a_1, a_2 constant where w_1, w_2 are solutions of the equation,

$$w_1(z_0) = 1 \qquad w_1'(z_0) = 0$$
$$w_2(z_0) = 0 \qquad w_2'(z_0) = 1$$

We call the pair $w_1(z)$, $w_2(z)$ a fundamental set of solutions. (Every solution is a linear combination of a fundamental set of solutions and $z = z_0$ is called an ordinary point for the equation.)

7.14   Let $p(z), q(z)$ be single valued and analytic in the annulus $0 < |z| < R$, and consider the differential equation

$$w''(z) + p(z)\, w'(z) + q(z)\, w(z) = 0$$

For any z_0 in the annulus, form $w_1(z, z_0)$ and $w_2(z, z_0)$, a fundamental set of solutions at z_0. Next, continue $w_1(z, z_0)$ around $z = 0$ by the usual process: that is, take z_1 in the circle of convergence of $w_1(z; z_0)$, form

$\sum_0^\infty w_1^{(n)}(z_1, z_0)(z - z_1)^n/n!$ a continuation of $w_1(z, z_0)$ and repeat the process along a path C, returning to a neighborhood of z_0. Let $w_1^*(z, z_0)$ denote such a continuation.

Show that

(a) at each stage of the continuation the series

$$\sum_0^\infty w_1^{(n)}(z_k) \frac{(z - z_k)^n}{n!}$$

is a solution of the differential equation.

(b) $w_1^*(z, z_0) = aw_1(z, z_0) + bw_2(z, z_0)$
$w_2^*(z, z_0) = cw_1(z, z_0) + dw_2(z, z_0)$

(c) there is a combination $W = Aw_1 + Bw_2$ so that $W^* = \lambda W$, where

$$\begin{vmatrix} a - \lambda & c \\ b & d - \lambda \end{vmatrix} = 0$$

7.15   In 7.14, show that

(a) The roots λ_1, λ_2 are independent of the choice of w_1, w_2 as fundamental solutions.

(b) $(z^\rho)^* = e^{2\pi i \rho} z^\rho = \lambda z^\rho$ for $\rho = (1/2\pi i) \log \lambda$

(c) $(W(z)/z^\rho)^* = W(z)/z^\rho$ is thus single valued in $0 < |z| < R$, and so possesses a Laurent expansion

$$W(z) = z^\rho \sum_{-\infty}^\infty a_n z^n$$

7.16   In 7.14, if λ_1, λ_2 are distinct, then $\sum_{-\infty}^\infty a_n z^{n+\rho_1}$, $\sum_{-\infty}^\infty b_n z^{n+\rho_2}$ form a fundamental set of solutions.

If $\lambda_1 = \lambda_2$, then choose

$$w_1^* = \lambda_1 w_1 \qquad w_2^* = cw_1 + dw_2$$

and show that $d = \lambda_1$; that is,

$$w_2^* = cw_1 + \lambda_1 w_2$$

$$\left(\frac{w_2}{w_1}\right)^* = \frac{w_2}{w_1} + \frac{c}{\lambda_1}$$

and hence

$$\frac{w_2}{w_1} - \frac{c}{\lambda_1 2\pi i} \log z \text{ is single valued}$$

and so

$$w_2(z) = w_1(z)\left(\frac{c}{\lambda_1 2\pi i} \log z + \sum_{-\infty}^{\infty} c_n z^n\right)$$

7.17   If, in 7.13, $p(z)$ has (at worst) a pole of order one, and $q(z)$ a pole of order not exceeding two, show that the principal parts of the Laurent expansions in 7.15 and 7.16 have only a finite number of terms. Hence the solutions can be found readily by taking $w(z) = \sum_0^\infty a_n z^{n+\rho}$, $a_0 \neq 0$, and using undetermined coefficients. Such a point ($z = 0$ here) is called a *regular singular point* for the differential equation.

7.18   Consider

$$z^2 w'' + z w' + (z^2 - \alpha^2) w = 0 \qquad \alpha = \text{const}$$

This is Bessel's equation of order α. Show that $z = 0$ is a regular singular point (7.17) and find a fundamental set of solutions for $\alpha = 0$, $\alpha = 1/2$.

7.19 $z(1 - z) w'' + [c - (a + b + 1) z] w' - abw = 0$ is   called   the hypergeometric differential equation. Show that $z = 0, 1, \infty$ are regular singular points, all others being ordinary points.

7.20   Show that one solution of 7.19, appropriately normalized, is

$$\frac{\Gamma(c)}{\Gamma(a)\Gamma(b)} \sum_0^\infty \frac{\Gamma(a + n)\Gamma(b + n)}{n! \,\Gamma(c + n)} z^n$$

This specific series is termed the hypergeometric function, and written $_2F_1(a, b; c; z)$.

Chapter
EIGHT | **HYDRODYNAMICS**

In this chapter we look briefly at some applications of the theory of analytic functions to a certain class of fluid motions. We shall see one way (there are also others) to give intuitive, concrete, and practically usable contents to the theory.

8.1 THE EQUATIONS OF HYDRODYNAMICS

We merely state the equations. A detailed derivation may be found, for example, in *Hydrodynamics* by H. Lamb.

(1)
$$\frac{\partial u}{\partial t} + u\frac{\partial u}{\partial x} + v\frac{\partial u}{\partial y} + w\frac{\partial u}{\partial z} = X - \frac{1}{\rho}\frac{\partial p}{\partial x}$$

$$\frac{\partial v}{\partial t} + u\frac{\partial v}{\partial x} + v\frac{\partial v}{\partial y} + w\frac{\partial v}{\partial z} = Y - \frac{1}{\rho}\frac{\partial p}{\partial y}$$

$$\frac{\partial w}{\partial t} + u\frac{\partial w}{\partial x} + v\frac{\partial w}{\partial y} + w\frac{\partial w}{\partial z} = Z - \frac{1}{\rho}\frac{\partial p}{\partial z}$$

(2)
$$\frac{\partial \rho}{\partial t} + \frac{\partial}{\partial x}(\rho u) + \frac{\partial}{\partial y}(\rho v) + \frac{\partial}{\partial z}(\rho w) = 0$$

(3)
$$\rho = f(p)$$

These equations govern the motion of a fluid, with velocity (u, v, w), pressure p, and density ρ, in the presence of external forces X, Y, Z (such as gravity).

Equations   (1) express the conservation of momentum.
                (2) is the conservation of mass.
                (3) is the so called equation of state.
(Already we have assumed a fluid model that is simpler than a more general case—we have neglected viscosity.)

265

We proceed to a progressively more specialized situation (flow model), finally obtaining a case for which the use of analytic functions is indicated. First, we restrict our attention to incompressible fluids (a good approximation is water): we consider only those cases for which $\rho \equiv$ const.

Next, we ignore the external forces and set $X = Y = Z = 0$.

We consider only steady motions: u, v, w, p are independent of t.

We further restrict our attention to two dimensional flows—flows parallel to the x, y plane. Then $w = 0$, all partial derivatives with respect to z (a space coordinate) vanish, and we have

(4)
$$u \frac{\partial u}{\partial x} + v \frac{\partial u}{\partial y} + \frac{1}{\rho} \frac{\partial p}{\partial x} = 0$$

$$u \frac{\partial v}{\partial x} + v \frac{\partial v}{\partial y} + \frac{1}{\rho} \frac{\partial p}{\partial y} = 0$$

(5)
$$\frac{\partial u}{\partial x} + \frac{\partial v}{\partial y} = 0$$

Equations 4 and 5, although a very special case of the general one, are still very difficult to treat. We recognize in (5) one of the Cauchy-Riemann conditions for $u - iv$ to be an analytic function of $z = x + iy$, and might well wonder if (4) is consistent with the second C $-$ R equation. Consider, as well as (4) and (5),

(6)
$$v_x - u_y = 0$$

There is a physical justification that we merely allude to: for the two-dimensional model, $v_x - u_y$ is the curl of the velocity vector, and it can be shown that a hydrodynamical flow [solution of (1), (2), and (3)] that starts out irrotationally (curl $= 0$ at $t = 0$) remains irrotational for all time (curl $\equiv 0$). Hence it is reasonable to expect that there are nontrivial flows satisfying (4), (5), and (6).

One last observation here: if u, v satisfy (5), (6), that is, $u - iv$ is an analytic function of $z = x + iy$, then the left sides of (4) yield

$$uu_x + vv_x + \frac{1}{\rho} p_x \qquad uu_y + vv_y + \frac{1}{\rho} p_y$$

respectively, and so (4) yields

(7)
$$\tfrac{1}{2}(u^2 + v^2) + \frac{1}{\rho} p = \text{const}$$

called Bernoulli's equation.

Let us summarize these results. The *two dimensional, steady, irrotational flow of an incompressible nonviscous fluid in the absence of external forces* obeys

(4), (5), and (6). That is, if $u - iv$ is any analytic function (of z), and p is determined from (7), we have a solution of the stated flow problem. There remains to adapt this general solution to particular cases, such as the flow in a channel, or around a given obstacle. We devote the following sections to such adaptation, and we shall refer to a flow, or fluid motion, without further qualification.*

8.2 THE COMPLEX POTENTIAL

It is now convenient to introduce the function $\chi(z)$ for which

$$w = u - iv = \chi'(z) = (\phi + i\psi)_x = \frac{1}{i}(\phi + i\psi)_y$$

or,

$$\phi_x = u \qquad \phi_y = v$$

$$\psi_x = -v \qquad \psi_y = u$$

We call the function χ the complex potential; indeed, ϕ is the potential function for the flow, with grad ϕ being the velocity vector (u, v):

$$\bar{w} = u + iv = qe^{i\sigma}$$

The function ψ is called the stream function for the flow, and the level lines, $\psi = $ const, are the paths traced out by the fluid particles. These results follow immediately from our interpretation of the complex line integral for an arbitrary sourceless and irrotational vector-field:

$$\int_a^b w\, dz = \int_a^b \chi'\, dz = \chi(b) - \chi(a)$$

$$= \phi_b - \phi_a + i(\psi_b - \psi_a)$$

$$= \text{work} + i\, \text{flux}$$

Thus, if we take a portion of a level curve $\psi = $ const for our path of integration, $\psi_b \equiv \psi_a$, and the flux of fluid (per unit height) is zero. That is to say, the flow is everywhere tangent to the level lines $\psi = $ const.

These observations are very useful in the construction of flows in specific channels, or past given bodies: the boundaries of the flow region are stream lines, and hence $\psi = $ const at such lines.

A sort of converse statement is also true: given any flow, we may visualize any stream line as the trace of a solid boundary and regard the flow on one side of this stream line as the flow past this boundary.

* Henceforth $z = x + iy$.

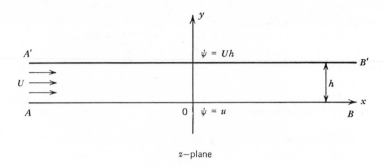

Figure 8.1

z−plane

As an example, consider the flow given by

(1)
$$\chi(z) = az + b$$
$$w = \chi'(z) = a$$

Thus the velocity of the fluid is everywhere constant, $\bar{w} = \bar{a}$. For convenience, let us consider a to be real, $a = U > 0$.

Then

$$\phi + i\psi = U(x+iy) + b$$
$$\phi = Ux + \text{const}$$
$$\psi = Uy + \text{const}$$

and the stream lines are $y = $ const. Thus $\chi(z) = Uz + $ const is the complex potential for uniform flow from left to right. If we restrict our attention to the upper half plane, $y > 0$, we may regard $y = 0$ as a boundary, and $\chi = Uz + $ const represents uniform flow in the upper half-plane. Or, further, we may restrict the vector field even further, to the channel bounded by $y = 0$, $y = h$. The same complex potential $\chi = Uz + $ const represents uniform flow in this channel.

Since $\chi(z)$ involves an arbitrary constant of integration, we may choose this constant at our convenience. Thus it is customary to choose $\psi = 0$ on the lower stream line, so that $\psi = Uh$ on the upper stream line of the channel of Figure 8.1; in this way (starting from $\psi = 0$ on one boundary), the value of $\psi = Uh$ on the upper boundary represents the flux of fluid through the channel.

For a general value of a, the complex potential $\chi = az$ yields identically the same type of flow except that the flow direction is given by $\bar{w} = \bar{a}$, making an angle arg $\bar{a}$ with the real axis.

As a second example, we consider

(2) $$\chi(z) = k \log z \qquad k > 0$$

$$w = \frac{k}{z} = u - iv = qe^{-i\sigma}$$

Thus

$$q = \frac{k}{r} \qquad \sigma = \theta$$

and, the flow is directed radially outwards from the origin, $q = |w|$ being constant on circles concentric with the origin. Let us compute the flux for this flow (Figure 8.2).

$$\text{work} + i\,\text{flux} = \int_C \frac{k}{z}\,dz = 2\pi i k$$

where C is any simple closed curve containing the origin. Thus, the flux is constant, flux $= 2\pi k$, so that fluid is emanating from the origin at a constant rate. The field is not sourceless at $z = 0$ since $z = 0$ is a singular point; in fact, we may regard $z = 0$, here, as a *source* of "strength" $2\pi k$; that is, a source providing a flux of $2\pi k$ units (per unit height). Clearly, a unit source at $z = a$ is provided by the complex potential

$$\chi(z) = \frac{1}{2\pi} \log (z - a)$$

If we take $k < 0$ in the above example, the flow direction is reversed, the flux across any curve C containing $z = 0$ is negative, and we regard $z = 0$ as a *sink*.

Figure 8.2

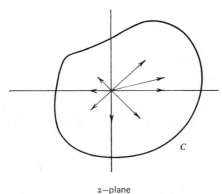

z–plane

Finally, let us consider

$$\chi(z) = ik \log z \qquad k > 0$$

so that

$$q = \frac{k}{r} \qquad \sigma = \theta - \frac{\pi}{2}$$

The stream lines are $r = $ const, the flux across a closed curve surrounding $z = 0$ is zero, and the flow turns around the origin in a *vortex motion*.

8.2.1 Examine the flow in the vicinity of a source and sink of equal strengths, located at $z = -h$ and $z = h$, respectively $(h > 0)$: $\chi(z) = k \log[(z - h)/(z + h)]$. Identify the stream lines and the equipotential lines.

8.2.2 For the complex potential of Example 8.2.1, take $k = k'/2h$, k' independent of h, let $h \to 0$, thus obtaining the complex potential for a *dipole*. Describe the flow in the vicinity of a dipole, and identify the stream lines. What is the flux produced by a dipole?

8.2.3 As an example of the superposition of two flows, consider the complex velocity potential for a uniform flow past a dipole

$$\chi(z) = U\left(z + \frac{1}{z}\right)$$

Find the stream lines, and interpret the flow as a flow uniform at infinity, around a cylinder.

8.3 FLOW IN CHANNELS: SOURCES, SINKS, AND DIPOLES

Let us take another look at the channel flow depicted in Figure 8.1 for which the complex velocity potential may be taken to be

$$\chi(z) = Uz$$

The leftmost portion of the channel, AA', extends to infinity, which we describe suggestively as "upstream infinity," while the exit end of the channel, BB', is at "downstream infinity." Fluid is supplied at AA', with constant flux Uh, and is removed at BB'. Thus, in some sense, there must be a source at AA', upstream infinity, and a sink at BB', downstream infinity. Actually, both AA' and BB' are at infinity, *the* point at infinity for the z-plane.

In order to see more clearly this source-sink (dipole?) combination, we may use the Schwarz-Christoffel mapping to advantage, and map the polygonal region consisting of the channel onto the upper half of an auxiliary complex plane, the τ-plane (Figure 8.3).

Figure 8.3

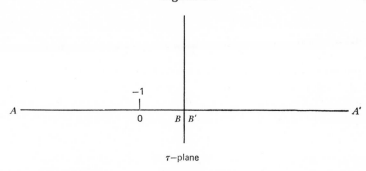

τ–plane

Let us arrange to leave AA' at infinity, and bring BB' to a finite point of the τ-plane. Since we may choose three points arbitrarily, let us map AA' to infinity, 0 to -1 and B to the origin of the τ-plane. Then we have (see Section 7.5)

$$\frac{dz}{d\tau} = \frac{C}{\tau}$$

$$z = C \log \tau + D = -\frac{h}{\pi} (\log \tau - i\pi)$$

while

$$\chi(z) = Uz = -\frac{Uh}{\pi} (\log \tau - i\pi)$$

Thus, in the upper half of the τ-plane, we have

$$\chi = -\frac{Uh}{\pi} (\log \tau - i\pi)$$

which represents the flow to a sink at $\tau = 0$. Observe that the strength of this sink must be Uh from the nature of the channel flow. On the other hand, we obtained previously that $\chi = -k \log \tau$ represents a sink of strength $2\pi k$, giving $2Uh$ instead of Uh for this example. However, we are concerned only with the fluid entering the channel or, in the τ-plane, with the contribution from the sink in the *upper half-plane*. Thus, the constant must be double ($2Uh$ instead of Uh) in order that half the contribution be Uh, as required.

In a similar way, we can convince ourselves that there is a source at AA' for the channel, while the source-sink combination is actually a dipole at infinity.

8.3.1   Map the channel of Figure 8.1 onto the upper half τ-plane so that BB' maps to $\tau = -1$, the origin maps to $\tau = \infty$, and AA' maps to $\tau = 1$, and verify that $\chi = h/\pi \log (\tau - 1)/(\tau + 1)$.

Flow in Channels: Sources, Sinks, and Dipoles   **271**

8.4 FLOW IN CHANNELS: CONFORMAL MAPPING

So far, in Section 8.2 and 8.3, we have proceeded in an *inverse* manner in constructing flows; that is, we have started from a given analytic function and considered the flow that results from the equation $u - iv = w(z)$. This approach can at best provide us with a dictionary of cases that we can search through in an attempt to solve a specific problem. What we need is a *direct* method for constructing a specific flow. Let us return once more to the channel flow of Figure 8.1. We require the analytic complex velocity potential,

Figure 8.4

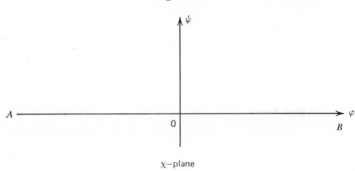

χ–plane

$\chi(z) = \phi + i\psi$. From the given data of the channel, we know certain properties that $\chi(z)$ must possess. Thus $\psi = $ const on AB, and also $\psi = $ const on $A'B'$. In fact, normalizing χ appropriately, we may take $\psi = 0$ on AB, $\psi = Uh$ on $A'B'$.

Furthermore, $d\chi/dz = w$, so that

$$\frac{d\chi}{dz} \to U \text{ as } z \to AA'$$

(z approaches upstream infinity) and also as z approaches downstream infinity. These statements become more suggestive if we consider the χ-plane, and the image of the channel of the z-plane that the flow induces in the χ-plane (Figure 8.4).

The boundary $y = 0$ of the z-plane maps into $\psi = 0$ of the χ-plane. Can we locate the images of the points A, 0, and B? We have normalized ψ by taking $\psi = 0$ on the lower stream line. We can equally well normalize ϕ so that $\phi = 0$ at $x = y = 0$, in this way, the origin of the z-plane maps onto the origin of the χ-plane.

Next, $d\chi/dz = \phi_x + i\psi_x \to U$ as z tends to infinity in the channel. That is, $\phi_x \to U$ as $x \to \pm\infty$, so that it is reasonable to expect ϕ to behave like Ux for large χ: $\phi \to -\infty$ as $x \to -\infty$, and $\phi \to \infty$ as $x \to \infty$.

Thus we expect A and B to map onto similarly situated points of the χ-plane.

A comparison of Figures 8.1 and 8.4 suggests that $\chi = Uz$. This equation certainly works here (it also agrees with the results of Section 8.2); since three points of one plane map to three specific points of the other plane, corresponding to the three parameters of the Riemann mapping theorem, we have an indication that the above solution is uniquely determined.

This point is worth dwelling on a little further, since the method of verification suggests a *direct technique* for the construction of various channel flows.

Consider now the z-plane channel of Figure 8.1, and its image under the flow, the channel of Figure 8.4 in the χ-plane. We introduce an auxiliary third complex plane, the τ-plane, and map both channels to the upper half of the τ-plane by the Schwarz-Christoffel mapping. We arrange to map A to infinity, the origin 0 to $\tau = -1$ and B to $\tau = 0$ (any three convenient boundary points of the τ-plane can be selected: simplicity should be our guiding principle). Then

$$\frac{dz}{d\tau} = \frac{C_1}{\tau} \qquad \frac{d\chi}{d\tau} = \frac{C_2}{\tau}$$

(see 7.5)

$$\frac{d\chi}{dz} = \frac{d\chi}{d\tau} \Big/ \frac{dz}{d\tau} = \frac{C_2}{C_1} = \text{const}$$

Thus $\chi = (\text{const}) \cdot z = Uz$, and the uniqueness of the Schwarz-Christoffel transformation indicates that our previous reasoning on uniqueness was valid.

Example 1.   *Flow in a corner.* We consider fluid occupying the first quadrant, moving vertically down at A and horizontally to the right at B (Figure 8.5). The image of the flow region in the χ-plane we may take to be

Figure 8.5

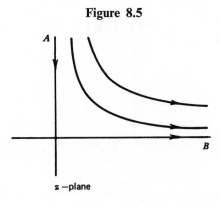

z —plane

the upper half-plane, $\psi = 0$ on the boundary stream line (and normalizing ϕ so that the origins correspond).

The mapping function $\chi = kz^2$ sends the z-plane onto the upper half χ-plane for arbitrary real positive k. Thus,

$$w = \frac{d\chi}{dz} = 2kz = u - iv$$

$$u = 2kx \qquad v = -2ky$$

The velocity is infinite at $z =$ infinity, and is zero at the corner, $z = 0$. A point in the flow at which $w = 0$ is called a *stagnation point*.

8.4.1 Find the stream lines and the pressure (from Bernoulli's equation) for the flow of Example 1.

8.4.2 Find the complex velocity potentials for the corner flows indicated (Figure 8.6). Locate any stagnation points, and the stream lines.

Figure 8.6

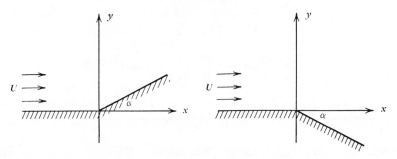

8.4.3 The uniform flow $w = U$ is disturbed by the presence of a "spike", $x = 0$, $0 \leq y \leq 1$, sticking into the flow region. Find the complex velocity · potential $\chi(z)$, and the velocity vector $\bar{w}$, as well as the pressure, p (Figure 8.7).

Figure 8.7

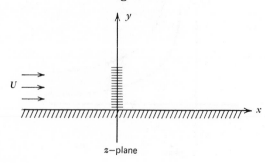

z—plane

Figure 8.8

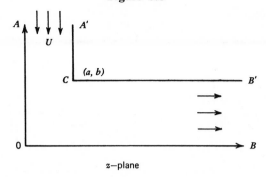

z—plane

Example 2. Consider now the flow problem for the right-angled channel of Figure 8.8. The velocity vector is to tend to $-iU$ as z approaches the entrance AA'. Clearly, the flux of fluid entering the channel is Ua since the channel width at AA' is a. Thus, $\psi = 0$ on AOB, and $\psi = Ua$ on $A'CB'$, and the image of the flow region in the χ-plane is the channel indicated in Figure 8.9.

In order to construct the specific $\chi(z)$ that maps the z-plane channel onto the χ-plane channel, we construct the auxiliary τ plane, and map each of the two polygonal channels onto the upper half of the τ-plane. Looking ahead, this procedure will yield

$$z = z(\tau), \qquad \chi = \chi(\tau)$$

expressing $\chi = \chi(z)$ in parametric form. We choose the allowable boundary points of the τ-plane in such a way as to make the computations as simple as possible. By choosing the point A to map to infinity of the τ-plane, the Schwarz-Christoffel formula is considerably simplified. Thus, we do so; let us map the origins onto the origin of the τ-plane, and points B onto

Figure 8.9

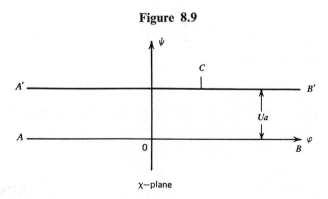

χ—plane

Figure 8.10

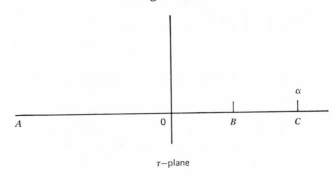

τ−plane

$\tau = 1$ (Figure 8.10). The location of the image of C will be determined by the mapping. We have thus

(1)
$$\frac{dz}{d\tau} = \frac{C_1\sqrt{\alpha - \tau}}{\sqrt{\tau}\,(1 - \tau)} \qquad\qquad \text{(see 7.5)}$$

(2)
$$\frac{d\chi}{d\tau} = \frac{C_2}{1 - \tau} \qquad C_1,\, C_2 = \text{constant}$$

(3)
$$\frac{d\chi}{dz} = \frac{d\chi/d\tau}{dz/d\tau} = \frac{C_2}{C_1}\sqrt{\frac{\tau}{\alpha - \tau}}$$

For concreteness, let us take the principal-value branches of the square roots in Section 8.4 (1). Then both $\sqrt{\tau}$, $\sqrt{\alpha - \tau}$ are real and positive for τ real and positive ($\tau < \alpha$).

Integrating from $\tau = 0$ along the positive real axis, we obtain

(4)
$$z = C_1 \int_0^\tau \sqrt{\frac{\alpha - t}{t}}\,\frac{dt}{1 - t}$$

(5)
$$\chi = C_2 \int_0^\tau \frac{dt}{1 - t}$$

We see immediately that

(6) C_1 and C_2 are real and positive.

Consider next Section 8.4(3), and rewrite it so that we may let $\tau \to \infty$, corresponding to $z = A$, where $d\chi/dz = w = u - iv = Ui$.

(7)
$$\frac{d\chi}{dz} = \frac{C_2}{C_1}\,i\sqrt{\frac{-\tau}{\alpha - \tau}}$$

the radical being real and positive for τ real and negative. Let $\tau \to -\infty$; we obtain $d\chi/dz = +i\,(C_2/C_1) = iU$

(8)
$$\frac{C_2}{C_1} = U$$

Next, if we let $\tau = 1$ in Section 8.4 (3), corresponding to the exit point of the channel, we have

(9)
$$\frac{d\chi}{dz} = \frac{C_2}{C_1} \frac{1}{\sqrt{\alpha - 1}} = \text{constant}$$

Thus the exit velocity is a constant, and from flux considerations we have

$$\frac{C_2}{C_1} \frac{1}{\sqrt{\alpha - 1}} \cdot b = Ua$$

or

$$\sqrt{\alpha - 1} = \frac{b}{a}$$

(10)
$$\alpha = 1 + \frac{b^2}{a^2}$$

We now have evaluated the constants in Sections 8.4 (4) and 8.4 (5) except for the determination of C_1 or C_2 itself. It is a simple matter to rewrite Section 8.4(5) as

$$\chi = -C_2 \log (1 - \tau)$$

and obtain the appropriate form valid for $\tau > 1$ and real. This corresponds to

$$\chi = \text{real} + iUa$$
$$= -C_2 \log (\tau - 1) + i\pi C_2$$

Thus

$$C_2 = \frac{Ua}{\pi} \qquad C_1 = \frac{a}{\pi}$$

and collecting our formulas

(11)
$$z = \frac{a}{\pi} \int_0^\tau \sqrt{\frac{1 + b^2/a^2 - t}{t}} \, \frac{dt}{1 - t}$$

$$\chi = -\frac{Ua}{\pi} \log (1 - \tau)$$

$$\frac{d\chi}{dz} = U \sqrt{\frac{\tau}{1 + \dfrac{b^2}{a^2} - \tau}}$$

COMMENTS

1. The evaluation of the constants in the Schwarz-Christoffel mapping of the channel in the z-plane onto the upper half τ-plane has been obtained previously [see Section 7.7 (4)]; at that time, considerable labor was involved that has been circumvented here. We have not actually obtained "something for nothing" however; one difficult computation has been replaced by several simpler ones.

2. If we recall the earlier discussion of sources and sinks, we can frequently reduce the amount of labor in determining the mapping functions from simple channels to half planes. Thus in order to solve for $\chi = \chi(\tau)$ as in Section 8.4(2) we might argue as follows. The channel of Figure 8.9 is to be mapped onto the upper half τ-plane of Figure 8.10, with the boundary points corresponding. Ignoring the fact that the χ-plane has any flow significance, imagine a uniform flow in the channel, with a source at AA', and a sink at BB'; in the τ-plane we have thus

$$\chi = -k \log (\tau - 1)$$

with $k = Ua/\pi$ as previously.

8.5 FLOWS PAST FIXED BODIES

Let us start with an example. Consider the superposition of a source of strength k at the origin, and a uniform flow parallel to the real axis. The complex velocity potential for the superimposed flow is then the sum of the complex potentials for the individual flows (Figure 8.11).

$$(1) \qquad \chi(z) = Uz + \frac{k}{2\pi} \log z \qquad k > 0$$

$$w = U + \frac{k}{2\pi z}$$

and

$$\psi(x, y) = Uy + \frac{k}{2\pi} \theta$$

From symmetry, we expect part of the x-axis to be a stream line. In fact, on $\theta = \pi$, $y = 0$, $\psi = k/2$, so that

$$(2) \qquad \psi = Uy + \frac{k\theta}{2\pi} = \frac{k}{2}$$

is a stream line, consisting of a portion of the x-axis and a pair of curves as indicated in Figure 8.11; we may regard Section 8.5 (1) as the complex

Figure 8.11

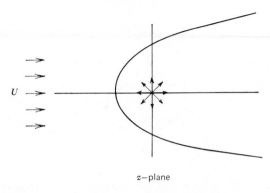

z—plane

potential for a flow past a blunt body, similar in shape to the cross-section of an airplane wing near the leading edge. We note that there is a stagnation point at $z = -k/2\pi U$, that is, $w = d\chi/dz$ vanishes at this point. The mapping from the z-plane to the χ-plane thus ceases to be conformal at $z = -k/2\pi U$ and, indeed, angles are doubled at this point.

Example 1.   Consider the flow whose complex potential is

$$\chi(z) = Uz + \frac{k}{2\pi} \log z - \frac{k}{2\pi} \log (z - 1)$$

that is, the superposition of a uniform flow with a source flow and a sink flow. Show that there is a stream line consisting of portions of the x-axis and a blunt shaped closed finite curve, similar to the shape of a ship. (First locate the stagnation points.)

Example 2.   Superposing a uniform flow with the flow from a dipole, for example,

(3)
$$\chi(z) = U\left(z + \frac{a^2}{z}\right) \qquad a > 0$$

$$w = U\left(1 - \frac{a^2}{z^2}\right)$$

$$\psi = U\left(r - \frac{a^2}{r}\right) \sin \theta$$

we have $\psi = 0$ on the lines $\sin \theta = 0$, $r = a$, so that the flow can be interpreted as flow past an infinite cylinder of radius a (Figure 8.12). There are

Flows Past Fixed Bodies   **279**

Figure 8.12

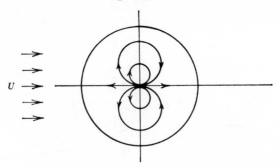

stagnation points at $z = \pm a$ and the velocity on the circular boundary is

$$\bar{w} = U\left(1 - \frac{a^2}{\bar{z}^2}\right)$$

$$= U(1 - e^{2i\theta}) = qe^{+i\sigma}$$

$q = 2\,|\sin\theta|\,U$, $\sigma = -(\pi/2) + \theta$ on the upper semicircle, $(\pi/2) + \theta$ on the lower.

These examples illustrate the *inverse* procedure described in Section 8.2; we start with a known complex potential, locate the stream lines, and carve out a flow between specific stream lines, or on one side of a certain stream line. If we are sufficiently ingenious, we can obtain reasonably close approximations to some regularly shaped profiles, and then have solved the precise flow problem past these profiles.

A more satisfactory procedure would be to start with a given body and construct the complex potential for the required flow past the body. Let us reconsider the flow of Example 2, and investigate a *direct* method for solving the problem.

Figure 8.13

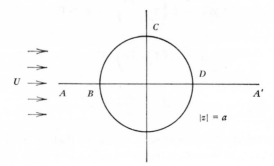

Figure 8.14

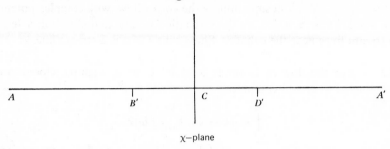

χ–plane

Example 3.   Flow past a *circular cylinder*. We suppose that the cylinder, whose cross-section in the z-plane is $|z| = a$, has fluid flowing past it, with $w = d\chi/dz \to U$ as $z \to \infty$ (Figure 8.13). By symmetry, the line $ABCDA'$ must be a stream line, which we take to be $\psi = 0$. Then, fixing our attention on the portion of the flow above this stream line, the image of the flow region in the χ-plane will be the full upper half-plane (Figure 8.14).

This mapping problem we have already solved (see Example 4.5.2) yielding again

$$\chi(z) = U\left(z + \frac{a^2}{z}\right)$$

Another approach to this mapping problem is useful, and is exploited in the next section. Instead of proceeding directly from the z-plane to the χ-plane, we introduce an auxiliary complex plane, $\tau = \log z/a$ (Figure 8.15).

The mapping function from the τ-plane to the χ-plane can be written down by means of the Schwarz-Christoffel mapping, giving

$$\chi(\tau) = Ua(e^\tau + e^{-\tau})$$

$$\tau = \cosh^{-1}\frac{\chi}{2Ua}$$

Figure 8.15

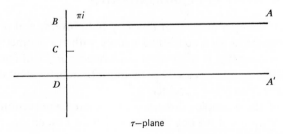

τ–plane

8.5.1   Consider the flow past a circular cylinder, with complex potential $\chi(z) = U(z + a^2/z)$, and superimpose the vortex flow, with complex potential $\chi = -ik \log z$, $k > 0$. Locate the stagnation points, and sketch in a few of the stream lines near the cylinder. Show that $r = a$ is a stream line.

8.5.2   For the flow in Example 8.5.1 let C be any simple closed curve encircling $|z| = a$, and compute

$$\int_C w \, dz = \text{work} + i \text{ flux}$$

The real part of this integral is called the *circulation* of the flow, and represents the work required to produce the *circulatory component* of the flow. We speak, here, of a flow with circulation.

8.5.3   Find the flow past a plate of finite width, $2d$, making an angle α with the real axis, symmetrically placed with respect to the origin (Figure 8.16).

$$w = \frac{d\chi}{dz} \to U \quad \text{as} \quad z \to \infty$$

(Map the exterior of the plate onto the exterior of a circle and use Example 3.) Compute the pressure, p, from Bernoulli's equation.

Figure 8.16

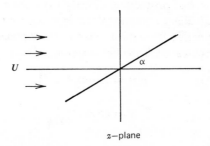

z—plane

8.6 FLOWS WITH FREE BOUNDARIES

We have examined a number of flows by both inverse and direct methods and, until now, have been concerned mainly with the geometry of the flow region, as this geometry is of greatest consequence in identifying the stream lines, and so, the complex velocity potential. The actual values of the velocity vector, $\bar{w}$, have been almost incidental.

A survey of the examples discussed in the previous sections shows that $q = |w| = 0$ at points of the boundary at which the flow encounters an abrupt change of direction through an angle less than π (into a corner), but around a sharp bend, at an angle greater than π, $q \to \infty$. Our senses, perhaps, tell us that some slow down into a corner, and some speed up around a bend, are natural. However, let us recall Bernoulli's equation, $\frac{1}{2}\rho |w|^2 + p = \text{const} = p_0$, where p_0 is the pressure at points for which $|w| = 0$ (the *rest* pressure). Solving for p, we obtain

$$p = p_0 - \frac{1}{2}\rho |w|^2$$

and as

$$|w| \to \infty \qquad p \to -\infty$$

Figure 8.17

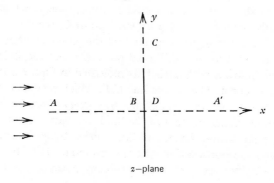

z–plane

Since p is the pressure in our fluid model, long before p reaches $-\infty$ we realistically would have the fluid changing state to a vapor. We have not included this possibility in our original equations, and if we wish to consider flow problems around sharp external corners for which q remains finite, we must accordingly alter our model of the flow.

Such modifications have been studied by various authors, and the interested reader should consult a text on hydrodynamics for such discussions.

We shall content ourselves by considering a specific example. Consider the flow past a plate of finite width, perpendicular to, and superimposed upon, a uniform flow (Figure 8.17). The plate is specified by $x = 0$, $-b \leq y \leq b$, and $w = d\chi/dz \to U$ as $z \to \infty$. From symmetry, $ABCDA'$ will be a stream line, for example $\psi = 0$, and since the fluid must turn through $180°$ at C, we expect $|w| = \infty$ there.

The flow region above this central stream line maps onto the upper half χ-plane (Figure 8.18), and from the Schwarz-Christoffel mapping, we have

$$\chi(z) = U\sqrt{z^2 + b^2}$$

$$w = u - iv = \frac{Uz}{\sqrt{z^2 + b^2}}$$

Flows with Free Boundaries   **283**

Figure 8.18

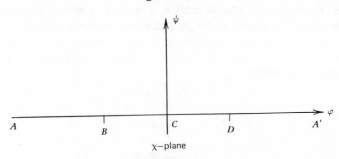

χ—plane

In order to modify our physical model appropriately, we suppose that, because of greatly reduced pressure near the point C, the fluid changes state to a vapor, and that the region occupied by the vapor extends downstream to infinity. Thus, we add the additional postulate that the fluid flow in the upper half-plane follow a path $ABCA'$ as indicated in Figure 8.19, and in the vapor region, $p = $ const. The stream line CA', which bounds the flow region and the vapor region, or *wake*, is called a *free stream line*, or *free boundary*, and the postulated change of state of the fluid to a vapor is called *cavitation*.

We no longer can locate the stream line $ABCA'$ by geometrical considerations alone, and the portion CA', the free boundary is to be found, along with the associated flow as part of the problem. From Bernoulli's equation

$$(2) \qquad \tfrac{1}{2}\rho(u^2 + v^2) + p = \text{const}$$

$$= \tfrac{1}{2}\rho U^2 + p_\infty$$

where p_∞ is the pressure at $z = \infty$, where $u = U, v = 0$. Since our postulated cavity (vapor region) extends to infinity, continuity considerations give us the condition

$$(3) \qquad p = p_\infty$$

Figure 8.19

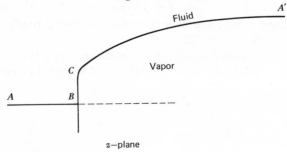

z—plane

for the constant pressure in the vapor region. Further, on the free stream line, continuity of the pressure in the fluid with the pressure in the cavity yield

$$\tfrac{1}{2}\rho(u^2 + v^2) + p_\infty = \tfrac{1}{2}\rho U^2 + p_\infty$$

on CA', the free boundary. This equation yields the "free boundary" condition

(4) $$u^2 + v^2 = U^2 \qquad q = U$$

To restate our problem, we seek a flow, or a complex velocity potential $\chi(z)$, such that $w = d\chi/dz = U$ at infinity, with the lines ABC, AED being stream

Figure 8.20

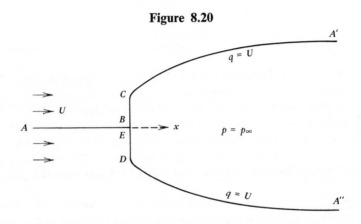

lines, CA' and DA'' being free stream lines (on which $q = U$). Taking $\psi = 0$ on the central stream line $ABCA'$, $AEDA''$, we observe that the image of the required flow region is the full χ-plane (Figure 8.20), except that it is cut along the real axis from the image of B to $+\infty$; the two "banks" of this cut correspond to two different stream lines.

Now, the central stream line is partially determined, by symmetry, to coincide with the negative real axis, and the plate DC, while the free boundaries are restricted by a velocity condition, $q = U$. This suggests that we consider the image of the flow region, not only in the χ-plane, but also in the w-plane, the "velocity plane," or the *hodograph plane*. For convenience, we consider the complex w/U-plane, $w/U = q/U\, e^{-i\sigma}$,

$$\tau = \frac{q}{U}\, e^{-i\sigma} \text{ plane}$$

The free boundaries map onto arcs of $|\tau| = 1$, the unit circle; the point at infinity, at which $q = U$, maps onto $\tau = 1$. On the real axis AB, τ is real, the corner at B is a stagnation point, $\tau = 0$, so that the line AB maps onto the segment $(0, 1)$ of the real τ-axis. Following the path BC, we have $\tau = -i$(real)

Figure 8.21

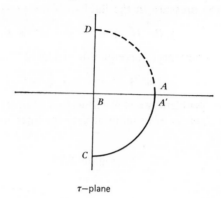

τ–plane

while at C, $\tau = -i$. Proceeding along the free boundary, CA', $|\tau| = 1$, and arg τ increases from $-\pi/2$ towards 0.

In a similar manner, we find that the image of the line $AEDA''$ in the τ-plane is the first quadrant of the unit circle, bounded by the coordinate axes.

What we have found is very important: the image of the required (unknown) flow region of the z-plane is completely determined in the hodograph plane, and this fact enables us to construct the complex velocity potential $\chi(z)$ for the flow.

Comparing Figures 8.21 and 8.20, we see that the image of the boundary of the cavity, $A'CDA''$, in the τ-plane is precisely the right half of the unit circle, bounded by the imaginary axis, the flow region corresponding to the interior of the semicircle (why?).

Let us map the quarter circle $ABCA'$ of the τ-plane (Figure 8.21) onto its image in the χ-plane normalizing χ so that $\chi = 0$ at B (Figure 8.22). The interior of the quarter circle maps onto the upper half of the χ-plane. The mapping is readily accomplished by observing that the quarter circle maps,

Figure 8.22

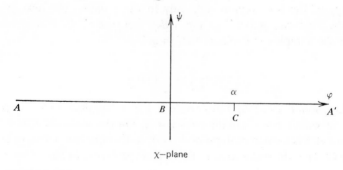

χ–plane

by log τ, to a polygonal region so that the Schwarz-Christoffel mapping gives us the solution directly. We obtain

$$(5) \qquad \tau^{-1} = \sqrt{\frac{\chi - \alpha}{\chi}} - i\sqrt{\frac{\alpha}{\chi}}$$

where α is the location of C in the χ-plane (to be determined).
Next,

$$\tau = \frac{w}{U} = \frac{1}{U}\frac{d\chi}{dz}$$

this result gives us

$$(6) \qquad U\,dz = \left(\sqrt{\frac{\chi - \alpha}{\chi}} + i\sqrt{\frac{\alpha}{\chi}}\right) d\chi$$

that is, the connection between the flow region and the complex velocity potential. Integrating Section 8.6 (6) along the central stream line,

$$(7) \qquad U\left(z - \frac{\pi i}{2}\right) = \int_\alpha^\chi \frac{\sqrt{t - \alpha} + i\sqrt{\alpha}}{\sqrt{t}}\,dt$$

Integrating to $\chi = 0$, at B, gives $z = 0$, or,

$$U\left(-i\frac{\pi}{2}\right) = \int_\alpha^0 \frac{i(\sqrt{\alpha - t} + \sqrt{\alpha})}{\sqrt{t}}\,dt$$

$$= -i\left(\frac{\pi\alpha}{2} + 2\alpha\right)$$

Thus,

$$(8) \qquad \alpha = \frac{\pi U}{\pi + 4}$$

and the parametric equations of the free stream line CA', from Section 8.6 (7), are

$$(9) \qquad Ux = \int_\alpha^\chi \sqrt{\frac{t - \alpha}{t}}\,dt$$

$$U\left(\frac{\pi}{2} - y\right) = \int_\alpha^\chi \sqrt{\frac{\alpha}{t}}\,dt$$

8.6.1 The following exercise illustrates another free boundary flow. Follow the indicated steps, and verify the mapping functions given (Figure 8.23). Fluid occupies the infinite reservoir above the real axis of the z-plane,

Figure 8.23

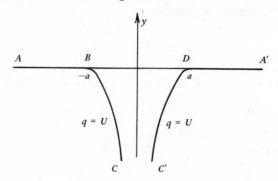

and emerges from the slit $-a < x < a$, forming two free stream lines BC, DC' extending to infinity, and on which $q = U$. Our aim is to find the ratio k of the limiting width between the free boundaries and the entrance width, $2a$. This is the problem of the *vena contracta*. See Figure 8.24.

(a) The image of the flow region is the strip in the χ-plane,

$$\psi = 0 \text{ corresponding to the stream line } x = 0$$

(b) The image of the flow region in the hodograph plane $\tau = (q/U) e^{-i\sigma}$, and in the $v = \log \tau$ plane is shown in Figure 8.24.

(c) Introduce an auxiliary complex plane, the t-plane, and map both the v-plane polygon, and the χ-plane channel, onto the upper half t-plane.

$$v = i \arc \sin t + \frac{\pi i}{2}$$

$$\chi = -2k \frac{Ua}{\pi} \log t + ikUa$$

$$\frac{d\chi}{dz} = \frac{d\chi}{dt}\frac{dt}{dz} = -\frac{2kUa}{\pi t}\frac{dt}{dz} = Uie^{i \arc \sin t}$$

$$dz = \frac{2ka}{\pi t} i(\sqrt{(1 - t^2)} - it)\, dt$$

Integrate from B, along BC,

$$x + a + iy = \frac{2ka}{\pi} \int_{-1}^{t} \left(1 + i\frac{\sqrt{1 - t^2}}{t}\right) dt$$

Figure 8.24

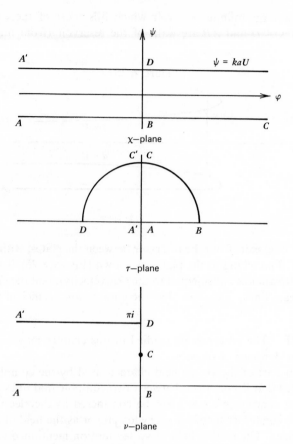

χ–plane

τ–plane

ν–plane

t–plane

Let

$$t \to 0 \qquad a - ka = \frac{2ka}{\pi}$$

$$k = \frac{\pi}{\pi + 2}$$

8.6.2 THE BORDA MOUTHPIECE

Fluid occupies an infinite reservoir which fills most of the z-plane. Two parallel plates $A'D$ and AB are walls of the reservoir (from infinity to the

Figure 8.25

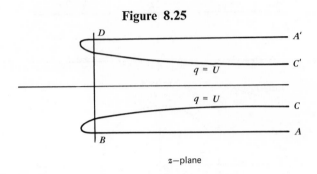

z—plane

y-axis), and fluid exits from the reservoir between the plates, with free stream lines DC', CB attaching to the plates as shown (Figure 8.25). Taking $q = U$ on the free stream lines (also equal to the exit velocity), find the equations for the free stream lines, and show that the contraction coefficient is $k = \frac{1}{2}$.

COMMENT. The vector fields studied in this chapter may be interpreted also in other domains of physics.

Thus, that part of the vector field characterized by the complex potential Section 8.2(2) that is contained between two equipotential lines (an annulus between two concentric circles) can be considered as the electrostatic field between two condenser plates of a Leyden jar, or as the field of heat flow in a high chimney. The related field of vortex motion mentioned at the end of Section 8.2 can be considered as the magnetic field generated by a stationary electric current flowing in an infinitely long straight wire.

There are other physical applications of the theory of analytic functions, e.g., to certain problems of elasticity, which differ more from those studied in this chapter.

In this chapter we study certain functions for large values of the variable. We shall develop the essential concept in the second section, starting from an example in the first section.

9.1 ASYMPTOTIC SERIES

Consider the function

$$(1) \qquad f(x) = \int_0^\infty e^{-xt} \frac{dt}{1 + t}$$

defined, as shown, by an improper integral, convergent for all $x > 0$. We wish to study $f(x)$ for large values of x, or, more sharply expressed, as $x \to \infty$.

Thus, $f(x) \geqq 0$, and since $1/(1 + t) \leqq 1$ for $0 \leqq t < \infty$, we have

$$f(x) \leqq \int_0^\infty e^{-xt} \, dt = \frac{1}{x}$$

It is clear that $f(x) \to 0$ as $x \to \infty$.

Suppose we ask for some finer detail: for example, does $\lim_{x \to \infty} x f(x)$ exist? The answer becomes clear when we integrate (1) by parts, obtaining

$$(2) \qquad f(x) = \frac{1}{x} - \frac{1}{x} \int_0^\infty e^{-xt} \frac{dt}{(1 + t)^2}$$

As shown above, we have

$$0 \leqq \int_0^\infty e^{-xt} \frac{dt}{(1 + t)^2} \leqq \int_0^\infty e^{-xt} \, dt = \frac{1}{x}$$

and consequently

$$f(x) = \frac{1}{x} - \left(\text{a quantity smaller than } \frac{1}{x^2} \right)$$

Thus

$$\lim_{x \to \infty} xf(x) = 1$$

We may repeat this procedure, and after n integrations by parts, we obtain

$$(3) \qquad f(x) = \frac{1}{x} - \frac{1!}{x^2} + \frac{2!}{x^3} - \frac{3!}{x^4} + \cdots + (-1)^{n-1}\frac{(n-1)!}{x^n} + R_n$$

$$R_n = (-1)^n \frac{n!}{x^n} \int_0^\infty \frac{e^{-xt}\,dt}{(1+t)^{n+1}}$$

with

$$|R_n(x)| \leqq \frac{n!}{x^n} \int_0^\infty e^{-xt}\,dt = \frac{n!}{x^{n+1}}$$

One of the first things we might investigate is the full infinite series suggested by (3),

$$(4) \qquad \sum_{n=0}^{\infty} (-1)^n \frac{n!}{x^{n+1}}$$

However, this series diverges for every x, no matter how large. For a long time, mathematicians, obsessed by a need for complete rigor, regarded such divergent series (4) as nonsense. However, in the sense that will be described in this section, this and similar series are completely meaningful, and of great importance in many parts of applied analysis.

Let us return to the finite series (3), and consider the result for some specific value of x, for example, $x = 10$. The series alternates in sign, and however many terms we take, the *magnitude of the "remainder, R_n" is less than that of the first omitted term.* Thus

$$f(10) = \frac{1}{10} - R_1 \qquad\qquad |R_1| \leqq \frac{1}{10^2}$$

$$= \frac{1}{10} - \frac{1}{10^2} + R_2 \qquad\qquad |R_2| \leqq \frac{2}{10^3}$$

$$= \frac{1}{10} - \frac{1}{10^2} + \frac{2!}{10^3} - R_3 \qquad |R_3| \leqq \frac{3!}{10^4}$$

and so on. The remainders decrease in size till $n = 10$, after which the $n!$ in the numerator grows faster than the powers of 10 in the denominator. If we stop at 10 terms, we have

$$f(10) = \frac{1}{10} - \frac{1}{10^2} + \cdots - \frac{9!}{10^{10}} + R_{10}$$

with $|R_{10}| \leqq 10!/10^{11}$. Thus if we neglect R_{10}, we have $f(10)$ correct to four significant figures ($|R_{10}| < 2.8 \times 10^{-4}$)

For a larger value of x, for example $x = 100$, three terms of the series leave an error term $|R_3| \leq 6/10^8$. We see that for large x, a few terms (but not too many) of the series (4) serve to approximate $f(x)$ with remarkable accuracy, even though the series diverges for all x. The series (4) is an example of an *asymptotic series*, and we write

$$(5) \qquad f(x) \sim \sum_0^\infty (-1)^n \frac{n!}{x^{n+1}}$$

This notation is suggestive, and carries the following precise meaning. We define $f(z) \sim g(x)$ as $x \to \infty$ to stand for the statement

$$\lim_{x \to \infty} \frac{f(x)}{g(x)} = 1$$

If no confusion can arise, and it is clear that $x \to \infty$ is implied, we may abbreviate even further, and write merely

$$f(x) \sim g(x)$$

with this notation, we have, for our example

$$f(x) \sim \frac{1}{x}$$

since

$$\lim_{x \to \infty} \frac{f(x)}{\dfrac{1}{x}} = 1$$

It is also true that

$$f(x) \sim \frac{1}{x} - \frac{1}{x^2}$$

since

$$f(x) = \frac{1}{x} - \frac{1}{x^2} + R_2 \qquad |R_2| \leq \frac{2}{x^3}$$

and so

$$\frac{f(x)}{\dfrac{1}{x} - \dfrac{1}{x^2}} = 1 + \frac{R_2}{\dfrac{1}{x} - \dfrac{1}{x^2}} \to 1 \text{ as } x \to \infty$$

Thus, when we write

$$f(x) \sim \sum_0^\infty (-1)^n \frac{n!}{x^{n+1}}$$

the notation implies that for each integer k, and each partial sum

$$S_k(x) = \sum_0^{k-1} (-1)^n \frac{n!}{x^{n+1}}$$

it is true that $f(x) \sim S_k(x)$. However, there is even more to be added to the strict concept of asymptotic series. We give a precise definition in the next section.

9.2 NOTATION AND DEFINITIONS

We digress, for the moment, from our study of asymptotic series in order to introduce some additional notation. We shall be faced with certain functions about which we shall repeatedly have to say "a function that grows, as $x \to \infty$, no faster than x ," or, "a function that, when divided by x , tends to zero, as $x \to \infty$." A very useful contraction of these unwieldy statements follows: in the first instance we write $f(x) = O(x)$, where "O" is a "capital O," or "big O," and read "$f(x)$ *is big O of* x"; in the second instance we write $f(x) = o(x)$; now "o" is a "lower case o," or "small o," and we read "$f(x)$ *is little o of x*" (as $x \to \infty$).

DEFINITION. If $|f(x)/g(x)|$ is bounded as $x \to \infty$, we write $f(x) = O(g(x))$, "big O"; if $f(x)/g(x) \to o$ as $x \to \infty$, we write $f(x) = o(g(x))$, "little o."

For example, from Section 9.1 (3), we can write a variety of statements:

$$f(x) = O\left(\frac{1}{x}\right) \qquad f(x) \sim \frac{1}{x} \qquad f(x) = o(1)$$

The first result merely states that

$$\frac{f(x)}{\frac{1}{x}} = xf(x)$$

is bounded; the second says that $\lim_{x \to \infty} x f(x) = 1$, and implies the first statement. The third result says $f(x)/1 \to 0$ as $x \to \infty$, that is, $\lim_{x \to \infty} f(x) = 0$. The symbols $O(1)$, $o(1)$, are "generic symbols" representing, respectively, any absolutely bounded function, and any function that tends to zero (as $x \to \infty$). We can, when desired, use the same notation to describe behavior as $x \to a$ rather than $x \to \infty$.

We can now present a precise definition of asymptotic series. From Section 9.1 (1),

$$f(x) = \frac{1}{x} - \frac{1}{x^2} + \frac{2!}{x^3} - \cdots + (-1)^{n-1}\frac{(n-1)!}{x^n} + R_n(x)$$

$$= S_n(x) + R_n(x)$$

$$|R_n| \leqq \frac{n!}{x^{n+1}}$$

Then,

$$|f(x) - S_n(x)| = |R_n(x)| \leq \frac{n!}{x^{n+1}}$$

so that

$$f(x) = S_n(x) + O\left(\frac{1}{x^{n+1}}\right)$$

$$x^n |f(x) - S_n(x)| = o(1)$$

are both correct statements. We adopt Poincaré's definition of asymptotic series: the series $\sum_0^\infty a_n/x^{n+1}$ is called an asympotic expansion of $f(x)$, as $x \to \infty$ if

(1) $f(x) \to 0$ as $x \to \infty$

(2) $\lim\limits_{x \to \infty} x^n[f(x) - S_n(x)] = 0$

for each integer n, and partial sum

$$S_n(x) = \frac{a_0}{x} + \frac{a_1}{x^2} + \cdots + \frac{a_{n-1}}{x^n}$$

and write

$$f(x) \sim \sum_0^\infty \frac{a_n}{x^{n+1}}$$

9.2.1 Show that $e^{-x} = o(x^{-n})$ for each integer n, as $x \to \infty$.

9.2.2 Let $a_1, a_2, \ldots a_k$

$\qquad b_1, b_2, \ldots b_k$

denote constants,

$$a_1 \neq 0, \qquad b_1 > b_2 > b_3 > \cdots > b_k$$

Then, for $x \to \infty$

$$a_1 e^{b_1 x} + a_2 e^{b_2 x} + \cdots + a_k e^{b_k x} \sim a_1 e^{b_1 x}$$

9.2.3

$$f(x) = \frac{1}{1+x} = \sum_0^\infty \frac{(-1)^n}{x^{n+1}} \qquad |x| > 1$$

Show that

$$f(x) \sim \sum_0^\infty \frac{(-1)^n}{x^{n+1}}$$

9.2.4 If

$$f(x) = \sum_0^\infty \frac{a_n}{x^{n+1}}$$

converges for $|x| > R$, show that also

$$f(x) \sim \sum_0^\infty \frac{a_n}{x^{n+1}} \qquad \text{as } x \to \infty$$

9.2.5 Show that $\sin x = O(1)$ as $x \to \infty$, but that $1 = O(\sin x)$ is false.

9.2.6 Show that $\int_\delta^\infty e^{-xt}/(1+t)\,dt = o(x^{-n})$ for all n (as $x \to \infty$), provided $\delta > 0$.

9.2.7 If $e^{-x} \sim \sum_0^\infty a_n/x^{n+1}$ as $x \to \infty$, show that

$$a_0 = a_1 = \cdots = a_n = \cdots = 0$$

9.2.8 If $f(x) \sim \sum_0^\infty a_n/x^{n+1}$ show that with

$$S_n(x) = \frac{a_0}{x} + \frac{a_1}{x^2} + \cdots + \frac{a_{n-1}}{x^n}$$

$$f(x) = S_n(x) + o\left(\frac{1}{x^n}\right)$$

$$f(x) = S_n(x) + O\left(\frac{1}{x^{n+1}}\right) \quad \text{for each } n = 1, 2, \ldots$$

9.3 MANIPULATING ASYMPTOTIC SERIES

The asymptotic series we are concerned with are of the form

$$\sum_0^\infty \frac{a_n}{x^{n+1}} \qquad \text{or} \qquad \sum_1^\infty \frac{b_n}{x^n}$$

they are "formal" power series, in powers of $1/x$. The term "formal" refers to the fact that we are not concerned with the convergence of these series; indeed, they may well diverge for all x. However, in using these series, we deal with a finite number of terms only, so that convergence is not required.

(1) If $f(x)$, $g(x)$ both tend to zero as $x \to \infty$, and if

$$f(x) \sim \sum_1^\infty \frac{a_n}{x^n} \qquad g(x) \sim \sum_1^\infty \frac{b_n}{x^n}$$

then

$$f(x) + g(x) \sim \sum_{n=1}^\infty \frac{a_n + b_n}{x^n}$$

and

$$f(x)g(x) \sim \sum_2^\infty \frac{c_n}{x^n} \qquad c_n = \sum_{r=1}^n a_r b_{n-r}$$

These rules are easy to remember: they are the same rules that apply to convergent power series (which are also asymptotic series). The proof is immediate. Let $S_n(x)$, $\Sigma_n(x)$ be the partial sums of the asymptotic series for $f(x)$, $g(x)$ respectively. Then

$$x^n[f(x) - S_n(x)] = o(1)$$

$$x^n[g(x) - \Sigma_n(x)] = o(1) \text{ for each } n = 1, 2, \ldots,$$

Adding,

$$x^n\{f(x) + g(x) - [S_n(x) + \Sigma_n(x)]\} = o(1)$$

which is to say,

$$f(x) + g(x) \sim \sum_1^\infty \frac{a_n + b_n}{x^n}$$

The product rule can be easily visualized if we write

$$f(x) = S_p(x) + O\left(\frac{1}{x^{p+1}}\right)$$

$$g(x) = \Sigma_q(x) + O\left(\frac{1}{x^{q+1}}\right)$$

Then,

$$f(x) \cdot g(x) = S_p(x) \cdot \Sigma_q(x) + O\left(\frac{1}{x^{r+1}}\right) \qquad r = \min(p, q)$$

multiplying the two polynomials S_p, Σ_q together, and ignoring terms $1/x^{n+1}$ and higher, we find

$$S_p(x)\Sigma_q(x) = \sum_2^n \frac{c_k}{x^k} + O\left(\frac{1}{x^{n+1}}\right)$$

$$c_k = \sum_{r=1}^k a_r b_{k-r}$$

Thus if $p, q > n$, we have

$$f(x) \cdot g(x) = \sum_2^n \frac{c_k}{k} + O\left(\frac{1}{x^{n+1}}\right)$$

for each n

$$x^n\left[f(x) \cdot g(x) - \sum_2^n \frac{c_k}{x^k}\right] = o(1)$$

(2) If $f(x) \sim \sum_2^\infty a_n/x^n$ (with the additional assumption that $f(x)$ behaves like x^{-2} as $x \to \infty$), then $\int_x^\infty f(t)\, dt$ exists, and

$$\int_x^\infty f(t)\, dt \sim \sum_2^\infty \frac{a_n}{(n-1)x^{n-1}}$$

That is, we can integrate term by term, obtaining the asymptotic series for the integrated function.

The proof is uncomplicated: we note that the expression

$$\int_x^\infty o\left(\frac{1}{t^{n+1}}\right) dt \text{ stands for } \int_x^\infty h(t)\, dt \qquad |h(t)| \le \frac{M}{t^{n+1}}$$

$$\left| \int_x^\infty h(t)\, dt \right| \le M \int_x^\infty \frac{dt}{t^{n+1}} \le \frac{M}{nx^n} = o\left(\frac{1}{x^n}\right)$$

Thus,

$$\int_x^\infty o\left(\frac{1}{t^{n+1}}\right) dt = o\left(\int_x^\infty \frac{dt}{t^{n+1}}\right) = o\left(\frac{1}{x^n}\right)$$

We now write

$$f(x) = \sum_2^n \frac{a_k}{x^k} + o\left(\frac{1}{x^{n+1}}\right)$$

$$\int_x^\infty f(t)\, dt = \sum_2^n \frac{a_k}{(k-1)x^{k-1}} + \int_x^\infty o\left(\frac{1}{t^{n+1}}\right) dt$$

$$= \sum_2^n \frac{a_k}{(k-1)x^{k-1}} + o\left(\frac{1}{x^n}\right)$$

which implies

$$x^n\left(\int_x^\infty f(t)\, dt - \sum_2^{n+1} \frac{a_k}{(k-1)x^{k-1}}\right) = o(1)$$

. (3) When they exist, asymptotic series are unique. That is, given $f(x) \to 0$ as $x \to \infty$, and

$$f(x) \sim \sum_1^\infty \frac{a_n}{x^n} \qquad f(x) \sim \sum_1^\infty \frac{b_n}{x^n}$$

then $a_n = b_n$, $n = 1, 2, 3, \ldots$.

The proof follows directly from the definition of asymptotic series. Thus

$$x\left[f(x) - \frac{a_1}{x}\right] = o(1) \qquad x\left[f(x) - \frac{b_1}{x}\right] = o(1)$$

or,

$$\lim_{x \to \infty} xf(x) = a_1 \qquad \lim_{x \to \infty} xf(x) = b_1 \qquad a_1 = b_1$$

In precisely the same way,

$$x^2\left[f(x) - \frac{a_1}{x} - \frac{a_2}{x^2}\right] = o(1) \qquad x^2\left[f(x) - \frac{b_1}{x} - \frac{b_2}{x^3}\right] = o(1)$$

yields

$$a_2 = b_2 = \lim_{x\to\infty} x^2\left[f(x) - \frac{a_1}{x}\right]$$

and so on.

(4) The converse statement to (3) is not correct. If $\sum_1^\infty a_n/x^n \sim f(x)$, then $f(x)$ is not uniquely determined. There are many functions, such as e^{-x}, which have asymptotic series, each of whose coefficients is zero. In fact,

$$e^{-x} \sim \sum_1^\infty \frac{a_n}{x^n} \qquad \text{if, and only if} \qquad a_n = 0,\ n = 1, 2, \ldots$$

(see Example 9.2.7).

Clearly, given $\sum_1^\infty a_n/x^n \sim f(x)$, then $\sum_1^\infty a_n/x^n \sim f(x) + e^{-x}$ is equally valid. Given an asymptotic expansion, there are many functions having this expansion.

(5) Very frequently, in specific problems, we content ourselves with just one term of an asymptotic expansion: with the first nonvanishing term that we call the *dominant term* of the expansion. This one term is often all that we need in a given problem; if ever we need more terms, the same ideas may lead us to their expressions.

Also, it may well occur that the function $f(x)$ we are studying does *not* have an asymptotic expansion of the form we are considering. For example $f(x) = e^x/(x-1)$.

However, $e^{-x}f(x) = 1/(x-1)$ does have an asymptotic expansion of our type,

$$\frac{1}{x-1} \sim \sum_0^\infty \frac{1}{x^{n+1}}$$

$$e^{-x}f(x) \sim \sum_0^\infty \frac{1}{x^{n+1}}$$

and the dominant term is $1/x$. In this case, we say that the dominant term, as $x \to \infty$, for $f(x)$ is e^x/x. (For other types of asymptotic series see Example 9.3ff.)

(6) Let us return to the example we studied in Section 9.1.

(1)
$$f(x) = \int_0^\infty e^{-xt}\frac{dt}{1+t}$$

and reexamine the construction of its asymptotic series, which we now look for in the form

$$f(x) \sim \sum_0^\infty \frac{a_n}{x^{n+1}}$$

Writing

$$\int_0^\infty e^{-xt} \frac{dt}{1+t} = \int_0^\delta \frac{e^{-xt}}{1+t} \, dt + \int_\delta^\infty e^{-xt} \frac{dt}{1+t}$$

and noting that

$$\left| \int_\delta^\infty e^{-xt} \frac{dt}{1+t} \right| \leq \int_\delta^\infty e^{-xt} \, dt$$

$$= \frac{e^{-x\delta}}{x}$$

$$= o(x^{-n}) \text{ for every } n.$$

we recognize that the full asymptotic series for $\int_\delta^\infty e^{-xt}/(1+t) \, dt$ is identically zero, for each $\delta > 0$. Thus

$$f(x) = \int_0^\infty \frac{e^{-xt}}{1+t} \, dt$$

$$\sim \int_0^\delta \frac{e^{-xt}}{1+t} \, dt$$

$$= \int_0^\delta \sum_0^\infty (-1)^n t^n e^{-xt} \, dt \qquad \text{when } \delta < 1$$

$$= \sum_0^\infty (-1)^n \int_0^\delta t^n e^{-x} \, dt$$

$$= \sum_0^\infty (-1)^n \left[\int_0^\infty t^n e^{-xt} \, dt - \int_\delta^\infty e^{-xt} t^n \, dt \right]$$

$$\sim \sum_0^\infty (-1)^n \int_0^\infty t^n e^{-xt} \, dt$$

$$= \sum_0^\infty (-1)^n \frac{n!}{x^{n+1}}$$

We have used $\int_\delta^\infty e^{-xt} t^n \, dt = o(x^{-m})$ for all m. We might see this result in the following way: $t = \delta + \tau$ gives

$$\int_0^\infty e^{-x\delta} e^{-x\tau} (\delta + \tau)^n \, d\tau = e^{-x\delta} \left(\text{polynomial in } \frac{1}{x} \right)$$

by expanding $(\delta + \tau)^n$ in the binomial series.

(7) The example we have been studying,

$$f(x) \sim \sum_0^\infty (-1)^n \, \frac{n!}{x^{n+1}}$$

suggests one further essential distinction between asymptotic series and convergent series. Thus, for a specific value of x, say $x = 10$, we get $f(10)$ to the best possible accuracy by taking 10 terms of the asymptotic series (here, the remainder has its least value). While this accuracy is good enough for most practical purposes, it is also the best we can do. On the other hand, in a convergent series we can come as close to the correct value as we desire merely by taking enough terms.

9.3.1   Let

$$f(x) = \int_0^\infty \frac{e^{-xt}}{1+t^2} \, dt$$

show that

$$f(x) \sim \sum_0^\infty \frac{(-1)^n (2n)!}{x^{2n+1}}$$

9.3.2   If $f(x)$ is bounded, $|f(x)| \leq M$, show that

$$\int_\delta^\infty e^{-xt} f(t) \, dt = O(e^{-x\delta}) \qquad \delta > 0$$

9.3.3   Given

$$f(x) = \int_0^\infty e^{-xt^2} \frac{dt}{1+t}$$

show that

$$f(x) \sim \int_0^\delta \frac{e^{-xt^2}}{1+t} \, dt \qquad \text{as } x \to \infty$$

and that

$$f(x) \sim \sum_0^\infty \frac{(-1)^n \Gamma\left(\dfrac{n}{2} + \dfrac{1}{2}\right)}{2x^{n/2+1/2}}$$

(In this example, we have series in powers of $1/\sqrt{x}$; this series is also called an asymptotic series; the substitution $x = y^2$ transforms it into the more usual form.)

Manipulating Asymptotic Series   **301**

9.3.4   Find the dominant term, as $x \to \infty$, of

$$f(x) = \int_0^1 e^{-xt^3}\, dt$$

9.3.5   If $f(x)$ is analytic for $|x| < R$, $f(x) = \sum_0^\infty a_n x^n$, and $f(x)$ is continuous on the x-axis, show that

$$\int_0^1 f(t)e^{-xt}\, dt \sim \sum_0^\infty a_n \frac{n!}{x^{n+1}}$$

9.3.6   If $f(x)$ is continuous, $\int_0^1 f(t)\, e^{-xt}\, dt \sim f(0)/x$ and

$$\int_0^1 f(t)e^{-xt^2}\, dt \sim \frac{f(0)}{2}\sqrt{\frac{\pi}{x}}$$

for the dominant terms.

9.4 LAPLACE'S ASYMPTOTIC FORMULA

In this section, we examine some definite integrals of the form

$$(1) \qquad\qquad I_n = \int_a^b \phi(x)[f(x)]^n\, dx$$

in their dependence on the parameter n, for n very large. Integrals of this form were studied extensively by Laplace, who came across them in his investigation of the probabilities in a very large number n of trials. In the cases he studied, the functions $\phi(x)$ and $f(x)$ are connected with probabilities and are nonnegative. For our purposes we shall retain the assumption $f(x) > 0$, and otherwise require only such continuity conditions on $\phi(x)$ and $f(x)$ as are useful for our purposes.

We are concerned with finding the dominant term, as $n \to \infty$, for the integral I_n of (1). Since $f(x) > 0$, we may write $h(x) = \log f(x)$, and take as our normalized form

$$(2) \qquad\qquad I_n = \int_a^b \phi(x)e^{nh(x)}\, dx$$

If we compare this integral (2) with the example in Section 9.1 (1) it should be clear that the largest contribution to the integral should be governed mostly by those values x at which $h(x)$ is largest. In order to perceive these ideas more clearly, let us first examine a special case. Let $h(x)$ have its maximum value at $x = a$, the left end point, and steadily decrease as x increases. For convenience (and by setting $x = a + t$), we may assume the interval to be $0 \leqq t \leqq c$, and

$$I_n = \int_0^c \phi(t)e^{nh(t)}\, dt$$

$$= e^{nh(0)}\int_0^c \phi(t)e^{n[h(t)-h(0)]}\, dt$$

$$h(t) < h(0) \qquad\qquad \text{for } 0 < t \leqq c$$

Thus

$$e^{-nh(0)}I_n = \int_0^c \phi(t)e^{n[h(t)-h(0)]}\, dt$$

We may anticipate a more general result: this integral has a dominant term which behaves like a power of n. Since $h(t) - h(0) = 0$ at $t = 0$, and decreases as t increases, then $h(t) - h(0) \leqq -\delta_1 < 0$ for $t \geqq \delta$. Thus

(3) $$e^{-nh(0)}I_n = \int_0^\delta \phi(t)e^{n[h(t)-h(0)]}\, dt + O(e^{-n\delta_1})$$

$$e^{-nh(0)}I_n \sim \int_0^\delta \phi(t)e^{n[h(t)-h(0)]}\, dt$$

and this result holds for every fixed $\delta > 0$, no matter how small. It is now clearer that the dominant term will depend only on the values of $\phi(x)$ and $h(x)$ near $x = 0$.

In order to proceed further with the asymptotic nature of (3), we require more information about $\phi(x)$ and $h(x)$ near $x = 0$. We shall not attempt to treat "the most general case"; instead, we consider two cases of most common occurrence. In fact, we consider only those $h(x)$ for which $h''(x)$ is continuous on $0 \leqq x \leqq c$, and $\phi(x)$ continuous on $0 \leqq x \leqq c$ [a more involved analysis is possible in which only $h'(x)$ need be continuous, instead of $h''(x)$].

(a) $h(x)$ has a maximum at $x = 0$

$$h'(0) = -k, \qquad k > 0$$

Then

$$h(x) - h(0) = -k\, x\psi(x)$$

defines

$$\psi(x), \qquad \psi(0) = 1$$

with $\psi'(x)$ continuous.

In this case, "$h(x)$ behaves linearly for small x."

The above assumptions permit us to take $t\,\psi(t) = x$, and regard t as a (single-valued) function of x for all t sufficiently small. (For this purpose, it is enough that $d/dt\,(t\,\psi(t)) = \psi(t) + t\,\dot\psi(t)$ be nonzero for all t small. This is guaranteed since $\psi(0) = 1$ and $\dot\psi(t)$ is continuous.) Equation 3 transforms into

(4) $\quad e^{-nh(0)}I_n \sim \displaystyle\int_0^{\delta\psi(\delta)} \phi[t(x)]e^{-nkx}\left(\frac{dt}{dx}\right)dx$

$$= \int_0^{\delta_1} g(x)e^{-nkx}\,dx \qquad \text{where } g(x) = \phi[t(x)]\frac{dt}{dx}$$

$$= \int_0^{\delta_1} [g(x) - g(0)]e^{-nkx}\,dx + g(0)\int_0^{\delta_1} e^{-nkx}\,dx$$

$$= \int_0^{\delta_1} [g(x) - g(0)]e^{-nkx}\,dx + g(0)\int_0^{\infty} e^{-nkx}\,dx + O(e^{-nk\delta_1})$$

$$\sim \frac{g(0)}{nk} + \int_0^{\delta_1} [g(x) - g(0)]e^{-nkx}\,dx$$

If $g(0) \neq 0$, the dominant term of (4) is $g(0)/nk$. This fact can readily be seen as follows.

Since δ_1 is any positive number, no matter how small, and since $g(x)$ is continuous, $|g(x) - g(0)| < \epsilon$, ϵ arbitrary, by taking x small. Thus

$$\left|\int_0^{\delta} [g(x) - g(0)]e^{-nkx}\,dx\right| \leq \int_0^{\delta} |g(x) - g(0)|\,e^{-nkx}\,dx$$

$$< \epsilon \int_0^{\delta} e^{-nkx}\,dx \qquad \text{for all } \delta \text{ small}$$

$$< \frac{\epsilon}{nk}$$

Hence

$$\left[e^{-nh(0)}I_n - \frac{g(0)}{nk}\right]n \sim 0 \qquad \text{and} \qquad I_n \sim e^{nh(0)}\frac{g(0)}{nk}$$

$$g(0) = \phi[t(0)]\frac{dt}{dx}\bigg|_{t=0} = \phi(0)$$

$$k = -h'(0)$$

Collecting these results, we have

(5) $\qquad\qquad I_n = \displaystyle\int_0^{c} \phi(t)e^{nh(t)}\,dt$

$$\sim \int_0^{\delta} \phi(t)e^{n[h(t)-h(0)]}\,dt\; e^{nh(0)}$$

$$\sim e^{nh(0)}\frac{\phi(0)}{-nh'(0)}$$

valid for $\phi(x)$ continuous, $\phi(0) \neq 0$,

$$h(x) < h(0) \quad \text{for} \quad x > 0, \qquad h'(0) = -k < 0$$

and $h''(x)$ continuous.

(b) In the previous case, $h(x)$ behaved like $h(0) - kx$ near $x = 0$; that is, $h(x)$ was essentially linear near $x = 0$. We now consider $h(x) - h(0) = -kx^2\psi(x)$, $\psi''(x)$ continuous, $\psi(0) = 1$, so that $h(x) - h(0)$ behaves quadratically near $x = 0$. Then

(6)
$$e^{-nh(0)}I_n = \int_0^c \phi(t)e^{n[h(t)-h(0)]}\, dt$$

$$= \int_0^\delta \phi(t)e^{n[h(t)-h(0)]}\, dt + O(e^{-n\delta})$$

$$\sim \int_0^\delta \phi(t)e^{-nkt^2\psi(t)}\, dt$$

$$= \int_0^{\delta\sqrt{\psi(\delta)}} \phi[t(x)]e^{-nkx^2}\left(\frac{dt}{dx}\right) dx$$

with

$$t^2\psi(t) = x^2 \qquad 0 \leq t \leq \delta$$

With reasoning analogous to the one in (a), we obtain the dominant term

(7)
$$e^{-nh(0)}I_n \sim \phi(0)\int_0^\infty e^{-nkx^2}\, dx$$

$$= \frac{\phi(0)}{2}\sqrt{\frac{\pi}{nk}}$$

$$= \frac{\phi(0)}{2}\sqrt{\frac{2\pi}{-nh''(0)}}$$

(c) A particular case involving (7) is worth noting:

(8)
$$I_n = \int_a^b \phi(x)e^{nh(x)}\, dx$$

where $h(x)$ has a unique maximum at $x = \xi$, $a < \xi < b$; $\phi(x)$ is continuous, $\phi(\xi) \neq 0$, and $h''(x)$ is continuous near $x = \xi$. The numbers a, b may be finite or infinite (if they are infinite, we assume the absolute convergence of the integral).

This special case can be treated by two applications of (7), translating $x = \xi + t$, $x = \xi - t$ to obtain two versions of (7). These yield

$$(9) \qquad I_n = \int_a^b \phi(x) e^{nh(x)} \, dx$$

$$\sim \phi(\xi) e^{nh(\xi)} \sqrt{\frac{2\pi}{-nh''(\xi)}}$$

Example.   Consider the problem of finding the *dominant* behavior of $n! = \int_0^\infty t^n e^{-t} \, dt$. As it stands, this integral is not of the form we have been studying. Here, $h(t) = \log t$, which increases from $-\infty$ to $+\infty$ as t runs from 0 to ∞.

Not every such integral we encounter will be in our standard form. We must be prepared to try, by various transformations, to bring them into the form we have been discussing. In this instance, the transformation $t = nx$ in the integral above yields

$$(10) \qquad n! = n^{n+1} \int_0^\infty (xe^{-x})^n \, dx$$

or

$$\frac{n!}{n^{n+1}} = \int_0^\infty e^{n[\log t - t]} \, dt$$

Now $h(t) = \log t - t$, as $h'(t) = (1/t) - 1$, has a unique (finite) maximum at $t = 1$, $h(1) = -1$. Equation 9 applies, and we find

$$(11) \qquad n! \sim \sqrt{2\pi n} \left(\frac{n}{e}\right)^n$$

This formula, called Stirling's formula, expresses the asymptotic value of $n!$ in terms of readily computed quantities.

9.4.1   Show that

$$I_n = \int_0^{\pi/2} \sin^n \theta \, d\theta \sim \frac{1}{2} \sqrt{\frac{2\pi}{n}}$$

9.4.2   Establish that

$$\Gamma(1 + x) \sim \sqrt{2\pi x} \left(\frac{x}{e}\right)^x \qquad \text{as } x \to \infty$$

9.4.3   Take $x = n + a$ in 9.4.2, and show that $\Gamma(n + a + 1) \sim n^a \Gamma(n + 1)$ or, what is the same thing,

$$\Gamma(n + a) \sim n^a \, \Gamma(n), \qquad \text{as } n \to \infty$$

9.4.4   Consider

$$I_n = \int_0^1 x^{\alpha-1}(1-x)^n \, dx = \frac{\Gamma(\alpha)n!}{\Gamma(n+\alpha+1)}$$

Find the dominant term of I_n, and establish the result of 9.4.3.

9.5   PERRON'S EXTENSION OF LAPLACE'S FORMULA

In this section we consider the problem of determining the dominant term (or the first few terms) of the asymptotic expansion of integrals of the form

$$(1) \qquad\qquad I_n = \int_0^{z_0} z^{\alpha-1} f(z) e^{nh(z)} \, dz$$

where $f(z)$, $h(z)$ are single-valued analytic functions in a neighborhood of $z = 0$, α is a fixed complex number, $\mathbf{R}\alpha > 0$, and z_0 a fixed limit of integration. We visualize that many problems of interest can be transformed into this form. (See, for example, Section 9.4 (3).) Finally, we assume that $h(0) = 0$. Thus

$$(2) \qquad\qquad h(z) = b_p z^p + b_{p+1} z^{p+1} + \cdots \qquad b_p \neq 0$$

In light of the results of the strictly real case of Section 9.4, we anticipate that the dominant term in (1) will be largely determined by those values z for which $|e^{nh(z)}| = e^{n\mathbf{R}h(z)}$ is greatest, that is, for z such that $\mathbf{R}h(z) = \max$.

In order to describe these concepts more clearly, we consider first a neighborhood of $z = 0$, and the curves $\mathbf{R}h(z) = 0$. These curves are easily visualized when we examine the first term of $h(z)$,

$$b_p z^p = |b_p| \, e^{i\beta} \, (re^{i\theta})^p$$

with real part $|b_p| \, r^p \cos(p\theta + \beta)$. The curves $|b_p| \, r^p \cos(p\theta + \beta) = 0$ are straight lines,

$$p\theta + \beta = \frac{\pi}{2} + n\pi \qquad n = 0, 1, \cdots 2p - 1$$

forming $2p$ rays from the origin, or p lines through the origin. Clearly, these p straight lines through the origin are tangents to p curves at the origin, the curves $\mathbf{R}h(z) = 0$ (Figure 9.1).

These curves separate regions in which $\mathbf{R}h(z) < 0$, $\mathbf{R}h(z) > 0$ alternately. We make one further assumption: the point z_0, of Section 9.5 (1), lies in a region for which $\mathbf{R}h(z) < 0$, and we take the path of integration, from $z = 0$ to $z = z_0$ to lie completely inside the same region (for which $\mathbf{R}h(z) < 0$). In this manner, the maximum of $|e^{nh(z)}|$ on the path of integration occurs at $z = 0$.

Figure 9.1

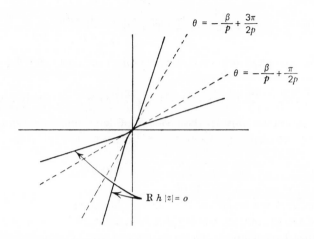

$$\theta = -\frac{\beta}{p} + \frac{3\pi}{2p}$$

$$\theta = -\frac{\beta}{p} + \frac{\pi}{2p}$$

R $h\,|z| = o$

We now deform the path of integration in a definite way (Figure 9.2). First, we proceed on the straight line segment, from $z = 0$ to $z = \rho e^{i\omega}$, which bisects the angle formed by the tangent lines bounding the region, and from $\rho e^{i\omega}$ to z_0 in any convenient way (provided, of course, that we remain inside our region for which $\mathbf{R}h(z) < 0$). We assume this latter path of integration to remain completely inside the region, so that $\mathbf{R}h(z) \leqq -\delta_1 < 0$ on this path, and we have

(3) $$\left| \int_{\rho e^{i\omega}}^{z_0} z^{\alpha-1} f(z) e^{nh(z)} \, dz \right| \leq M e^{-n\delta_1}$$

or $$= O(e^{-n\delta_1})$$

Figure 9.2

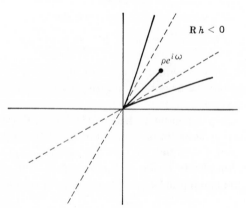

R $h < 0$

$\rho e^{i\omega}$

Accordingly,

$$(4) \qquad I_n = \int_0^{\rho e^{i\omega}} z^{\alpha-1} f(z) e^{nh(z)} \, dz + O(e^{-n\delta_1})$$

$$\sim \int_0^{\rho e^{i\omega}} z^{\alpha-1} f(z) e^{nh(z)} \, dz$$

and (4) remains valid for all $0 < \rho$ sufficiently small.

On the ray from $z = 0$ to $z = \rho e^{i\omega}$, the bisector of the rays on which $\cos(p\theta + \beta) = 0$, we have $\cos(p\theta + \beta) = -1$, $\sin(p\theta + \beta) = 0$. Thus

$$(5) \qquad h(z) = b_p z^p + b_{p+1} z^{p+1} + \cdots$$

$$= -|b_p| \, r^p + b_{p+1} r^{p+1} e^{i(p+1)\theta} + \cdots$$

with leading coefficient real and negative. Rewriting $h(z)$, we have

$$h(z) = z^p(b_p + z \, b_{p+1} + \cdots)$$

so that $h(z)$ behaves like $b_p z^p$ for all z close to the origin, and in particular on the ray from 0 to $\rho e^{i\omega}$. What we shall show is easy to remember and to manipulate: The *dominant term of*

9.5.1

$$I_n = \int_0^{z_0} z^{\alpha-1} f(z) e^{nh(z)} \, dz$$

$$\sim \int_0^{\rho e^{i\omega}} z^{\alpha-1} f(z) e^{nh(z)} \, dz$$

$$\sim e^{i\alpha\omega} \int_0^{\rho} r^{\alpha-1} f(re^{i\omega}) e^{-n|b_p| r^p} \, dr$$

that is, the dominant term of I_n comes from the first nonzero term of $h(z)$, and a path of integration can be chosen so that this first term is real.

The evaluation of this dominant term is straightforward.

$$(6) \qquad I_n \sim e^{i\alpha\omega} f(0) \int_0^{\rho} r^{\alpha-1} e^{-n|b_p| r^p} \, dr$$

$$= e^{i\alpha\omega} \frac{f(0)}{p} \int_0^{\rho_1} t^{(\alpha/p)-1} e^{-n|b_p| t} \, dt$$

$$\sim e^{i\alpha\omega} \frac{f(0)}{p} \frac{\Gamma\!\left(\dfrac{\alpha}{p}\right)}{(n|b_p|)^{\alpha/p}}$$

(We can, with no loss of generality, take $f(0) \neq 0$, as the term $z^{\alpha-1}$ can absorb any zeros of $f(z)$ at $z = 0$.)

The detailed proof of the result given in (6) does not involve any new ideas. Starting with

$$I_n \sim e^{i\alpha\omega} \int_0^\rho r^{\alpha-1} f(re^{i\omega}) e^{n[-|b_p| \, r^p|+\cdots]} \, dr$$

we define a new variable of integration:

$$r^p - \frac{b_{p+1}}{|b_p|} r^{p+1} e^{i(p+1)\theta} - \cdots = t^p$$

or, more precisely,

(7)
$$t = r\left[1 - \frac{b_{p+1}}{|b_p|} re^{i(p+1)\theta} - \frac{b_{p+2}}{|b_p|} r^2 e^{i(p+2)\theta} + \cdots\right]^{1/p}$$

$$= r\psi(r) \qquad r = t\chi(t) \qquad \chi(0) = 1$$

where the pth root is the principal-value branch; that is, the pth root approaches 1 as $r \to 0$.

Then

(8)
$$I_n \sim e^{i\alpha\omega} \int_0^{t(\rho)} r^{\alpha-1} f(e^{i\omega}r) e^{-n \, |b_p| \, t^p} \left(\frac{dr}{dt}\right) dt$$

$$= e^{i\alpha\omega} \int_0^{t(\rho)} t^{\alpha-1} [\chi(t)]^{\alpha-1} f[e^{i\omega} t\chi(t)] \left(\frac{dr}{dt}\right) \cdot e^{-n \, |b_p| \, t^p} \, dt$$

$$= e^{i\omega\alpha} \int_0^{t(\rho)} t^{\alpha-1} F(t) e^{-n \, |b_p| \, t^p} \, dt$$

with $F(t) = [\chi(t)]^{\alpha-1} f[e^{i\omega} t \, \chi(t)] \, dr/dt$, analytic near $t = 0$ (even if quite messy looking). The upper limit of integration, $t(\rho)$ is no longer necessarily real, and the real r-axis is mapped, by $t = r \, \psi(r)$, onto a curve near the real t-axis. We now deform the path of integration in (8) in the customary way, to the real point $t = \delta_2$, and thence to $t(\rho)$, obtaining

(9) $\quad e^{i\omega\alpha} \displaystyle\int_0^{\delta_2} t^{\alpha-1} F(t) e^{-n \, |b_p| \, t^p} \, dt + O(e^{-n\rho_1})$

$$\sim e^{i\omega\alpha} \int_0^\delta t^{\alpha-1} F(t) e^{-n \, |b_p| \, t^p} \, dt$$

$$\sim e^{i\omega\alpha} \frac{F(0)\Gamma\left(\dfrac{\alpha}{p}\right)}{p(n \, |b_p|)^{\alpha/p}}$$

We evaluate

$$F(0) = [\chi(0)]^{\alpha-1} f(0) \frac{dr}{dt}\Big|_0$$

$$= f(0)$$

Hence

(10)
$$I_n = \int_0^{z_0} z^{\alpha-1} f(z) e^{nh(z)} \, dz$$

$$\sim e^{i\omega\alpha} \frac{f(0)\Gamma\left(\frac{\alpha}{p}\right)}{p(n\,|b_p|)^{\alpha/p}}$$

as stated.

COMMENT. The method used here suggests much more. Since $F(t)$ is analytic near $t = 0$, $F(t) = \sum_0^\infty F^{(k)}(0)\, t^k/k!$, we obtain from (9) and (10) the full expansion

(11)
$$I_n \sim e^{i\omega\alpha} \sum_{k=0}^\infty \frac{F^{(k)}(0)}{k!} \frac{\Gamma\left(\frac{\alpha+k}{p}\right)}{(n\,|b_p|)^{\frac{\alpha+k}{p}}}$$

an asymptotic expansion in powers of $(1/n)^{1/p}$. The relation between $F^{(k)}(0)$ and the coefficients of $f(z)$ is somewhat messy, and involves the inversion of a power series (see Section 6.39) among other things. The first few terms are readily accessible by expansions about $t = 0$ ($z = 0$) and equating coefficients.

Example 1. We take another look at the example of Section 10.4.

(12)
$$\frac{n!}{n^{n+1}} = \int_0^\infty (te^{-t})^n \, dt$$

$$= \int_0^\infty e^{n[\log t - t]} \, dt$$

On the path of integration, the real t-axis, $\log t - t$ has a maximum at $t = 1$. Setting $t = 1 + z$, we have

$$e^n \frac{n!}{n^{n+1}} = \int_{-1}^\infty e^{n[\log(1+z) - z]} \, dz$$

where now

$$h(z) = \log(1 + z) - z = -z^2/2 + z^3/3 \cdots$$

We have $p = 2$, $b_2 = -\frac{1}{2}$.

The curves $\mathbf{R}h(z) = 0$ have as their tangents

$$\mathbf{R}\left(\frac{-z^2}{2}\right) = 0 \qquad \mathbf{R}(x^2 - y^2 + 2ixy) = 0$$

or

$$x^2 - y^2 = 0$$

These tangents bound four regions, in which (near $z = 0$) $\mathbf{R}h(z)$ is alternately positive and negative, being negative on the real axis (Figure 9.3). The real

Figure 9.3

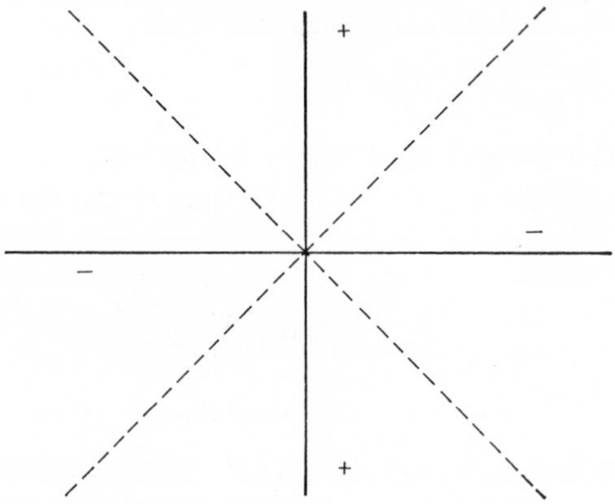

axis already bisects the appropriate angle. To obtain the dominant term, we have

$$\frac{e^n n!}{n^{n+1}} \sim \int_{-\delta}^{\delta} e^{n[\log(1+z)-z]}\, dz$$

$$\sim \int_{-\delta}^{\delta} e^{-nz^2/2}\, dz$$

[retaining the first term only in $h(z)$]

$$\sim \int_{0}^{\delta} e^{-nz^2/2}\, dz + \int_{-\delta}^{0} e^{-nz^2/2}\, dz$$

$$\sim 2\int_{0}^{\delta} e^{-nz^2/2}\, dz$$

$$\sim 2\int_{0}^{\infty} e^{-nz^2/2}\, dz$$

(why?)

$$= \sqrt{\frac{2\pi}{n}}$$

and again

$$n! \sim \sqrt{2\pi n}\left(\frac{n}{e}\right)^n$$

Example 2.   We consider the integral

$$\int_{-1}^{1}\frac{e^{ixt}}{\sqrt{1-t^2}}\,dt = 2\int_{0}^{1}\cos xt\,\frac{dt}{\sqrt{1-t^2}}$$

for real x, and look for the dominant term as $x \to \infty$. This integral occurs frequently in certain applied problems and represents a Bessel function:

(13) $$J_0(x) = \sum_{0}^{\infty}\frac{(-1)^n}{(n!)^2}\left(\frac{x}{2}\right)^{2n} = \frac{1}{\pi}\int_{-1}^{1}\frac{e^{ixt}}{\sqrt{1-t^2}}\,dt$$

On the path of integration, the real t-axis, $h(t) = it$, and $\mathbf{R}h(t) \equiv 0$. Now, $h(t) = i(t_1 + it_2)$, $\mathbf{R}h = -t_2$. We see that the region in which $\mathbf{R}h < 0$ is characterized by $t_2 = \mathbf{I}(t) > 0$. If, near $t = -1$, $t = 1$, we deform the contour in order to bisect the angle formed by the curves $\mathbf{R}h(t) = 0$, and at the same time enter the region $\mathbf{R}(h) < 0$, we are led to the vertical segments $(-1, B)$, $(1, A)$ (Figure 9.4). Since $\oint e^{ixt}\,dt/(1 - t^2)^{1/2} = 0$, from Cauchy's theorem, we may write

$$J_0(x) = \frac{1}{\pi}\int_{-1}^{1}\frac{e^{ixt}}{\sqrt{1-t^2}}\,dt$$

$$= \frac{1}{\pi}\int_{-1}^{-1+i\delta}e^{ixt}\frac{dt}{\sqrt{1-t^2}} - \frac{1}{\pi}\int_{1}^{1+i\delta}e^{ixt}\frac{dt}{\sqrt{1-t^2}}$$

$$+ \frac{1}{\pi}\int_{-1}^{1}e^{ix(t+i\delta)}\frac{dt}{\sqrt{1-(t+i\delta)^2}}$$

$$= \frac{1}{\pi}\int_{0}^{\delta}\frac{e^{ix(-1+i\tau)}i\,d\tau}{e^{i\pi/4}\sqrt{(2-i\tau)\tau}} - \frac{1}{\pi}\int_{0}^{\delta}\frac{e^{ix(1+i\tau)}i\,d\tau}{e^{-i\pi/4}\sqrt{(2+i\tau)\tau}} + O(e^{-\delta x})$$

$$\sim \frac{1}{\pi}e^{-ix}e^{i\pi/4}\int_{0}^{\delta}\frac{e^{-x\tau}\,d\tau}{\sqrt{\tau(2-i\tau)}} + \frac{1}{\pi}e^{ix}e^{-i\pi/4}\int_{0}^{\delta}\frac{e^{-x\tau}\,d\tau}{\sqrt{\tau(2+i\tau)}}$$

$$\sim \frac{1}{\pi}e^{-ix+(i\pi/4)}\frac{1}{\sqrt{2}}\sqrt{\frac{\pi}{x}} + \frac{1}{\pi}e^{ix-(\pi i/4)}\frac{1}{\sqrt{2}}\sqrt{\frac{\pi}{x}}$$

$$= \sqrt{\frac{2\pi}{x}}\cos\left(x - \frac{\pi}{4}\right)$$

Figure 9.4

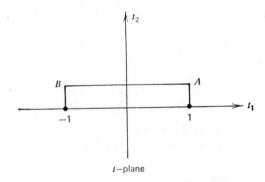

t—plane

COMMENT.   The process of selecting the path of integration to bisect the angle formed by successive curves $\mathbf{R}h(z) = 0$, into a region $\mathbf{R}h(z) < 0$ can be visualized picturesquely in the following manner: we regard $\mathbf{R}h(z) = h_1(x, y)$ as the altitude of a mountain and are located at the point $x = y = 0$ (with $h_1(0, 0)$ normalized to be zero). The paths $h_1(x, y) = 0$ are constant altitude paths. We wish to proceed down hill from our location $x = y = 0$. Among all possible directions pointing down (between two level paths) the one bisecting the directions of the level paths is the steepest. The method we have described is accordingly called the "method of steepest descent."

9.5.1   Find the dominant term, as $x \to \infty$, for

$$K_0(x) = \int_1^\infty \frac{e^{-xt}}{\sqrt{t^2 - 1}}\, dt$$

[$K_0(x)$ is another Bessel function: $y = K_0(x)$ satisfies $xy'' + y' - xy = 0$.]

9.5.2   Find the dominant term, as $x \to \infty$, for

$$f(x) = \int_0^1 e^{ixt}\, \frac{dt}{\sqrt{1 - t^2}}$$

9.5.3   Find the dominant term, as $n \to \infty$, for

$$I_n = \int_0^{\pi/2} e^{n\cos\theta}\, d\theta$$

9.6   THE SADDLE-POINT METHOD

In this section, we give several examples to illustrate further possibilities using asymptotic methods.

Figure 9.5

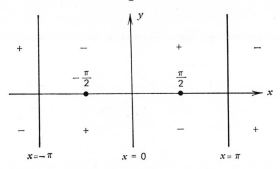

Example 1.

$$(1) \qquad I_n = \int_{-\pi/2}^{\pi/2} e^{in \cos t} \, dt$$

Here, $h(z) = i \cos z$, $\mathbf{R}h(z) = \sin x \sinh y$ and the full path of integration lies on one of the curves $\mathbf{R}h(z) = 0$. Let us plot the curves $\mathbf{R}h(z) = 0$ in the vicinity of the real axis, near the interval $(-\pi/2, \pi/2)$, and label the regions that they bound plus or minus accordingly as $\mathbf{R}h > 0$ or $\mathbf{R}h < 0$ in these regions. By using Cauchy's theorem, we may deform the path of integration for (1) from the real axis to any desired curve whose end points are at $-\pi/2$, $\pi/2$ (Figure 9.5).

$$\int_{-\pi/2}^{\pi/2} e^{in \cos z} \, dz = \int_{-\pi/2 \atop C}^{\pi/2} e^{in \cos z} \, dz$$

Near $z = -\pi/2$, the directions for which $\mathbf{R}h = 0$ are the real directions, with $\mathbf{R}h(z) < 0$ above the point $(y > 0)$. The steepest descent path is along $z = -\pi/2 + iy$, $0 < y$ (bisecting the angle at $-\pi/2$). At $z = \pi/2$, the steepest descent path is along $z = \pi/2 - iy$, $y > 0$. We discard the prospect of using curves C such as in Figure 9.6, since near $z = \pi/2$, $\mathbf{R}h(z) > 0$ on the curves; instead, we try to choose a new path of integration that permits us to enter

Figure 9.6

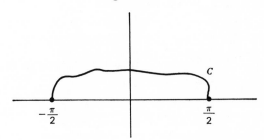

Figure 9.7

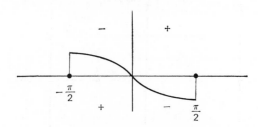

regions in which $\mathbf{R}h(z) < 0$, and on steepest descent paths if possible. In order to achieve this result, we are forced to select a curve that passes through the origin (Figure 9.7). In fact, the steepest descent paths at the origin similarly bisect the curves $\mathbf{R}h = 0$ there, and so make an angle of $\pi/4$ with the axes. We have

$$(2) \qquad I_n = \int_{-\pi/2}^{\pi/2} e^{in\cos z}\, dz$$

$$= \int_{-\pi/2}^{0} e^{in\cos z}\, dz + \int_{0}^{\pi/2} e^{in\cos z}\, dz$$

$$= J_n + K_n$$

$$J_n = \int_{0}^{\delta_1} e^{in\cos z} i\, dy - \int_{0}^{\delta_2} e^{in\cos z} e^{3\pi i/4}\, d\rho + O(e^{-n\delta})$$

with

$$z = -\pi/2 + iy \qquad z = e^{3\pi i/4}\rho$$

respectively.

$$J_n \sim \int_{0}^{\delta_1} e^{-n\sin y} i\, dy - \int_{0}^{\delta_2} e^{in[1-(z^2/2)+\cdots]} e^{3\pi i/4}\, d\rho$$

$$\sim \int_{0}^{\delta_1} e^{-ny} i\, dy - e^{in} \int_{0}^{\delta_2} e^{-n\rho^2/2} e^{3\pi i/4}\, d\rho$$

$$\sim \frac{i}{n} - e^{3\pi i/4} \frac{e^{in}}{2} \sqrt{\frac{2\pi}{n}}$$

Similarly,

$$K_n = \int_{0}^{\delta} e^{in[1-(z^2/2)+\cdots]} e^{-i\pi/4}\, d\rho + \int_{0}^{\delta} e^{in\cos z} i\, dy + O(e^{-n\delta})$$

with

$$z = e^{-\pi i/4}\rho \qquad z = \frac{\pi}{2} - iy$$

respectively.

$$K_n \sim e^{-i\pi/4}e^{in}\int_0^\delta e^{-n\rho^2/2}\,d\rho + \int_0^\delta e^{-n y}i\,dy$$

$$\sim e^{-i\pi/4}\frac{e^{in}}{2}\sqrt{\frac{2\pi}{n}} + \frac{i}{n}$$

(3) $$I_n \sim \frac{2i}{n} + \sqrt{\frac{2\pi}{n}}\,e^{in}e^{-i\pi/4}$$

The dominant term is clearly

$$\sqrt{\frac{2\pi}{n}}\,e^{i[n-(\pi/4)]}$$

which comes, not from the vicinity of the end points $\pm\pi/2$ (as in previous examples), but from the vicinity of the origin, an interior point on the original path of integration.

COMMENT. This example, of Section 9.6, and Example 2 of Section 9.5, both involved integrals of the form

(4) $$I_n = \int_C^{b}{}_a\ z^{\alpha-1}f(z)e^{nh(z)}\,dz$$

for which $\mathbf{R}h(z) \equiv 0$ along all of C. In the example of Section 9.5, $\mathbf{R}h(z) > 0$ on one side of C, $\mathbf{R}h(z) < 0$ on the other side of C. In such cases, the steepest descent lines, aP and bQ are normals to C, entering the region in which $\mathbf{R}h(z) < 0$ (Figure 9.8). On the segment PQ, $\mathbf{R}h(z) \leqq -\delta < 0$, and

$$\left|\int_P^Q z^{\alpha-1}f(z)e^{nh(z)}\,dz\right| = O(e^{-n\delta})$$

The second example is quite different. On the path of integration there is a point at which $h'(z) = 0$, a point at which angles are not preserved. The

Figure 9.8

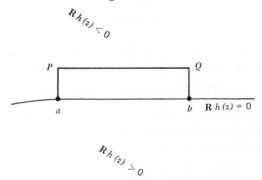

level lines $\mathbf{R}h(z) = $ const, and in particular, the path of integration, $\mathbf{R}h(z) = 0$, of the h-plane have z-plane images that intersect at angles less than π (at $\pi/2$ for this example). This situation is typified by $h(z) = z^2$ at $z = 0$;

$$\mathbf{R}h(z) = x^2 - y^2 = 0: \qquad x + y = 0 \quad x - y = 0$$

The regions formed by these curves have $\mathbf{R}h < 0$ and $\mathbf{R}h > 0$ alternately (Figure 9.9). Viewing $\mathbf{R}h(z)$ as the altitude of a mountain, the point $z = 0$ is a

Figure 9.9

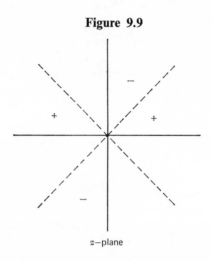

z-plane

col or *pass*. Alternately the shape of the surface, whose height is given by $\mathbf{R}h(z)$, looks like a saddle, and the point $z = 0$, where $h'(z) = 0$, is called a *saddle-point*. (The surface looks like an ordinary saddle for $h(z) = z^2$, while for $h = z^3$, or generally $h = z^p$ for $p > 2$, there are p plus regions, and p minus regions. The case $p = 3$ is sometimes referred to as a "monkey saddle," but the term saddle-point is used in all instances.)

Example 2. These ideas are further illuminated by the following integral

(5)
$$I_n = \int_{-\pi}^{\pi} e^{in(x - \sin x)} \, dx$$

In this instance, $h(z) = i(z - \sin z)$, $\mathbf{R}h(z) = -y + \cos x \sinh y = 0$ on the path of integration. Saddle-points are located at the points

$$h'(z) = i(1 - \cos z) = 0 \qquad z = 0 \qquad \pm 2\pi \cdots$$

Near $z = 0$, $h(z) = i(z^3/6 - z^5/120 + \cdots)$; the regions $\mathbf{R}h(z) > 0$, < 0 are marked plus and minus respectively (Figure 9.10); the tangents to the curves $\mathbf{R}h(z) = 0$ at the origin are given by

$$\mathbf{R}(iz^3) = 0 = \mathbf{R}(ir^3 e^{3i\theta}) = \mathbf{R}(r^3 e^{i(3\theta + \pi/2)})$$

Thus

$$\cos\left(3\theta + \frac{\pi}{2}\right) = 0 \qquad \sin 3\theta = 0 \qquad \theta = 0, \frac{\pi}{3}, \frac{2\pi}{3}$$

In order to deform the contour to a steepest descent path, with $\mathbf{R}h(z) \leq 0$ everywhere on the path, we start at $z = -\pi$, along $z = -\pi + iy, 0 \leq y \leq \delta_1$. Next, we arrange to enter the saddle-point at $z = 0$ along $\arg z = 5\pi/6$ (the

Figure 9.10

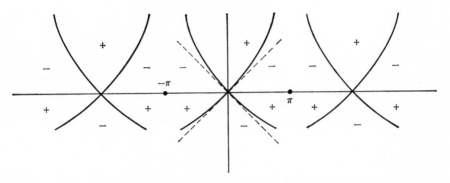

steepest descent path, bisecting the rays $\arg z = 2\pi/3$, $\arg z = \pi$). To reach $z = \pi$, we must leave $z = 0$ along the ray $\arg z = \pi/6$ and return to $z = \pi$ along $z = \pi + iy, 0 \leq y \leq \delta_1$.

9.6.1   Complete the details of the above example, and find the dominant terms (a) from the vicinity of the end points $z = -\pi$, $z = \pi$ and (b) from the saddle-point $z = 0$.

Example 3.   As our final example, we consider the following specific problem: given the formula

(6) $$J_n(x) = \frac{\left(\dfrac{x}{2}\right)^n}{\sqrt{\pi}\,\Gamma(\tfrac{1}{2} + n)} \int_{-1}^{1} e^{ixt}(1 - t^2)^{n-1/2}\, dt$$

for the appropriately normalized Bessel function of order n, $J_n(x)$, we investigate the asymptotic behavior, as $n \to \infty$, of $J_n(n \sin \alpha)$, for a fixed α, $0 < \alpha < \pi/2$.

(7) $$J_n(n \sin \alpha) = \left(\frac{n \sin \alpha}{2}\right)^n \frac{1}{\sqrt{\pi}\,\Gamma(\tfrac{1}{2} + n)} \int_{-1}^{1} e^{n[it\sin\alpha + \log(1-t^2)]} \frac{dt}{\sqrt{1 - t^2}}$$

$$h(z) = i(\sin \alpha)z + \log(1 - z^2)$$

The Saddle-Point Method   **319**

On the path of integration, $\mathbf{R}h(z) = \log(1 - x^2) \leqq 0$, while $\mathbf{R}h(z) = -y \sin \alpha + \log |1 - z^2| < 0$ near the path of integration, for $y > 0$. Thus, concentrating on the integral of (7) all by itself,

$$I_n = \int_{-1}^{1} e^{nh(z)} \frac{dz}{(1 - z^2)^{1/2}}$$

we note that for any curve such as C in Figure 9.11, we have $\mathbf{R}h(z) \leqq -\delta$, $I_n = O(e^{-n\delta})$. Thus it is clear that I_n decreases exponentially, and we must determine this behavior precisely before we can proceed. In order to achieve

Figure 9.11

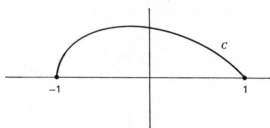

this result, we require the nature of $\mathbf{R}h(z)$ at all points of the z-plane through which we might conceivably deform our path of integration. In fact, we wish to find a new path of integration on which the *maximum of* $\mathbf{R}h(z)$ *is as small as possible.*

Let us examine the saddle-points of $h(z)$:

$$h'(z) = i \sin \alpha - \frac{2z}{1 - z^2} = 0$$

(8)
$$z = \frac{i}{\sin \alpha}(1 \pm \cos \alpha)$$

$$= i \tan \frac{\alpha}{2}, \qquad i \cot \frac{\alpha}{2}$$

The geometry of the level curves $\mathbf{R}h(z) = \text{const}$ is quite specific at these points and aids us considerably in forming our picture of $\mathbf{R}h(z)$.

At $z = i \tan \alpha/2$,

$$\mathbf{R}h(z) = -2 \sin^2 \frac{\alpha}{2} + 2 \log \left(\sec \frac{\alpha}{2} \right)$$

$$= -2 \sin^2 \frac{\alpha}{2} - \log \left(1 - \sin^2 \frac{\alpha}{2} \right) = -p < 0$$

the level curve $\mathbf{R}h(z) = -p$ consists of two branches intersecting orthogonally at $z = i \tan \alpha/2$, and can be traced back to the real axis at points P, Q (Figure 9.12). Below the level curve PSQ, $\mathbf{R}h(z) > -p$, as is also the case immediately above the portion TSW. The other two regions formed at S have $\mathbf{R}h(z) < -p$.

In order to reach $z = +1$ from $z = -1$, our path must cross the positive y-axis (if we require $\mathbf{R}h(z) \leqq 0$ on the path). If we cross at S on the steepest descent (ascent) path, then, clearly, max $\mathbf{R}h(z)$ occurs at S. Crossings at

Figure 9.12

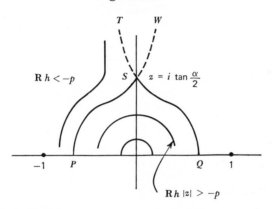

points other than S must invoive $\mathbf{R}h(z) > -p$ on the contour. Thus, max $\mathbf{R}h(z)$ is least on the steepest descent path through the lower saddle-point, $z = i \tan \alpha/2$. We find, near $z = i \tan \alpha/2$,

$$h(z) = h\left(t + i \tan \frac{\alpha}{2}\right)$$

$$= h\left(i \tan \frac{\alpha}{2}\right) + \frac{t^2}{2} h''\left(i \tan \frac{\alpha}{2}\right) + \cdots$$

$$= -2 \sin^2 \frac{\alpha}{2} - 2 \log \cos \frac{\alpha}{2} - t^2 \cos \alpha \cos^2 \frac{\alpha}{2} + \cdots$$

Hence

(9)

$$J_n(n \sin \alpha) \sim \left(n \frac{\sin \alpha}{2}\right)^n \frac{1}{\sqrt{\pi}\, \Gamma(\frac{1}{2} + n)} \exp\left[n\left(-2 \sin^2 \frac{\alpha}{2} - 2 \log \cos \frac{\alpha}{2}\right)\right]$$

$$\cos \frac{\alpha}{2} \int_{-\delta}^{\delta} \exp\left[-nt^2 \cos \alpha \cos^2 \frac{\alpha}{2}\right] dt \sim \left(\frac{\sin \alpha e^{\cos \alpha}}{1 + \cos \alpha}\right)^n \frac{1}{\sqrt{2\pi n \cos \alpha}}$$

Additional Examples and Comments on Chapter Nine

9.1   The complementary error function

$$\text{erfc } x = \frac{2}{\sqrt{\pi}} \int_x^\infty e^{-t^2}\, dt$$

Show that

$$\text{erfc } x \sim \frac{2}{\sqrt{\pi}} e^{-x^2} \left(\frac{1}{2x} - \frac{1}{2^2 x^3} + \frac{1.3}{2^3 x^5} - \cdots \right)$$

9.2   Show that

$$J_n(n) \sim \frac{\Gamma(\tfrac{1}{3})}{\pi 2^{2/3} 3^{1/6} n^{1/3}}$$

[see Section 9.6 (6)].

9.3   Show that

$$\int_0^\infty e^{-nx} \log x\, dx = -\frac{\log n}{n} - \frac{\gamma}{n}$$

where

$$-\gamma = \Gamma'(1) = \frac{d}{dz} \Gamma(z) \Big|_{z=1}$$

(Start with $\Gamma(\alpha)/n^\alpha = \int_0^\infty x^{\alpha-1} e^{-nx}\, dx$.)

9.4   Show that

$$\int_0^\pi e^{-n \sin x} \log x\, dx \sim -\frac{\gamma}{n} - \frac{\log n}{n}$$

(see 9.3).

9.5   Find the dominant term in the asymptotic expansion of

$$f(x) = \int_{-\infty}^\infty e^{-t^2 + ixt}\, dt \qquad x \to \infty$$

(Let $t = xz$ first.)

9.6   Find the dominant term of the asymptotic expansion of

$$f(x) = \int_{-\infty}^\infty e^{-t^4 + ixt}\, dt$$

(see 9.5.)

(Show that there are two saddle-points in the upper half-plane, and that the steepest descent path contributions at these two points are of the same order.)

9.7   (A problem of A. Wintner.)

$$f(x) = \int_{-\infty}^{\infty} e^{-t^{2n}+ixt}\, dt \qquad n = 2, 3, 4, \ldots$$

We require the dominant term of the asymptotic expansion of $f(x)$ as $x \to \infty$. Establish the following results, and so find the dominant term.

(a) Let $t = x^{1/2n-1}\, z$. Then $h(z) = iz - z^{2n}$.

(b) Of the $2n - 1$ saddle-points, ζ, $h'(\zeta) = 0$, n lie in the upper half-plane and $n - 1$ in the lower half—plane.

(c) $\mathbf{R}h(\zeta) = $ max at the two saddle-points in the upper half-plane nearest the real axis, for example,

$$\zeta_0 \qquad \zeta_{n-1} \qquad \text{and} \qquad h(\zeta_{n-1}) = \overline{h(\zeta_0)}$$

(d) With $P = x^{2n/2n-1}$,

$$f(x) \sim x^{1/2n-1}(e^{Ph(\zeta_0)} + e^{\overline{h(\zeta_0)}P})$$

$$\left(\int_{-\infty}^{\infty} \exp \frac{-|h''(\zeta_0)|\, Pu^2}{2}\, du \right) \exp -\frac{\pi}{4}\frac{2n-2}{2n-1} i$$

9.8   Obtain the following generalization of Laplace's formula 9.4 (10):

$$\int_a^b g(z)e^{nh(z)}\, dz \sim \sum g(\zeta_\nu)e^{nh(\zeta_\nu)} \sqrt{\frac{-2\pi}{nh''(\zeta_\nu)}}$$

where $\sqrt{-2/nh''(\zeta)}$ is in the direction of passing from valley to valley in the direction of integration, and the points ζ_ν are saddle-points. (See also 9.6 and 9.7.)

INDEX

Essential singularity, 203
  behavior near, 220
  residue at, 210
Euler's formula, 39
Exponential function, 36
  inverse of, 46
  mapping by, 129

Fixed point, 126
Fluid flow, in channels, 275
  about cylinder, 280
Flux, 143
Fourier series, 202
Free boundaries, 282
Functions, analytic, 75
  beta, 239
  entire, 187
  gamma, 235
  inverse, 46
  limits of, 20
  multiple valued, 54, 140
Fundamental theorem of algebra, 188

Gamma function, 235
Gauss' theorem on zeros of polynomials, 70
Geometric series, 21
Green's identity, 175

Harmonic functions, 96
Hodograph plane, 285
Hydrodynamics, 265
Hyperbolic functions, 69

Identities, 43
Image, 85
Imaginary axis, 4
  part, 4
Indefinite integral, 162
Independence of path, 159
Integrals, contour, 146

line, 147
  principal value, 182
Inverse point, 127
Inversion of series, 227
Irrotational flow, 90
  vector field, 93
Isolated point, 206
  singular point, 206

Jensen's theorem, 229
Joukowski airfoil, 139

Lagrange's formula, 228
Laplace's equation, 95
  asymptotic formula, 302
Laurent series, 195
Level curves, 57
Limits, 20
Line integral, 147
Linear fractional transformation, 120
Liouville's theorem, 187
Logarithm, 46
  inverse of, 46
  mapping by, 131
  principal branch of, 47

Machin's formula, 24
Maclaurin's series, 188
Magnification, 105
Majorant, 262
Mapping, conformal, 87
  Mercator, 131
  Ptolemy (stereographic), 113
  Schwarz-Christoffel, 250
  univalent, 110
Maximum modulus theorem, 187
Meromorphic function, 99

nth roots of a complex number, 50
Natural boundary, 235

One to one function, 110
Order, of poles, 203
of zeros, 205

Phase (argument), 4
Point at infinity, 114
Poisson's integral formula, 219, 24:
Poles, 203
order, 205
Polynomial, 36
Powers, 49
Power series, 63
algebra of, 192
convergence of, 63
differentiation of, 78, 101
integration of, 192
Principal part of Laurent expansion,
203
Punctured plane, 55

Quaternions, 29–33

Real axis, 4
part, 4
Reflection principle, 240
Removable singularity, 203
Residue theorem, 206
Residues, 204
computation of, 208
Riemann mapping theorem, 243
Rouche's theorem, 227

Saddlepoint method, 314
Schwarz-Christoffel formula, 250
Schwarz' inequality, 205, 220
Series inversion, 227
Singular points, 202
Sink, 270
Source, 91, 270
Stagnation point, 274
Stereographic mapping, 113
Stream function, 267
streamlines, 267

Taylor's series, 188
Triangle inequality, 7, 25
Trigonometric functions, 38
inverse, 52
identities, 40, 43

Univalent, 110, 244

Vector(s), 2, 6
fields, 88
Vectorial operations, 17
Velocity potential, 267
Vena contracta, 288

Weierstrass, 220
Work, 143

Zeros, order, 205